Georg Christian Wittstein, Ferdinand von Mueller

# The Organic Constituents of Plants and Vegetable Substances and

## their Chemical Analysis

Georg Christian Wittstein, Ferdinand von Mueller

**The Organic Constituents of Plants and Vegetable Substances and their Chemical Analysis**

ISBN/EAN: 9783337375065

Printed in Europe, USA, Canada, Australia, Japan

Cover: Foto ©berggeist007 / pixelio.de

More available books at **www.hansebooks.com**

THE

# ORGANIC CONSTITUENTS

OF

# PLANTS AND VEGETABLE SUBSTANCES

AND

## THEIR CHEMICAL ANALYSIS.

BY

### DR. G. C. WITTSTEIN.

AUTHORISED TRANSLATION FROM THE GERMAN ORIGINAL, ENLARGED
WITH NUMEROUS ADDITIONS;

BY

### BARON FERD. VON MUELLER,

C.M.G., M. & PH. D., F.R.S.

MELBOURNE:

M'CARRON, BIRD & CO., 37 FLINDERS LANE WEST.

MDCCCLXXVIII.

MELBOURNE:

M'CARRON, BIRD AND CO., PRINTERS,

37 FLINDERS LANE WEST.

# INDEX.

## - PART I.

### DIVISION I.

### DIVISION II.

### DIVISION III.

## PART II.

### DIVISION I.

# PREFACE OF THE TRANSLATOR.

At the time when Dr. Wittstein's "*Anleitung zur chemischen Analyse von Pflanzentheilen auf ihre organischen Bestandtheile*" made its appearance, I was honoured by its distinguished author, whose friendship I have enjoyed for many years, by bringing this important work under my notice. Recognising its value in my own laboratory, I became eager to render it accessible also to chemical workers in the great British Empire, as well as in the North American States, through an English version. The consent for a translation was granted with equal liberality and disinterestedness both by the author and Mr. C. H. Beck, of Noerdlingen, the publisher of the work. Circumstances, over which I could exercise no control, and such as not readily occur except in the earlier phases of a young colony, have retarded, to my deep regret, for a series of years the issue of this translation, though the manuscripts were mostly prepared long ago; thus I redeem but late the obligation so far devolving on me, and this to the disadvantage of the author. I feel great pleasure in acknowledging the co-operation of Mr. L. Rummel, who for some years conducted many phyto-chemic and technic operations in my laboratory, in aiding me throughout in the work of translation, and also in its revision, while it was passing through the press. Thus I effected a large saving of my time, heavily taxed already by multifarious professional, departmental and special literary duties, and often even sacrificed in defending the dignity of a scientific position, or sustaining my means for further progressive researches. These circumstances may also plead excuse, should I not have realised the expectations, raised in reference to this translation by the meritorious author or by the Pharmaceutic Society of Victoria, which more particularly through its President, Jos. Bosisto, Esq., M.P., and its Hon. Secretary, C. R. Blackett, Esq., promoted the issue of this English edition. The delay, which arose in the publication, has had however one advantage; it is this, that I was enabled to supplement the original work with many additional notes on new and well authenticated phyto-chemic data, which transpired during the last few years, some claiming local originality here. These additions, for which I myself must take the responsibility, are distinguished by the marks of parenthesis, and have met with Dr. Wittstein's approbation. The "*Zeitschrift des allgemeinen œsterreichischen Apotheker-Vereins*" was one of the main-sources of the additional data, obtained for this translation. Moreover, in one respect I have effected alterations in the original, for which the author's concession was also obtained; they consist in my re-writing the two chapters on the systematic names and arrangement of all the plants, to chemical substances of which allusion is made in the work. Researches on extended material, even since Professor Wittstein issued his volume, have not only modified in many cases the systematic limits of the orders, genera and species of these plants, but have also shed light on the origin of many medicinal and other vegetable substances, coming within the scope of this work, and formerly more or less involved in obscurity. Furthermore, I have preferred in the enumerative chapter of plants a systematic arrangement to an alphabetic sequence, and for this I have adopted the Candollean (or reversed Jussieuan) system, with such a change, as enabled me to distribute the monochlamydeous orders (Coniferæ and Cycadeæ excepted) among the other dicotyledonous ordinal groups, according to their greatest mutual affinities. The atomic formulas have been left unaltered, as given in the original; but a tabular exposition is appended, demonstrating the symbols of molecules according to the modern doctrine, adopted in most of the recent chemical works. Added are also as new for convenience tabular comparisons of English with metric weights and measures; furthermore, comparative tables of

Celsius' and Fahrenheit's thermometers, and finally calculations of the specific gravity of alcohol, according to the degrees of its dilution.

The translation is effected not without a certain freedom, extended even to the etymologic construction of the chemical appellations; yet I have endeavoured to adhere to the original text, so far as the different idioms of the two languages permitted. If I have failed in fusing the translation into the best expressions and forms, then I may frankly concede, that unless under remarkable circumstances of exception we never will be able to wield fully a language, which has not been that of our youth. If, after much toil and expenditure (the print of this translation being effected at my own private expense), I could wish, to reap any reward for my aid in diffusing the knowledge of methods, adopted by a leading masterly operator in phyto-chemistry, it would be, that local observers in these southern colonies, as well as in any other countries, teeming with an almost endless number of yet novel objects for phyto-chemic inquiry for additional resources, may be armed with auxiliary means for extending not only in abstract the science of chemistry, but also the precincts of therapeutics, and the realms of technology in reference to vegetable products.

Melbourne, May, 1878.

# PREFACE OF THE AUTHOR.

W ITH the exception of a work* published ten years ago, literature has as yet been wholly unrepresented by a manual of phyto-chemic analysis. For this reason alone the above work met with a very favourable reception, which moreover was doubtless enhanced by the name of the author, who had for a long time been engaged in researches of that particular direction. My expectations on receiving it were great. But although I found a considerable treasure of experiences deposited therein, I had soon to arrive at the conclusion, that the course recommended by the author renders the performance of a phyto-chemical analysis an extremely tiresome and slow process, and one requiring a large share of patience; consequently, instead of animating to attempt analyses of such kind, it rather deters therefrom. Besides, the course of the analysis indicated therein did not meet with my approval.

I decided therefore, not being inexperienced myself in these kinds of operations, to publish the method, which I have followed for many years, and which after numerous repetitions and corrections has been proved most practical, since it is considerably shorter than any other, and, I venture to say, is not less accurate. By this I will not assume, that my work is not capable of improvement; indeed, I myself am striving for that incessantly still, and shall gladly acknowledge any aid from anyone, who may pursue the same objects.

* "*Anleitung zur Analyse von Pflanzen und Pflanzentheilen*," von Friedr. Rochleder, Wuerzburg.

But to restrict this work merely to the methodical course of investigation appeared to me too incomplete; on the contrary, I considered it necessary to submit a short review of the apparatus and reagents required for the investigation, as well as of the whole of the proximate constituents hitherto known, in order that my work might contain nearly everything most urgently required for any phyto-chemical analysis, while dispensing with the necessity of consulting other works, involving loss of time and often producing unsatisfactory results.*

WITTSTEIN.

Munich, February, 1868.

* All temperatures given in this book refer to the centesimal (Celsius') thermometer.

# INTRODUCTION.

THE chemical analysis of a plant is distinguished from that of a mineral in several ways.

1. With a mineral the object of the analysis is the investigation of the nature of its elements or of their simplest binary combinations.

With a plant it is otherwise, because the elements, constituting the same,* are few in number and always the same, namely, carbon, hydrogen, nitrogen and oxygen.†

It is true, all organic bodies do not contain these four elements together; but to all of them belongs carbon, with which in most cases is combined oxygen and hydrogen, in some hydrogen, oxygen and nitrogen, in others hydrogen and nitrogen, in others again only hydrogen, and in a very small number only oxygen. According to the nature of their elements, the organic bodies form therefore the following five groups:

First group, carbon combined with oxygen,
Second ,,            ,,            ,,            ,, hydrogen,
Third  ,,            ,,            ,,            ,,            ,, and oxygen,
Fourth ,,            ,,            ,,            ,,            ,,            ,, nitrogen,
Fifth   ,,            ,,            ,,            ,,            ,,            ,,   ,, and oxygen,

of which the third group in regard to members greatly excels the other four groups.

---

* Once for all be it here remarked, that the investigation of plants in regard to their inorganic constituents—the ash-analysis—is excluded from this book, since I have published regarding this branch of science special instructions already several years ago.

† Sulphur, present in a very few proximate constituents of plants, may here remain unconsidered.

General, perspicuous and readily applicable distinguishing characteristics of these five groups are at present unknown; only the determination, whether an organic body belongs to any of the three first or any of the last two groups, that is, whether it is free of nitrogen or nitrogenised, is as a rule easy, because nitrogenised bodies mostly burn with a so-called horny odour and liberate ammonia on application of alkalies. In dubious cases ignite the body with sodium, treat the mass with water, add to the solution subsulphate of iron and over-saturate afterwards with hydrochloric acid, when in the presence of nitrogen a floccous turbidity of prussian blue is produced.

The investigation, to which of these five groups an organic body belongs and in what proportions the elements stand to each other, is moreover the object of the so-called elementary analysis; consequently the latter often extends into the province of special phyto-chemic analysis, but, like ash analysis, needs not here to be considered, since for their execution most text-books on organic chemistry afford all the requisite instruction.

2. With a mineral the object pointed out before can only be attained by complete dissolution. The solution takes place either in one or in several operations, in the latter case under change of the solvent. As solvents serve mostly acids, not so frequently alkalies, seldom water, never alcohol or ether.

With a plant the determination of its constituents is likewise ascertained by dissolution, but always with leaving a considerable residue; and therefore this kind of solution belongs to those operations, described by the term of extraction. The most important extracting agents are here ether, alcohol and water. Of less importance are acids and alkalies.

3. In a mineral analysis, the single constituents of the object of examination are obtained either as such or in compounds of accurately known constitution, but always in proportions of weight expressing most accurately the composition of the mineral.

In a phyto-chemic analysis means for a precise separation of the constituents are often still wanting. Quantitative determinations often encounter invincible difficulties, can only, in comparatively few cases, be executed satisfactorily, and a qualitative result has mostly to satisfy the operator.

4. To fathom the chemical constitution of a mineral, one or a few grams weight are almost always quite sufficient.

On the contrary, to ascertain qualitatively, and even superficially, all the constituents of a plant, at least a hundred times more material is required, but which, if a profound study of the single constituents is aimed at, has still to be increased ten or even a hundred fold.

5. It belongs to the rarest cases, generally only occurring after intervals of years, to meet, in the analysis of a mineral, with a constituent previously quite unknown, the discovery of which means at the same time that of a new element.

In a phyto-chemic analysis, on the other hand, it is not uncommon to obtain constituents unknown before. It therefore offers still a very fertile field for discoveries, though it ought not to be overlooked, that the accurate decision in regard to the correctness of such a discovery is mostly no easy task, because it depends on the purity and quantity of the material, which is not always so obtainable.

6. During the analysis of minerals a decomposition of the constituents, as regards their elements, need of course not be apprehended; the occurring changes consist only in the absorption or liberation of oxygen, sulphur, a few other elements—and volatile acids; the relations of these to the object of examination, whether they are constituents of the same or not, are already answered satisfactorily by other operations of the analysis.

The phyto-chemic analysis, on the contrary, is never safe against irretrievable losses, and this is the more impeding, because those constituents are mostly lost, on the determination of which the success of the whole operation mainly depends. The causes of such losses are either the easy decomposibility, or the volatility, or the solubility of many organic substances. To avoid such losses must be the incessant endeavour of the analytic operator, if he will not run the risk of losing the fruits of perhaps months of toil.

In the preceding six paragraphs, on which it would be easy to enlarge, I have explained sufficiently, that the execution of a phyto-chemic analysis is one of those operations, the just performance of which demands in a high degree circumspectness and accuracy. But the real value of such work is only obtained, when the necessary chemical elementary knowledge is previously gained; and as such I must designate the complete theoretical and practical acquaintance with the qualitative and quantitative inorganic analysis, with the principal laboratary operations, with the characteristics of the most important groups of organic bodies, and with their elementary analysis. He, who undertakes a phyto-chemic analysis without being furnished with these treasures of knowledge, enters helpless the field of research, and has, irrespective of loss of time and means, no chance of results, while he may introduce dubious or incorrect data into science. Ancient phyto-chemic analysts have sometimes in these respects committed themselves enormously; but the modern operators ought not on that account to relax, but ought to remember, that we are justified in demanding from them something better than from their predecessors, and that perhaps the limits of their own exertions may in not too distant a time be considerably surpassed. At all events, the endeavours of the chemical worker must be incessantly directed towards the goal of perfection, and in this spirit I shall welcome also every contribution to the improvement of this work.

# PART I.

## DIVISION I.

THE PROXIMATE CONSTITUENTS OF PLANTS AND
VEGETABLE SUBSTANCES AS FAR AS HITHERTO
KNOWN; THEIR PROPERTIES; THEIR MODE OF
PREPARATION AND QUANTITATIVE ESTIMATION.

My original intention was to bring this work into a kind of
systematic form—into a number of natural groups; but I relin-
quished this purpose on account of very many great difficulties.
A very large number of the vegetable substances, treated on in
the following pages, are as yet so imperfectly investigated, that it
is impossible to determine their real constitution; consequently
they could not have been brought into any system at all. Moreover, not a few of the better-known substances have properties,
which leave it doubtful to which of the groups they belong. For
instance, there are many dyeing substances, which possess all the
properties of resins or acids, and therefore can be placed just as
well among the resins as among the acids, and had to be looked
for sometimes amongst the resins, sometimes amongst the dyeing
substances, and sometimes amongst the acids. A special index
might have alleviated this difficulty; still, the system would not
have been better for all that. I preferred, in order to follow out the
practical object, to bring all the names into one alphabetical order;
and to facilitate consultation by tables of synonymes.

The nature of a substance will be taught by its description. A
thorough investigation is wanting for a satisfactory result, and
herein lies an invitation to fill out gaps, and sometimes very considerable ones, in our knowledge of vegetable compounds hitherto
considered peculiar. Whoever devotes his time to the solution
of such problems deserves more praise, I firmly believe, than he
who engages in the examination of vegetables not analysed before.

[**Abieten.** Liquid hydrocarbon, obtained by distillation from the resin of Pinus Sabiniana. It is a colourless oil of a penetrating orange-like odour, of 0·694 sp. gr. at 15°, boiling at 101°. Dissolves very little in water, in five parts alcohol of 95%. It forms no compound with hydrochloric acid gas, and is slowly decomposed by warm nitric acid. It absorbs a great quantity of chlorine, and becomes thick. The A. is an excellent solvent for fats and volatile oils, except castor oil, though the latter dissolves two-thirds of its own volume of Abieten. Canada balsam dissolves two parts Abieten, Peru balsam one-fifth its volume. Wenzel.]

**Abietic Acid.** The constitution of colophony, *i.e.*, of the resinous substance exuded from coniferous trees by incisions of the stem, and which is fused afterwards in order to volatilise the essential oils, is, according to Maly, materially different, and especially much less complicated than it was heretofore considered. It is not a mixture of several isomeric acids (sylvic, pinic, pimaric acids = $C_{40} H_{30} O_4$ ), besides indifferent resins, but consists in the main (more than 90%) of a peculiar acid in the anhydrous form, named by the author Abietic acid, and composed according to the formula $C_{88} H_{62} O_8$. It is a bibasic acid, and crystallises slowly from a solution of colophony in common alcohol (of about 70%), as hydrate = $C_{88} H_{\overline{62}} O_8 + 2 HO$ in crusty masses. The anhydrous acid fuses at 100°, the hydrate not under 165°, without losing water; and even in higher temperatures it does not lose more water, until, through long-continued heating, it assumes a yellow or brown colour and becomes decomposed. The so-called sylvic acid was imperfectly purified abietic acid, the pinic acid nothing but genuine colophony, and pimaric acid appears to be nothing else but abietic acid.

**Absinthin** = $C_{40} H_{28} O_8$. The bitter ingredient of Artemisia Absinthium. Precipitate the decoction with tannic acid, mix the washed precipitate with oxyd of lead, dry, treat with alcohol, digest the solution with animal charcoal and evaporate. The remaining Absinthin has the appearance of slightly yellow drops of oil, but solidifies to a hard, indistinctly crystallised mass. It is friable to a powder, not influenced by the atmosphere, neutral, not aromatic, of a very bitter taste, fuses at 120° to 125°; dissolves very little in cold and sparingly in hot water, readily in alcohol, ether and alkalies; yields no sugar with diluted acids, and dissolves in concentrated sulphuric acid with a brown colour, which changes slowly to a green-blue and becomes dark-blue by a few drops of water, but separates grey flakes by an excess of the latter.

**Acetic Acid** = $C_4 H_3 \dot{O}_3 + HO$. Discovered in the juice of many plants, and especially of trees. Its properties are well known. The salts formed by it are all soluble in water, most of them readily so; they can be recognised with certainty by the blood-red colour they acquire with salts of oxyd of iron, by the smell of acetic acid they emit when heated with sulphuric acid and alcohol, and by the odour of alkarsin (oxyd of kakodyl), when heated dry with arsenious acid. Its production and quantitative estimation can, as it belongs to the volatile acids, only be effected by distilling the liquid in question with sulphuric acid until the distillate no longer shows any acid reaction. Should the liquid contain much organic matter, phosphoric instead of sulphuric acid has to be used, since towards the end of the operation, the latter acid is decomposed by the organic matter into sulphurous acid, which, in passing over, would contaminate the acetic acid. The distillate has to be saturated with soda, dried, heated until it is fused and weighed. One hundred parts by weight of this anhydrous salt contain 62·00 parts acetic acid. The result will be more exact by digesting the acid distillate with an excess of carbonate of baryta, finely ground in water, and filtering after the acid reaction has disappeared. The filtrate has to be precipitated with sulphuric acid, and the resulting sulphate of baryta has to be weighed. One hundred parts of the latter salt represent 43·78 parts acetic acid.

**Achilleic Acid.** Peculiar acid of Achillea Millefolium. In order to obtain it, precipitate the decoction of the herb by acetate of lead, decompose the deposit by sulphuret of hydrogen, over-saturate the acid liquid (containing green colouring matter and lime) with carbonate of potash, filter off from the lime, digest with animal charcoal, decompose the potassium salt with acetate of lead and the lead compound with sulphuret of hydrogen. A colourless liquid, without smell and of very acid taste; its density is 1·0148 in the utmost concentration, not volatile at 80°, is assuming a darker colour by protracted heating; crystallises in colourless prisms, soluble in two parts water, when exposed to the atmosphere; the solution is not precipitated by the acetate, but becomes turbid by the subacetate of lead; yields crystallisable salts with alkalies.

**Achillein.** The preparation designated by this name is nothing but an alcoholic extract of Achillea Millefolium, which has been treated with animal charcoal and evaporated to dryness. It has a yellow-brown colour, a peculiar smell and a bitter, not disagreeable taste similar to milfoil, the plant yielding it. It becomes moist by exposure to the atmosphere, and dissolves easily in water and in alcohol; in ether only when mixed with a few drops of acetic acid.

**[Achillein** (*see* IVAIN). The evaporated, aqueous extract, on treating with alcohol, yields up Achillein, Moschatin and organic acids. Moschatin is thrown down by addition of water, and the acids by subacetate of lead. Brittle, brown-red, hygroscopic mass of a bitter taste, easily soluble in water, less so in alcohol; insoluble in ether. Contains nitrogen. Planta Reichenau.]

**Acids.** All organic acids as yet investigated are combinations of carbon with hydrogen and oxygen—seldom without hydrogen, and more seldom still with nitrogen; they have an acid reaction and mostly an acid taste, and form salts with the bases. They are very widely diffused, and it is highly probable that there exists no vegetable organism, which does not contain one or the other; indeed, experience has shown that many plants contain more than one. They exist very seldom in the free state in the organism, but nearly always either partly or entirely saturated by anorganic or by organic bases (alkaloïds); in the event of the basic access being incomplete, they are the cause of the acid reaction of vegetable extracts. The substance found combined with fat acids, and which serves as substitute for bases, is oxyd of glyceryl.

Acids are either volatile or not, as in the case of alkaloïds. The volatile acids are always obtained by distillation with a fixed mineral acid, preferably phosphoric acid. For fuller information see section ix., Division III., Part II., where the way of their estimation is also indicated. The particulars for ascertaining, preparing and estimating non-volatile acids will be found in sections ii. and viii. The discovery of fat acids coincides in section ii. with the examination of fats.

**Acolyctin.** The root of Aconitum Lycoctonum contains no aconitin, but in its stead two other organic bases, the one of which has been termed Acolyctin, and the other, present in less quantity, Lycoctonin. In order to isolate the Acolyctin, evaporate or distil the alcohol from the tincture (treated beforehand as under "Aconitin," according to Geiger and Hesse, first with lime and afterwards with sulphuric acid), remove from the residue—diluted with water, if necessary—all resinous matters, decolourise with animal charcoal, add carbonate of soda in sufficient quantity for a decidedly alkaline reaction and bring to dryness. Grind the substance, extract with chloroform or absolute alcohol, filter, add some water, evaporate to a syrup, shake repeatedly with ether (which dissolves the Lycoctonin), dry and pulverise. Whitish powder of a bitter, not acrid taste, readily soluble in water, in diluted and concentrated alcohol and in chloroform, but not in ether; of alkaline reaction. When ether is added to the alcoholic solution, the whole of it will either be converted into a paste, or the Acolyctin will fall down as a white substance when more diluted. It separates in drops on the bottom of the vessel when ether is added to its solution in chloroform. The aqueous solution of the pure Acolyctin and of its salts is precipitated by the carbonates of alkalies. Ammonia has no effect on it at first, but after it has been kept for some time the whole solidifies to a colourless jelly. Tannic acid precipitates white, acetate of lead likewise, but an excess of it re-dissolves the precipitate. Subacetate of lead occasions no turbidness in the alcoholic solution, but a considerable one in the aqueous solution. Molybdate of ammonia produces a strong white turbidness with sulphate of Acolyctin. Chloride of gold yields a pale yellow precipitate; concentrated sulphuric acid causes no change of colour.

**Aconellin** = NARCOTIN.

**Aconitic Acid** = $C_{12} H_3 O_9 + 3HO$. In the herb of species of Aconitum and Delphinium Consolida, according to Baup also in Equisetum (*see* Equisetic Acid). Extract with water, saturate with carbonate of soda, precipitate the sulphuric and phosphoric acids with acetate of baryta, filter, precipitate with acetate of lead, wash the precipitate and decompose it in the moist state by sulphuret of hydrogen, separate the liquid by filtration from the sulphide of lead, concentrate by evaporation, and leave it to crystallise. It appears in white warts and needles, without smell, of a pleasant acid taste, fuses at 140° and decomposes in higher temperatures, dissolves readily in water, alcohol and ether. The salts of the alkalies and alkaline earths are readily soluble, most of the other salts only with difficulty.

**Aconitin** = $C_{60} H_{47} NO_{14}$. Specific alkaloïd of the root of Aconitum Napellus, A. ferox and probably of other species of this genus of Ranunculaceæ. Extract with hot water, acidulated by sulphuric acid, press out, evaporate to a syrupy consistence, saturate not quite sufficiently with carbonate of soda and add a sufficient quantity of calcined magnesia to give it a faint alkaline reaction, shake several times with ether, mix the ethereous solutions and distil off the solvent, heat the remnant with water and add diluted sulphuric acid by drops until it is dissolved, drive off the last traces of ether, leave to cool, filter, precipitate the filtrate with ammonia, collect the white flocky precipitate occasioned thereby, wash and dry without application of heat. White, voluminous, dull, amorphous, electric powder without odour, but even in the most minute particles of remarkably irritating effect upon the mucous membranes of the nose; of at first faintly bitter, afterwards long-lasting acrid, burning and lastingly harsh taste. It fuses at 80–85° under loss of 25 per cent. (water) and decomposes in higher temperatures, while emitting vapours of acid reaction. It dissolves in 1316 parts cold and in 43 parts boiling water, and these solutions have a neutral reaction; very readily soluble in alcohol, ether, benzol, chloroform and sulphide of carbon, and these solutions have a faint alkaline reaction when concentrated;

it also dissolves in diluted acids, but without yielding crystallisable salts. It dissolves colourless in concentrated sulphuric acid; and chromate of potash originates in the solution pale purple-red zones, similar to those which are produced by subjecting strychnin to the same process, but much paler. Its solution in concentrated nitric acid is gold-yellow. Caustic potash, ammonia and carbonate of potash give a white precipitate; carbonate of ammonia, bicarbonate and phosphate of soda none; chloride of mercury, sulphocyanide of potassium and tannic acid a white one, picric acid a yellow one, tincture of iodine a red-brown, chloride of gold a yellow-white, and chloride of platinum none.

**Acorn Sugar** = QUERCIT.

**Adansonin** = $C_{48} H_{36} O_{33}$. The bitter ingredient of the bark of Adansonia digitata and A. Gregorii. In order to obtain it in a pure state, evaporate the alcoholic extract to dryness, boil with water, mix with finely ground oxyd of lead, filter, evaporate nearly to dryness, shake with ether and leave the ethereous liquids to evaporate without heat. Fine, white needles of a smell similar to that of aloes or gentian, of extremely bitter taste, fusible by heat, but becoming afterwards carbonised; dissolves in six parts cold and three parts boiling ether, also readily soluble in absolute and in common alcohol and perceptibly in water. The solutions do not become turbid by alkalies and by metallic salts. Chloride of iron imparts to the alcoholic solution a greenish tinge.

**Æsculin** = $C_{42} H_{24} O_{26}$ + 3 HO. In the bark of Æsculus Hippocastanum; its occurrence in other plants (in the quassia, sandalwood, &c.) has to be further proved yet. Exhaust with alcohol of 80°, distil off the bulk of the alcohol and leave the rest to stand in the cold for a few weeks. The Æ., which will then have separated, has to be washed with ice-cold water and recrystallised repeatedly in ether-alcohol. Snow-white, fine crystalline needles, often globularly arranged and of the appearance of a loose powder, inodorous, of slightly bitter taste and of acid reaction. It fuses at 160° under loss of the water, and is decomposed in higher temparatures; dissolves in 672 parts cold, already in $12\frac{1}{2}$ parts boiling water. The cold saturated solution is colourless and of a faint blue fluorescence, which becomes more marked after the addition of well-water; it loses this property by acids, but recovers it through alkalies and alkaline earths. It dissolves in diluted acids or alkalies more readily than in water; the alkaline solution appears blue in the reflected and yellow in the transmitted light. Dissolves in 120 parts absolute alcohol, in 100 parts alcohol of 82%, in 80 parts rectified alcohol, and in 24 parts boiling absolute alcohol. It changes into sugar and æsculetin ($C_{18} H_6 O_3$) when boiled with diluted acids.

[**Agaric Acid** together with **Agaric Resin** have been prepared by G. Fleury from Polyporus officinalis by extracting with ether. Agaric resin is brown-red, pulverised light-brown; insoluble in water, easily soluble in ether and in absolute alcohol, less in alcohol of 70%, also in methylic alcohol, chloroform, acetic acid, insoluble in benzol and sulphide of carbon; easily soluble in ammonia and diluted potash-ley. It is only slightly bitter and fuses at 89·7°.

Agaric Acid crystallises in fine, white needles, fuses at 145·7°, not sublimable, dissolves easily in strong alcohol, less in chloroform, very little in ether and acetic acid, still less in sulphide of carbon and benzol. Water dissolves very little of it, but assumes an acid reaction. Its centesimal composition is C = 63·44, H = 9·75, O = 26·81.]

**Agaricin.** Solid, crystallisable fat, contained in mushrooms (Agaricus campestris and many other species), fusing between 148–150°, is not affected by caustic alkalies.

**Agrostemmin.** Alkaloïd alleged to exist in the seeds of Lychnis Githago. Obtained by extracting with alcohol of 40 per cent. containing acetic acid, and by precipitating with calcined magnesia. The precipitate to be treated with alcohol and left to crystallise. Yellowish white, minute scales, fusible by heat and slowly soluble in water, of perceptibly alkaline reaction and yielding crystallisable salts with acids.

[**Ailantic Acid,** prepared by Narajan-Dagi from the bark of Ailantus excelsa. The decoction of the bark is freed from lime by oxalic acid, from gum and colouring matter by subacetate of lead; the liquid is then evaporated after treating with sulphuret of hydrogen. Reddish brown, very bitter, deliquescent mass of wax-consistence, very easily soluble in water, less in alcohol and ether, insoluble in chloroform and benzol.]

**Albumin.** The widest distributed of the protein-substances and found in the sap of all vegetables, but occurring also in the solid or curdled state. When these juices are heated to the boiling point or only to 75°, the albumin loses its solubility and separates in almost white flakes, frequently coloured green by chlorophyll. By treating the coagulated mass successively with alcohol, ether and water containing hydrochloric acid, the albumin remains after drying as a yellowish or grey-white transparent mass, which swells up in water without dissolving in it, and by its behaviour coincides with protein (*see* Protein-substances). In order to determine the quantity of albumin contained in a liquid, collect after the latter has been purified in the above manner in a weighed filter, and dry at 120°.

**Alismin.** An acrid and bitter substance, occurring in the roots of Alisma Plantago, but only obtained in the extractive form, and has as yet not been closely examined.

**Alizarin** $= C_{20} H_6 O_6 + 4\ HO.$  In the root of Rubia tinctorum. Boil the pulverised root (madder) with water, precipitate the decoction with sulphuric acid, and boil the well-washed precipitate with chloride of aluminium, in order to dissolve the dyeing substances.  The solution, when mixed with a little hydrochloric acid, separates red flocks consisting of Alizarin, purpurin and resin.  These flocks have to be dissolved in alcohol, or in diluted liquor of ammonia under addition of freshly precipitated alumina, which combines with the dyeing matters.  This compound has to be boiled with concentrated solution of carbonate of soda, whereby purpurin is dissolved with deep-red colour, while Alizarin-alumina remains in the residue.  The latter has to be freed from resinous matter by washing with warm ether, and is then decomposed by boiling with diluted hydrochloric acid.  The remaining Alizarin has to be recrystallised in alcohol.  It appears in long translucent, dark-yellow prisms of great lustre, of neutral reaction, and of bitter taste; loses the water at 100° to 120°, turns opaque, darker red, fuses afterwards, and sublimates at 215° in orange-red needles; dissolves little in water, with purple colour in alkalies, with blood-red colour in concentrated sulphuric acid, and is precipitable by water without change, with red-yellow colour in sulphide of carbon, and with gold colour in alcohol and in ether.

[From the tubers of various Australian Droserae a substance has been obtained, which in physical properties closely resembles, and probably is identical with, Alizarin.—Baron von Mueller and Rummel.]

**Alkaloids.** Compounds of carbon, hydrogen and nitrogen, with or without oxygen, mostly of alkaline reaction, and able to form salts with acids.  Considering the short time (about fifty years), which has elapsed since their existence has been discovered, their number is rather considerable, and becomes greater continually, though only the minor portion of the plants as yet treated for such have given positive results, perhaps because either too little raw material has been employed, the quantity of alkaloids in vegetables being comparatively small, or because the proper method has not been adopted, or was not known, as is partly the case even at present.  Investigations in this branch of phyto-chemistry are nevertheless so far progressing, that I may venture some remarks regarding the approximate number and distribution of alkaloids.

The classification of plants into natural groups satisfies not only the botanist to a high degree, but also the chemist, for, just as the former comprises all the plants that show a certain harmony in regard to morphology and anatomy into one separate group, so the latter observes that the individuals of such a group resemble each other by containing certain widely-distributed substances in exceedingly large quantities (for instance, tannic acid, starch), or by yielding certain compounds of peculiar character in regard to smell, taste or effect on the animal organism, substances which are either confined to one family or only present in but few others. Of such substances, which appear to link together different families, each group contains either one or more, and amongst them are of high consequence the alkaloïds. The number of orders or suborders of plants amounts to about 400; when we admit that on the average each of them contains two or three specific alkaloïds, it follows that the whole vegetable kingdom produces about 1000, of which I need scarcely say only the minor portion (about one-fifth) has become accurately known. In making this computation, it has not been left out of regard, that some of the vegetable families (Labiatæ, Compositæ) are at least largely free of alkaloïds, that some have only one in common (for instance, Berberin belongs, as far as our knowledge goes, alike to Anonaceæ, Berberideæ, Cassieæ, Menispermeæ, Papaveraceæ, Ranunculaceæ, Rutaceæ), and that some families contain more than two or three alkaloïds (Cinchoneæ, Papaveraceæ, Solanaceæ, &c.).

Whereas the presence of an essential oil in vegetables is immediately recognised by the smell, general indications for the presence of alkaloïds are wanting. Indeed, all the alkaloïds known at present possess a very perceptible taste, which is mostly bitter and acrid, both to the highest degree; and those among them that are volatile have also a specific odour; but these characteristics of taste and smell are shared by many other substances devoid of alkaloïd properties. On the other hand, it would be unwarrantable to conclude from little or no taste the absence of alkaloïds in a plant, as the alkaloïd, when only present in minute quantities, may not be detected by means of tasting. In this case, the isolation of the alkaloïd is often surrounded by the greatest difficulties.

Before examining any vegetables on alkaloïds, it must be decided first, if the latter are volatile or not. When a herb is endowed with a strong smell, that becomes more striking by adding a solution of caustic potash, the presence of a volatile alkaloïd is evident. The non-volatile alkaloïds are not so easily found out, since they may not be recognised by the taste; consequently, an alkaloïd, before it can be considered as such, must be isolated in almost a pure form. Their estimation and isolation become therefore simultaneously necessary. This process of isolation serves, at the same time, for determining the quantity.

The fact that every aqueous extract of vegetables reddens the litmus-paper shows, that the akaloïds are not present in the free state in the organism, but bound to acids as salts. Generally, these salts dissolve readily in water, except the tannates, which dissolve slowly or not at all. In the latter case, *i.e.*, when tannic acid is present, the extraction, in order to be exhaustive, has to be carried on by means of a very diluted mineral acid (100 parts water to about two parts of sulphuric or hydrochloric acids). Alkalies and alkaline earths decompose all these salts easily by combining with the acid and leaving the alkaloïd free, whereby the latter, according to its nature, is either volatilised or remains in the solution or is precipitated. By these characteristics is also indicated the way for obtaining the alkaloïds in general.

The volatile alkaloïds are at ordinary temperature liquid, colourless, of a strong specific odour of the plant used for their preparation, mostly heavier than water (of those which are known, only coniin is lighter than water), of alkaline reaction, readily soluble in water, alcohol, ether and acids.

The non-volatile alkaloïds are mostly white, seldom yellow, without odour, mostly of an eminently bitter or acrid taste, of amorphous or crystalline structure, heavier than water, fusible or not, slowly or not at all soluble in water, but readily so in acids and mostly in alcohol too, and partly in ether, of feeble alkaline reaction even in the saturated alcoholic solution.

**Alkanna-red** = ANCHUSIN.

**Alchornin.** Peculiar substance, obtained from the bark of Alchornea latifolia by extracting with alcohol, treating the extract with ether, and evaporating the ethereous solution. White, pointed crystals; readily soluble in alcohol, ether and oil of turpentine, but not in alkalies and diluted acids. Its existence has been repeatedly questioned.

**Aloïn** = $C_{34} H_{18} O_{14} +$ HO. In the aloes (the dried juice of the leaves of various species of the genus Aloe). In some sorts of aloes it exists in the amorphous form, and can therefore not be obtained pure. Best adapted for its isolation is the Barbadoes aloe. Extract with cold water, evaporate in vacuo to a syrup, leave to cool, collect the separated crystalline grains, press and recrystallise in water of not more than 65°. It appears in sulphur-yellow grains when crystallised in water, and in concentrically radiated needles when crystallised in alcohol, of at first sweetish, afterwards intensely bitter taste, without odour and of neutral reaction. It loses the water on the waterbath, and when left there for a longer time it loses more water and becomes partly converted into a brown, amorphous resin. Afterwards it fuses and decomposes gradually. Dissolves in 600 parts of cold water, readily in alcohol, ether and alkalies. When any of these solutions are boiled, the Aloin becomes uncrystallisable.

**Aluchi-Resin.** From an unknown tree of Madagascar. Almost white on the outside, of black marble appearance on the inside, opaque, solid, friable, of strongly aromatic, pepperlike smell and of bitter taste. Contains volatile oil, a resin easily dissolving in alcohol, another resin more difficult to dissolve in alcohol (the latter crystalline to about 20%), an acid in the free state, an amorphous bitter substance and a salt of ammonia.

**Alyxia-Stereopten.** Exudates on the inner surface of the bark of Alyxia Reinwardti.* White, hair-shaped crystals of the odour of cumarine, of faintly aromatic taste and of neutral reaction. It sublimates at 75° to 87°, fuses in higher temperatures and becomes brown afterwards; dissolves little in cold, better in warm water, readily in alcohol, ether, acetic acid, oil of turpentine, caustic alkalies and carbonates of alkalies.

**Amanitin.** The alleged poisonous ingredient of the Fly-Agaric (Agaricus muscarius), obtained as yet only as extract, consequently impure.

[**Ambrosin.** A fossil resin, exudation of probably coniferous trees of South Georgia. Resembles amber, yields on melting succinic acid and a fragrant volatile oil. Dissolves copiously in oil of turpentine, alcohol, ether, chloroform and carbonate of potash, in less quantity and without decomposition in concentrated mineral acids.]

**Ammoniacum.** Gum-resinous exudation of Dorema ammoniacum. Yellowish white, half-transparent lumps, friable at low temperature, of concheous fracture, disagreeable smell, faintly bitter and acrid taste. Contains about 70% resin, soluble in alcohol and solution of alkalies, 18 gum, 4 bassorin, and a light volatile oil.

**Amygdalin** = $C_{40} H_{27} NO_{22} + 6 HO$. Found in many plants belonging to Rosaceæ, but its presence has been partly deduced from the fact, that these vegetables produce hydrocyanic acid, when distilled with water. Amygdalin has been obtained in two different modifications, viz., (*a*) in the crystalline form from the seeds of Prunus Amygdalus, Prunus Persica, Prunus domestica, Prunus Laurocerasus, Prunus Padus, and from the leaves, flowers and bark of the latter; (*b*) in the amorphous form from the leaves of Prunus Persica and Prunus Laurocerasus and from the seeds of Prunus Cerasus. Its presence has been deduced from the hydrocyanic acid in the distillate of the following plants:—Prunus capricida (leaves), Pr. spinosa (flowers and seeds), Pr. virginiana (bark); Amelanchier vulgaris, Cotoneaster vulgaris, Cratægus Oxya-

---

* Doubtless also in others of the many Australian, Indian, and Polynesian species known in the root of A. ruscifolia.—F. v. M.

cantha, Pyrus aucuparia, hybrida and torminalis (all flowers); Spiræa Aruncus, Japonica, sorbifolia (leaves).

The best material for the production of Amygdalin are bitter almonds. These have to be separated from the bulk of fat-oil by pressing; they are then extracted with strong alcohol and the latter removed by distillation. The residue is separated from the oil floating upon it, and mixed with half its volume of ether. The Amygdalin subsides, and has to be pressed and recrystallised in alcohol. It crystallises in alcohol with four equivalents water in colourless scales of mother-of-pearl lustre (in water with six equivalents water in prisms), is without odour, of at first sweetish, afterwards bitter taste, and of neutral reaction. It loses the whole of the water at 120°, liquefies at 200°, turns brown and decomposes, while evolving the odour of burnt sugar (caramel), afterwards of hawthorn and at last of animal empyreumatic substances. It dissolves at 8 to 12° in 15 parts water, in boiling water in unlimited quantity; at 8 to 12°, in 904 parts alcohol of 0·819, in 148 parts alcohol of 0·939, in 12 parts boiling alcohol of 0·939; not in ether. Its aqueous solution separates by adding dissolved emulsin into hydrocyanic acid, oil of bitter almonds ($C_{14} H_6 O_2$) and grape sugar (Dextrose).

**Amylum** = STARCH.

**Amyrin.** *See* ARBOLABREA RESIN.

**Anacahuit Tannic Acid** = $C_{16} H_{12} O_{10}$. In the Anacahuit wood (from Cordia Boissieri). Precipitate the aqueous infusion with acetate of lead, treat the precipitate with acetic acid, filter, precipitate the filtrate with ammonia, wash the precipitate and decompose under water by sulphuret of hydrogen, filter and evaporate. The solution is of a faint astringent taste, precipitates chloride of iron with black-green and glue with brown colour.

**Anacardic Acid** = $C_{44} H_{32} O_7$. Obtained in combination with lead in the preparation of Cardol. Wash with alcohol, decompose with hydrosulphide of ammonia, filter off from the sulphide of lead and decompose the filtrate by sulphuric acid. White, crystalline substance without smell in low temperature, in higher of a peculiar odour and of faintly aromatic, afterwards burning taste. It remains unaffected by heat at 150°, decomposes at 200°, leaves greasy spots on paper; its solution in alcohol has a decidedly acid reaction; dissolves readily in alcohol and in ether, with faint blood-red colour in concentrated sulphuric acid. Yields with bases partly crystalline, partly amorphous salts.

**Anchusin** = $C_{36} H_{20} O_8$ (Alkanna-red). The red dyeing matter of Anchusa tinctoria. Remove foreign colouring matters by extracting with cold water, dry, exhaust with alcohol, distil the latter off from the tincture (after adding a few drops of hydrochloric acid, in order to prevent the red matter from turning green), evaporate on the water-bath to a thickish mass, shake with ether, which assumes a dark-red colour, and with water; remove the aqueous stratum and repeat the operation with fresh water until the ethereous stratum has so far diminished as to become of a thick fluidity, and evaporate the ether. Dark red-brown, resinous, brittle mass of neutral reaction, fuses below 60°, volatilises by careful heating in violet-red vapours similar to those of iodine and condenses in voluminous flocks under partial decomposition. Not soluble in water, but readily in alcohol and better still in ether and oils with red, in concentrated sulphuric acid with amethyst-colour, in alkalies with blue colour, and precipitable in the latter solution by acids in brown-red flocks. The alcoholic solution decomposes and becomes green with ammonia.

**Andirin.** Bitter substance of the wood of Andira anthelmintica, yellow-brown, soluble in water, alcohol and ether; only known in the impure state.

**Anemonic Acid** = $C_{30} H_{14} O_{14}$. *See* ANEMONIN. White, amorphous powder, without taste, of acid reaction, insoluble in water, alcohol, ether, oils and diluted acids, combines with alkalies in yellow colour.

**Anemonin** = $C_{30} H_{12} O_{12}$. In the herb of Anemone nemorosa, pratensis, Pulsatilla, Ranunculus bulbosus, Flammula, sceleratus*. The aqueous distillate of the above herbs, when kept for some time in contact with the volatile oil, obtained by the same operation, separates crystals of Anemonin and a white pulverulent substance (Anemonic acid). Of the two substances only Anemonin dissolves in alcohol, therefore it can be easily separated. It appears in colourless, glossy prisms of the klinorhombic form, without odour, of highly poignant and burning taste when fused, of neutral reaction; softens at 150° and decomposes afterwards (formerly believed to be volatile), dissolves very little in cold water, dissolves in hot water and separates in crystals on cooling; dissolves little in cold, readily in boiling alcohol; not in cold, little in boiling ether; in chloroform; in fixed and volatile oils ; in aqueous alkalies under decomposition.

**Anethol** = $C_{20} H_{12} O_2$. Constitutes almost entirely the volatile oils of Foeniculum officinale, of Artemisia Dracunculus, of Pimpinella Anisum and of Illicium anisatum. It occurs in the solid and in the liquid state. The former is obtained by pressing the oil of the first, third or fourth of the above plants at 0° and by re-crystallising; the liquid is obtained from the oils of the first or second herb by distilling, collecting the distillate of

* And numerous other plants of the order.—F. v. M.

206–225° and rectifying until of a constant boiling point. The solid form appears in white laminæ, smells more faintly and pleasantly than anis-oil, fuses at 16°, boils at 220°, has at 12° a density of 1·044, at 25° a density of 0·984. The liquid form from fennel-oil does not congele at − 10° and boils at 225°; the liquid from Tarragon-oil boils at 206°.

**Angelic Acid** = $C_{10} H_7 O_3$ + HO. Ingredient of the root of Angelica Archangelica, of the Sumbul root (from Euryangium Sumbul) and also of the essential oil of the flowers of Anthemis nobilis, the less volatile part of which in boiling with alcoholic solution of caustic potash secedes into angelate and valerate of potassa. It volatilises with the steam by distilling the roots of Angelica with water, but may be obtained more completely by boiling the roots with milk of lime, percolating, concentrating the liquid and distilling with sulphuric acid. The distillate has to be saturated with carbonate of soda; is then evaporated, again distilled with sulphuric acid and kept in the cold for some days. Collect the crystals, wash with cold water and re-crystallise. It forms translucent, colourless prisms and needles of peculiarly aromatic smell, and very acid, burning and aromatic taste; fuses at 45°, and boils at 190°; dissolves slowly in cold, most readily in hot water, alcohol and ether. Its salts are mostly soluble in water; the lead, silver and copper salts slowly; the oxyd of iron salt is insoluble.

**Angelicin.** Crystalline resin of the root of Angelica Archangelica. The alcoholic tincture of the above root separates in evaporating into two liquids of different density, the denser one being aqueous of light-yellow colour, and containing much sugar; the lighter supernatant one brown and resinous. The latter has, after washing with water, to be saponified by caustic potash; this is dissolved in alcohol, subjected to carbonic acid, evaporated and treated with ether, which dissolves Angelicin and leaves it pure after evaporating. Fine, colourless needles, without smell; of at first imperceptible, afterwards burning and aromatic taste, easily fusible, not volatile, soluble in alcohol and in ether. [According to the latest researches of B. Brimmer, Angelicin has been found to be identical with Hydrocarotin.]

**Angusturin.** In the genuine Angustura bark from Galipea officinalis and G. Cusparia. Obtained by extracting with alcohol and evaporating. Fine, white crystals of a bitter and faintly acrid taste, little soluble in water, more in alcohol and in acids, not in ether and volatile oils; is precipitable by tannic acid.

**Anime.** Exudation of the stem of Bursera gummifera and Trachylobium Hornemanni. Yellow, transparent, of pleasant smell, especially on warming, and of mastic-like taste; softens in the mouth. Contains volatile oil and two crystalline resins, the one ot which dissolves readily, the other slowly in alcohol.

**Annatto Red** = $C_{16} H_{13} O_2$. The resinous dyeing matter of Annatto (the pulp of the fruit of Bixa orellana). Extract with water, and remove the aqueous solution containing yellow dyeing matter and impurities, dry the residue and extract with alcohol, evaporate the tincture, treat with ether and bring the solution to dryness. Red, amorphous, soluble in alcohol, ether and ley of potash; turns blue with concentrated sulphuric acid.

**Anthocyan.** Exhaust with alcohol, evaporate and treat with water; precipitate the blue solution with acetate of lead, decompose the green precipitate by sulphuret of hydrogen, filter and evaporate; extract with absolute alcohol and precipitate the solution by ether, which throws down the Anthocyan in flocks. Of amorphous form, soluble in water and in alcohol, turns red with acids, green with alkalies, yields with alkaline earths and oxyd of lead green compounds insoluble in water.

**Anthoxanthein.** Extract with alcohol, evaporate, exhaust with water, evaporate again, treat with absolute alcohol, dilute the solution with water, precipitate with acetate of lead and decompose the deposit with sulphuric acid; the Anthoxanthein remains dissolved, and is obtained by evaporating as an amorphous mass, soluble in water, alcohol and ether; becomes brown by alkalies, and reassumes a pale colour with acids.

**Anthoxanthin.** Extract with hot alcohol, filter while hot, and leave to stand in the cold; the A. subsides but mixed with fat, removable by heating with a little alkali, decomposing by acids and extracting with cold alcohol, which dissolves the fat acid. Amorphous, resinous substance of a beautiful yellow colour, insoluble in water, dissolves with gold-colour in alcohol, ether and oils, little in alkalies.

**Antiarin** = $C_{28} H_{20} O_{10} + 4\ HO$. In the sap of the Upas-tree (Antiaris toxicaria) which forms an ingredient of the Javanese arrow-poison. The above sap, mixed with alcohol to prevent decomposition, has to be concentrated and exhausted with boiling alcohol, the filtrate is evaporated to honey-consistence, and boiled with water. The A. crystallises in the hot solution, and is purified by rinsing and re-crystallising. Beautiful, silvery leaflets, similar to the malate of lime, losing the water at $112°$, fusing at $220°$, and decomposing afterwards, of neutral reaction, without smell; dissolves in 254 parts cold and in 27 parts boiling water, in 70 parts alcohol and in 2792 parts ether, more readily in diluted acids and alkalies than in water; not precipitable by tannic acid.

**Antirrhinic Acid** = DIGITALIC ACID, VOLATILE.

**Apiin** $= C_{24} H_{14} O_{13}$. In the leaves of Apium graveolens and Carum Petroselinum. Boil the green herb (gathered before the floral season) three times with water, percolate, wash the dark-green jelly obtained on cooling with cold water, dry, treat several times with boiling alcohol, mix the tinctures with water, distil the alcohol, percolate, press the remaining thickish mass, edulcorate with alcohol and boiling ether, and dry. Delicate, white powder, without taste or smell, fuses at 180° without loss of weight, decomposes in higher temperatures, dissolves readily in boiling water and congeals to a jelly on cooling or when mixed with cold water (even one part Apiin in 1500 parts water yields on cooling a thin jelly); dissolves in 390 parts cold alcohol, more readily in boiling alcohol, not in ether. The solution in boiling water assumes, even when highly diluted, a deep blood-red colour with sub-sulphate of iron. Yields sugar when boiled with diluted acids; dissolves readily in caustic alkalies and their carbonates with yellowish colour; precipitable by acids as a jelly.

**Aporetin, Erythroretin** and **Phæoretin.** Brown or black resins, obtained in analysing the root of rhubarb; they appear to be products of decomposition.

**Apyrin.** Alleged alkaloïd of the seeds of Attalea funifera. Precipitable by oversaturating with ammonia the extract prepared with diluted hydrochloric acid. White powder without smell or taste, little soluble in water.

**Arachidic Acid** $= C_{40} H_{39} O_3 + HO$. In the fat-oil of Arachis hypogæa. Saponify the above oil by a solution of caustic soda, decompose the soap with hydrochloric acid, macerate the fat acids with alcohol, percolate, press the remnant and dissolve it in boiling alcohol, collect the laminæ formed on cooling and recrystallise until they fuse at 75°. Minute, glossy laminæ, assuming a porcelain appearance by keeping, fusible at 75°; not soluble in water, scarcely in cold, readily in hot alcohol, very easily in ether. Its salts are similar to stearates and palmitates, and dissolve as a rule sparingly.

**Arbolabrea Resin.** Presumptively from Canarium commune. Soft, grey-green, of a strong smell similar to turpentine, cubebs and fennel, behaves similar to elemi. Contains a light green-yellow volatile oil, a readily soluble and a sparingly soluble crystalline resin. By treating successively with alcohol of different strengths four different crystalline resins have been extracted, named Amyrin, Brein, Breidin and Bryoidin.

**Arbutin** $= C_{24} H_{16} O_{14} + HO$. In the leaves of Arctostaphylos Uva ursi. Precipitate the decoction with subacetate of lead and evaporate the filtrate, freed from lead by sulphuret of hydrogen, to the point of crystallisation; the crystals have to be purified by recrystallising with animal charcoal. Long, colourless tufts of needles of silky lustre, losing the water at 100°, of bitter taste, fusing at 170°; slowly soluble in cold, most readily in boiling water, sparingly in alcohol, scarcely in ether, not precipitable by metallic salts; do not reduce the salts of copper; separate when boiled with diluted sulphuric acid into grape-sugar and hydro-kinon ($=C_{12} H_6 O_4$, Kawalier's Arctuvin); become transformed into kinon and formic acid by heating with superoxyd of manganese and sulphuric acid.

**Arctuvin.** *See* ARBUTIN.

**Aribin**$=C_{46} H_{20} N_4 + 16$ HO. In the bark of Pinckneya pubens. Extract with water and sulphuric acid, concentrate, remove the gypsum, neutralise almost completely with carbonate of soda, precipitate with acetate of lead, filter, treat with sulphuret of hydrogen, filter, precipitate with carbonate of soda and shake repeatedly with ether. Add hydrochloric acid to the ethereous solution, collect the chloride of A. precipitated thereby, purify by recrystallising, shake its aqueous solution with carbonate of soda and ether, and leave the ethereous solution to crystallise. Colourless, quadrangular, flat columns; effloresce when exposed to the atmosphere, turn white and opaque at 100° under loss of all the water; may also be obtained crystallised anhydrous in colourless rhombic pyramids and columns of great lustre; of alkaline reaction and of remarkably bitter taste. A. fuses at 229°, sublimates by careful heating below the fusing-point in very fine long needles (empyreumatic products appear only by quickly heating); dissolves in 7762 parts cold, more abundantly in hot water, readily in alcohol, less in ether, also in amyl-alcohol. Yields with acids easily crystallisable salts, precipitable by caustic alkalies and their carbonates.

**Aricin**$=C_{46} H_{26} N_2 O_8$ (isomeric with Brucin); according to Pelletier: $C_{20} H_{12} NO_3$. In the Quina de Cusco (from Cinchona pubescens). Extract with acid water, treat the liquid with milk of lime, wash and dry the precipitate and treat with alcohol, filter hot, and purify the A. formed after cooling by recrystallising in alcohol under aid of animal charcoal. Rigid needles, without taste at first, afterwards of aromatic and acrid, and, when dissolved in acids, of very bitter taste; of alkaline reaction, unalterable at 150°; fuses at 188° without loss; decomposes in higher temperatures; dissolves sparingly in water, more readily in alcohol than cinchonin, less than quinin, also in ether, in nitric acid with green colour. Its salts are easily crystallisable and precipitable by caustic alkalies and their carbonates; the precipitates dissolve a little in ammonia.

**Arnicin**$=C_{40} H_{30} O_8$. The bitter ingredient of Arnica montana, obtained from all parts of the plant. From the root: Boil with water, press, exhaust the remnant with alcohol, digest the tincture with oxyd of lead, remove the dissolved lead from the tincture by means of sulphuret of hydrogen, distil off the alcohol, bring the remnant to dryness and extract the A. by ether. The ethereous solution, when mixed and agitated with a solution of caustic potash, delivers up resin, fat and dyeing matter. Separate from the ley, treat with animal charcoal and evaporate to dryness. Dissolve the remnant in weak alcohol and evaporate the filtrate or precipitate with water. From the flowers: Exhaust with ether, distil off the solvent, extract from the remnant the A. by alcohol of 0·850, and purify the solution by animal charcoal. Gold-coloured amorphous mass of bitter taste; dissolves little in water, readily in alcohol, ether and alkalies; yields no sugar on treating with diluted acids.

**Arnotta Red**=ANNATTO RED.

**Arthanitin**=CYCLAMIN.

**Asafœtida.** Gum-resinous exudation of Ferula Asafœtida. Conglutinated grains of white colour, turned rose-red, violet and brown by the atmosphere; friable in the cold; of nauseous garlic odour and of acrid, bitter taste. Contains about 50% resin, 20 gum and $4\frac{1}{2}$ volatile oil. The resin is partly soluble in alcohol and ether, and partly insoluble in the latter.

**Asarabacca Camphor** = ASARON.

**Asaron** = $C_{40} H_{26} O_{10}$. In the root of Asarum Europæum,* seems to be identical with Asarit. Distil the dried root with eight parts water until three parts have distilled over. The A. will be found partly in the neck of the retort and at the bottom of the distillate in little white grains, partly crystallising in the distillate when left to stand cold. Pellucid, quadrangular, tabular crystals of pearly lustre and of 0·95 density, without smell or taste; fuse at 40°, sublimate in small quantities mostly undecomposed in fumes of strong odour that provoke coughing, dissolve little in hot water, readily in alcohol, ether and volatile oils, begin to boil at 280°, but become decomposed while the temperature rises to 300°, without distilling in the least.

**Asclepiadin.** Emetic substance of the root of Vincetoxicum officinale, obtained by extracting with strong alcohol. Pale yellow, bitter, amorphous, hygroscopic matter, non-nitrogenised, readily soluble in water, alcohol and ether-alcohol; without alkaline properties.

**Asclepion** = $C_{40} H_{34} O_6$. In the milky juice of Asclepias Syriaca. Warm the juice, treat the coagulated mass with ether, evaporate the extract and purify by recrystallising in absolute ether. White, cauliflowerlike mass or tufts of needles when

* And doubtless in other species.—F. v. M.

slowly evaporated; without smell or taste; fuses at 104°, decomposes in higher temperatures, not soluble in water and in alcohol, readily in ether, less in acetic acid and volatile oils.

**Asparagin** $= C_8\ H_8\ N_2\ O_6 + 2\ HO$. Widely diffused in germs and young shoots, as yet specially in Liliaceæ, Boragineæ, Malvaceæ (plants rich in mucilaginous sap), Gramineæ and Leguminosæ, also occurring in beet-roots, potato-sprigs and hop-shoots. The best material for its preparation is the juice or the aqueous extract of the respective vegetable substances, but as A. can not be obtained by precipitation and as its quantity is generally inconsiderable, the liquid has to be concentrated and kept in the cold for several days. The Asparagin separates in small crystals which have to be purified by recrystallisation. It forms colourless, hard, recto-rhombic prisms without odour and of insipid cooling taste, soluble in 40 parts cold and four parts boiling water, also in weak alcohol, not in absolute alcohol and ether, loses the water of crystallisation at 100° and fuses in higher temperatures, while swelling considerably and emitting ammonia and a faint horny odour.

**Aspartic Acid** $= C_8\ H_6\ NO_7 + HO$. Lermer observed the occurrence of asparagin in the decoction of the germs of barley-malt, when evaporated to syrup consistence; but when the above syrup had been kept for some time, no asparagin could be observed, but in its stead aspartic acid combined with magnesia. Since the asparagin ($= C_8\ H_8\ N_2\ O_6$) behaves like the amide of aspartic acid, and when boiled by itself in aqueous solution, but more readily under co-operation of acids or bases, is converted into aspartate of ammonia ($C_8\ H_3\ N_2\ O_6 + 2\ HO = NH_4\ O + C_8\ H_6\ NO_7$) from which the ammonia is instantly expelled by stronger bases; it is easily explained how the asparagin disappears by-and-bye in the syrup of malt and an aspartate takes its place. Though it follows herefrom that originally no aspartic acid but only asparagin was contained in the malt, I thought it advisable not to pass by the above acid, as it might occur in phytochemical analyses, no matter if pre-existing or originated in the course of the analytic process.*

The aspartic acid would have to be looked for in the precipitates occasioned by neutral or basic acetates of lead. After decomposing the acid precipitates by sulphuret of hydrogen and concentrating the liquid, it would separate, as it is little soluble in water. It forms a white, shining crystalline powder without smell and of acidulous afterwards broth-like taste, becomes decomposed by heat, while evolving ammonia and a horny odour, and swelling considerably; dissolves in 128 parts cold, readily in hot water, still

* According to Scheibler, the asparagin of beet-roots reappears as aspartic acid in the molasses obtained in the manufacture of beet-root sugar.—F. v. M.

less in weak alcohol, insoluble in strong alcohol. Most of its salts dissolve in water.

**Asperula Tannic Acid**$=C_{14} H_8 O_8$. In the herb of Asperula odorata and Galium Mollugo\*. Precipitate the aqueous extract with acetate of lead, treat the precipitate with acetic acid, filter, precipitate the filtrate with ammonia, wash the precipitate, decompose under water with sulphuret of hydrogen, filter and evaporate. Faintly brownish, amorphous mass of acidulous, acerb taste, dissolves in water and alcohol, slowly in ether, imparts a dark-green tinge to chloride of iron, does not precipitate glue and tartarated antimony.

**Athamantin**$=C_{24} H_{15} O_7$. In Peucedanum Oreoselinum. Desiccate the alcoholic extract of the roots or seeds, treat with ether, decolourise the ethereous tincture by animal charcoal and leave to evaporate spontaneously. Dissolve the crystalline remnant in alcohol, dilute with much water and recrystallise the precipitate, obtained after some time, in alcohol. Colourless, long needles, when heated of a peculiarly rancid, saponaceous odour, of at first somewhat bitter and rancid, afterwards faintly harsh taste, fusible at 79°, decomposes in higher temperatures while yielding valerianic acid; insoluble in water, easily soluble in weak alcohol and ether, more copiously in volatile and fixed oils and without decomposition in sulphuric acid; separates by diluted sulphuric acid and by alkalies into valerianic acid and oreoselon. (*See* PEUCEDANIN).

**Atherospermin** $= C_{30} H_{20} NO_5$. Alkaloïd of the bark of Atherosperma moschatum. Extract with warm water, acidified by sulphuric acid, press and precipitate with carbonate of soda. Wash and dry the precipitate and extract with sulphide of carbon. Distil with water containing sulphuric acid, precipitate the remaining liquid with ammonia, wash and dry the deposit. A white, voluminous, highly electric powder of crystalline appearance under the microscope, and of a pure and lasting bitter taste. It assumes a yellowish colour in the direct sunlight; fuses at 128°, and decomposes in higher temperatures, while emitting a smell of putrid meat, and afterwards a faint odour of herrings. Water dissolves only traces of it, but acquires a bitter taste; ether dissolves at 16° $1/1000$, when boiling $1/100$; alcohol of 93% at 16° $1/32$, at the boiling point half its weight. The cold alcoholic solution shows a decidedly alkaline reaction. Of greater solving power are chloroform, sulphide of carbon, oil of turpentine and other essential oils and diluted acids. Chlorine-water effects a yellow solution, not changeable by ammonia. Iodic acid shows towards A. the same reaction as towards morphin and oxyacanthin, viz.,

---

\* And many congeneric plants.— F. v. M.

it becomes deoxygenised, and iodine is set free. The neutral solution of the chloride of A. gives a white precipitate with caustic alkalies and the carbonates thereof, and with the iodide, ferrocyanide and sulphocyanide of potassium, and with chloride of mercury; a yellowish white with bi-iodide of potassium; a lemon-yellow with picric acid; a sulphur-colour with ferrocyanide of potassium; a dirty-yellow with phospho-molybdic acid; an ochre yellow with chloride of gold; a pale greenish-yellow with chloride of platinum, and a yellow or orange precipitate with nitrate of palladium.

**Atherosperma Tannic Acid** $= C_{20} H_{14} O_4$. In the bark of Atherosperma moschatum. Precipitate the decoction of the bark with acetate of lead, treat the precipitate with acetic acid, precipitate the filtrate by ammonia, decompose the precipitate under water by sulphuret of hydrogen, and evaporate the filtrate. Yellow liquid of faintly acid and astringent taste; greens 'the salts of oxyd of iron.

**Atropin** $= C_{34} H_{23} NO_6$. In all parts of Atropa Belladonna, Datura Stramonium, D. arborea, and very likely also in the other species of this genus. Bruise the whole plant just when it begins to blossom, under addition of a little water, press, boil the liquid, strain, evaporate to syrup consistence, add soda-ley in excess, shake, add twice its volume of alcohol of $90\%$, agitate repeatedly for two days, leave to stand, decant the spirituous liquid, acidify by sulphuric acid, distil off the alcohol, render the remnant alkaline by soda-ley, shake with ether, decant the ethereous liquid, distil off the ether, dissolve the remnant in alcohol, filter and leave to evaporate slowly. When coloured still, it has to be redissolved in alcohol and treated with animal charcoal. Fine white needles without odour (when moist and imperfectly purified of a nauseous, somewhat tobacco-like smell), of nauseous and lasting bitter taste; fuses at $92°$ without loss of weight, decomposes in higher temperatures for the greater part under emitting vapours of alkaline reaction, while a small part sublimates unchanged; dissolves in 300 parts of cold and in 50 parts boiling water, in 8 parts cold and in equal parts boiling alcohol, in 60 parts cold and in 40 parts boiling ether; the alcoholic solution shows a decidedly alkaline reaction. Caustic alkalies and the carbonates dissolve it also, but decompose it on heating. Dissolves easily in chloroform, oils, glycerin and diluted acids. Concentrated nitric acid effects a pale-yellow solution, which becomes of an orange-yellow colour on heating. Concentrated sulphuric acid dissolves it without colour, but becomes brown on heating while emitting an odour of orange and sloe flowers.

**Avenin.** Peculiar protein substance of oats (Avena sativa). To prepare it, grind the grains with water, dilute the pasty mass with water, strain after twelve hours, filter the liquid, precipitate with acetic acid, dissolve the precipitate in diluted liquor of ammonia, precipitate again with acetic acid and purify the precipitate by means of alcohol and ether. The Avenin is greyish-white, dissolves readily in water, does not coagulate by heat, dissolves also in an excess of acetic and hydrochloric acids.

**Azulen** $= C_{16} H_{12} + HO$. Ingredient of volatile oils and causes the blue and the brown or yellow-green colour of them, in the latter cases when mixed with a yellow resin. It distils with difficulty and can be obtained in the pure state by repeated fractional distillations and rectification. It boils constantly at $302°$, has a density of $0·910$; its vapour is also of blue colour. In the blue oil of chamomil there is scarcely 1 per cent. of Azulen; the patchouly oil (from Pogostemon Patchouly) with 6 per cent. and the wormwood oil with 3 per cent. Azulen are not of a blue colour, because they contain a comparatively large quantity of yellow resin.

**Balsams.** Natural combinations of resins with volatile oils, viscid or fluid at ordinary temperatures, becoming thicker and and often solid by age.

**Balsam of Copaiva** $=$ Copaiva Balsam.

**Balsam of Mace** $=$ Mace Balsam.

**Balsam of Mecca** $=$ Mecca Balsam.

**Balsam of Nutmeg** $=$ Nutmeg Balsam.

**Balsam of Peru** $=$ Peru Balsam.

[**Balsam of Sindor.** Three varieties have been examined, two of which are the exudations of the stem of Sindora species, the third probably originating from a kind of Dipterocarpus. The latter balsam is light-brown, thickish, of the odour of Copaiva balsam, and of $0·9221$ density at $27·5°$. By distillation with water it yields a beautiful red-brown, translucid resin, and a volatile oil of light-yellow colour, thin fluidity and $0·914$ spec. gr., boiling at $246°$ to $255°$, soluble in 4 to 5 parts cold, and 1 to $1\frac{1}{2}$ parts warm alcohol, of acid properties. The two other kinds of balsam are distinguished from the first by not yielding up their volatile oil to the vapours of water. The volatile oils, obtained by heating the balsams to about $255°$, are of a yellowish or greenish-yellow colour, $0·904$–$907$ spec. gr. and soluble in alcohol. The resin was in one case brittle, yellow-brown, soluble in alcohol, chloroform, ether and oil of turpentine, producing with alcohol a brilliant varnish on glass; in the other case the resin was transparent, brittle, dark-brown, and soluble in hot alcohol only with difficulty, but easily soluble in ether and chloroform.]

## Bases, Organic = Alkaloids.

**Basilicum Stearopten** $= C_{20} H_{22} O_6$. Obtained by the distillation of Ocimum Basilicum with water. The oil floating on the water solidifies almost entirely to a white, crystalline mass. When recrystallised in alcohol it appears in quadrangular prisms of a faint odour of the oil; when recrystallised in water, in tetrahedrons almost devoid of taste; it is neutral, dissolves little in cold, readily in hot water or alcohol, in six parts of ether, also in acids and alkalies. Isomeric or identical with the hydrate of oil of turpentine.

**Bassia Fat**, from the seeds of Bassia butyracea, B. longifolia and B. latifolia. Yellowish, slowly decolourised by light, of the consistence of butter and of 0·958 spec. grav., fuses at 27° to 29°, dissolves little in alcohol, readily in ether; contains olein, myristin, palmitin and stearin. (The last-mentioned was erroneously distinguished as Bassic acid).

**Bassorin** $= C_{12} H_{10} O_{10}$. Ingredient of Bassora-gum, Tragacanth and similar gummous exudations of plants (cherry-gum, anacardia gum), insoluble in water and swelling in it; can only likely be object of phytochemical analyses in exudations of the above and similar kinds. When such an exudation is treated with cold water, it swells up considerably and dissolves partially; by straining and repeatedly treating with fresh water, the soluble part is removed, but the remaining portion contains, like the vegetable mucus, always more or less lime-compounds which can only be removed by repeatedly treating with water containing hydrochloric acid. When dry, the Bassorin is yellowish-white, solid, brittle, transparent, without taste, swells in cold water to a transparent jelly, without dissolving, but dissolves by continued boiling to a gummous liquid, yields with diluted sulphuric acid gum and sugar, with nitric acid, mucic and a little oxalic acids.

**Bay Oil**, obtained by distilling the berries of Laurus nobilis with water. Greenish-yellow, of a thickish consistence, of the odour of bay-berries and turpentine, of faintly acid reaction, and 0·932 density. It consists of two polymeric hydrocarbons, $C_{20} H_{16}$, boiling at 164° and of 0·908 density, and $C_{30} H_{24}$, boiling at 250° and of 0·925 density, and of lauric acid $= C_{24} H_{24} O_4$.

**Bay Oil from Guiana.** Obtained by incisions, from the stem of an unknown tree. When rectified and desiccated, colourless, of the smell of oil of turpentine and lemons, of aromatic pungent taste, of 0·864 density, boils at 150° to 163°.

**Bdellium.** Exudation of the stem of Balsamodendron Africanum and B. Roxburghii. Red-brown, more or less transparent, viscous or hard, of a myrrhlike odour, and of bitter taste. Contains about 60% resin, fusing at 55° to 60°, 10% gum, 30% bassorine, and a volatile oil.

**Bebiric Acid.** In the fruit of the Bebir-tree (Nectandra Rodiei). Concentrate the cold aqueous extract of the fruit, filter when cold, precipitate by ammonia bebirin and siperin and mix the filtrate with nitrate of baryta. The impure precipitate has to be washed with cold water, is dissolved in boiling water, and left to crystallise. The crystals, purified by recrystallisation, are dissolved in boiling water, and precipitated by acetate of lead; the precipitate is washed and decomposed by sulphuret of hydrogen, and the filtrate is evaporated over sulphuric acid. At last it has to be purified by dissolving in ether and evaporating in vacuo. Deliquescent, white, crystalline mass of wax-lustre, fuses at 150°, and sublimates somewhat above 200° undecomposed in tufts of needles. Its combinations with potash and soda are deliquescent, and soluble in alcohol; those with baryta, lime and magnesia dissolve very little in water; the lead compound is little soluble even in boiling water.

**Bebirin** $= C_{38} H_{21} NO_6$. In the bark and fruit of the Bebir-tree (Nectandra Rodiei), besides perhaps a second alkaloïd (Sipirin) and a peculiar acid; also in the bark and leaves of Buxus sempervirens. Exhaust with boiling water containing sulphuric acid, concentrate, leave to cool, separate from the deposit containing tannic acid and sulphate of lime and precipitate the filtrate with ammonia. The dark-green precipitate is washed, dried at the atmosphere (whereby it becomes black through tannic acid) and dissolved in diluted sulphuric acid. The solution is treated with animal charcoal and again precipitated by ammonia, which occasions a white precipitate. Dry, dissolve in alcohol, evaporate and treat the remnant with absolute ether, which dissolves the Bebirin and leaves behind the sipirin. Both substances have to be purified by treating their alcoholic solutions with animal charcoal. White, highly electric powder of strong and lasting bitter and faintly resinous taste, loses nothing of its weight up to 120°, fuses at 180°, decomposes in a higher temperature, dissolves in 6650 parts cold and in 1466 parts boiling water, in 5 parts absolute alcohol, also readily in weak alcohol, in 13 parts ether; of decidedly alkaline reaction, saturates acids completely and forms amorphous salts, separates iodine from iodic iodic acid. Its salts have a very bitter and somewhat astringent taste, are precipitated by caustic alkalies and the carbonates thereof; the precipitates redissolve in liquids of potash or of ammonia a little more copiously than in those of other precipitating agents.

**Belladonnin.** Second alkaloïd of Atropa Belladonna; it is the yellow resin-like substance that prevents the crystallisation of the atropin. Mode of isolation : Dissolve the crude atropin in water by means of an acid, neutralise with carbonate of soda in order to remove a fluorescent substance of bluish colour, filter and add to the filtrate small quantities of carbonate of soda as long as, according to the temperature and concentration of the liquid, a conglutinating resinous or oily substance is formed. The precipitation of a pulverulent body, occurring afterwards, has to be avoided. Collect the precipitate on a linen cloth, rinse with water, dissolve again in acid water, decolourise as much as possible with animal charcoal, filter and, in order to prevent contamination by atropin, precipitate as carefully as before with carbonate of soda, collect, dissolve in absolute ether and evaporate.' A colourless or in thicker layers yellowish, gum-like mass, drying with difficulty, of not very bitter, but burning-acrid taste, fuses by heat and decomposes afterwards under emission of heavy white fumes of the odour of burning hippuric acid; dissolves readily in ether and in alcohol, little in water; of strongly alkaline reaction; dissolves also readily in acids while saturating them completely, but is less basic than atropin. Its sulphate yields with ammonia a white pulverulent precipitate that becomes soon glutinous, a property in which it resembles hyoscyamin. Tannic acid also precipitates the sulphate of B. white. The solution of B. in weak alcohol is precipitated by nitrate of silver, chloride of gold and bi-iodide of potassium.

**Benic Acid**$=C_{44} H_{43} O_3 + HO$. In the oil of Behen, from Moringa oleifera. The fat acids separated by hydrochloric acid from the soap obtained by means of soda-ley are pressed and the remnant crystallised in alcohol. Shining, white needles, similar to stearic acid; fuse at 76°.

**Benzoic Acid**$=C_{14} H_5 O_3 + HO$. Contained in large quantities in the benzoin and other aromatic resins and balsams, mostly accompanied by cinnamic acid; in small quantities in different odoriferous seeds and roots; often confounded with cumarin. It is obtained best and without much loss by boiling the respective substance (finely contused when in a dry state) with milk of lime and water, filtering, evaporating the filtrate, oversaturating when cold with hydrochloric acid, collecting the crystalline deposit, pressing, dissolving in the least possible quantity of hot water, percolating, crystallising, collecting and drying. White laminæ and needles of mother-of-pearl lustre, mostly of faintly benzoic odour and of slightly acid taste, which fuse at 120°, boil at 239°, but begin to volatilise already at 145°, while emitting vapours which irritate the eyes considerably and provoke coughing. The acid dissolves in 200 parts cold and in 24 parts boiling water, in two parts of cold and in equal parts of boiling absolute alcohol, in 25 parts ether, also readily soluble in fixed and in volatile oils, and without change and readily in concentrated sulphuric acid. Almost all its salts dissolve in water and mostly readily so; the lead salt and some other metallic salts dissolve sparingly, the salt of oxyd of iron is insoluble.

**Benzoin.** Exudation of the stem of Styrax Benzoin and S. officinale. Yellowish or brownish, often undermixed with white almond-shaped masses, brittle, of peculiar pleasant smell and acrid, balsamic taste, fuses easily while emitting vapours of benzoic acid, yields to water only traces of benzoic acid, dissolves readily in alcohol and in acetic acid and partly in ether. It contains, besides 18 to 20 per cent. of benzoic acid and a little volatile oil, four resins, distinguishable by the different degrees of solubility. Some kinds contain also cinnamic acid.

**Berberin** $= C_{40} H_{17} NO_8 + 9HO$. Discovered (1824) in the bark of Geoffroya inermis and called Jamaicin; afterwards (1826) found in the bark of Xanthoxylum Clava Herculis and called Xanthopicrit; 1835 obtained from the bark of the root of Berberis vulgaris and called Berberin, but only lately recognised in its true nature. Occurs, according to recent observations, also in the following plants and appears to be very widely distributed: in the flowers of Berberis vulgaris, in an Indian and Mexican species of Berberis, in the bark of Xylopia polycarpa, in the root of Jateorrhiza Columbo and in the wood of Coscinium fenestratum; in Jeffersonia diphylla, Leontice thalictroides and Podophyllum peltatum; in Coptis Teeta, C. trifolia, Hydrastis Canadensis and Xanthorrhiza apiifolia.—Boil the bark of the root of Berberis with water, evaporate to honey-consistence, treat with boiling alcohol, add one-eighth water, distil off the alcohol and leave the remaining liquid to stand for some days in the cold. Strain the mass, hardened by fine yellow needles, press and recrystallise in hot water. This product being chloride of Berberin, has to be converted into the sulphate, its solution mixed with solution of baryta until alkaline, subjected to carbonic acid and evaporated. Draw out the B. with alcohol and precipitate the solution with ether or recrystallise in water. Fine, yellow needles of a pure and lasting bitter taste; loses the water at 100°, fuses at 120° to a red-brown resin and decomposes in higher temperatures, has a neutral reaction, dissolves slowly in cold water and alcohol, readily in both when hot, not in ether, readily in alkalies with brown colour; forms with acids crystallisable salts of mostly gold-yellow colour, neutral and bitter.

**Betulin**$=C_{50} H_{40} O_4$. In the epidermis of the bark of Betula alba. Exhaust the dry bark first with boiling water and, after it has been dried again, with boiling alcohol and filter while hot; when cold the B. separates and has to be recrystallised in ether. Voluminous white flocks or warty masses, without smell or taste; fuses at 200° with an odour of heated birch-bark; may be sublimated in a current of air; insoluble in water; dissolves in 120 parts cold and in 80 parts hot alcohol, also in ether, oils and alkalies and precipitable from the latter by acids; soluble in concentrated sulphuric acid and precipitable by water.

**Betuloretic Acid**$=C_{72} H_{66} O_{10}$. Covers as a white resin the young shoots and the upper surface of the young leaves of Betula alba, and is obtained by removing it mechanically. It has to be purified by dissolving in hot alcohol; evaporating; dissolving in ether, which leaves behind a black substance; evaporating; dissolving the residue by solution of carbonate of soda, and precipitating by an acid. White flakes or white friable mass softening in the mouth, fusing at 94°, in alcoholic solution of very bitter taste and acid reaction; dissolves also in ether and alkalies, and with beautiful red colour in concentrated sulphuric acid.

**Bicahyba Fat**, from Myristica Bicahyba, similar to Nutmeg balsam.

**Birch-Stearopten**$=$Betulin.

**Boheic Acid**$=C_{14} H_8 O_{10} + 2HO$. Found as yet only in small quantities in the black tea (from Thea Chinensis), besides much iron-bluing tannic acid. The aqueous extract of tea is precipitated by acetate of lead; filtered; the filtrate (containing acetate of lead) saturated by ammonia, the yellow precipitate collected, washed, mixed with absolute alcohol, pervaded with sulphuret of hydrogen; filtered, and the filtrate evaporated in a vacuum. Pale yellow, very hygroscopic substance, similar to gallo-tannic acid; fuses at 100°; dissolves in every amount in water and alcohol; the solution imparts a brown colour to chloride of iron without precipitating it.

**Boletic Acid**$=$Fumaric Acid.

**Borneen**$=C_{20} H_{16}$. Constitutes (contaminated with a little resin) the camphoric oil of Dryobalanops Camphora, and forms the non-oxygenised constituent of the oil of valeriana. The portion of the latter oil which, in rectifying, distils first, has to be distilled again with recently-melted caustic potash, whereby valerol remains as valerate of potash and borneol and Borneen distil over, from which all borneol may be separated by repeated fractional distillations, retaining only the first distilling portions. The Borneen is a colourless oil of turpentine-like odour, lighter than water; boils at 160°.

**Borneol** or **Solid Borneo Camphor**$=C_{20} H_{18} O_2$. In the excavations of the stem of Dryobalanops Camphora, also occurring in the crude oil of valerian. White, pellucid, easily friable, small crystals, lighter than water (according to other authorities, heavier than water), of the smell of camphor and pepper, and of burning taste, boils at 212°, and is converted by heating with nitric acid into ordinary camphor$=C_{20} H_{16} O_2$.

**Botany Bay Gum Resin.** Exudation of the stem of various species of Xanthorrhœa. Red-yellow, often with a green-grey rind, brittle, of shining fracture, of pleasant balsamic odour and acerb aromatic taste, fuses easily and burns with the smell of storax. Contains in the main a resin, soluble in alcohol, ether, alkalies, alkaline earths and oils; some volatile oil, a little benzoic acid and bassorine. Yields by treating with nitric acid a large quantity of picric acid.

**Brasilin**$=C_{44} H_{20} O_{14}$. The dyeing substance of the Brasil-wood and the Sappan-wood (from Peltophorum Linnæi, Cæsalpinia Crista and C. Sappan). Cannot be obtained directly from these woods or only with difficulty; has been obtained from the crystalline deposit occurring in a cask filled with extract of sappan-wood, by dissolving in absolute alcohol and crystallising under exclusion of light and air. Amber-yellow or brownish rhombo-hedra or klino-rhombic short prisms, obtained in straw or gold coloured needles with 3 at. water from a weak alcoholic solution, which lose the water at 90° while turning brown, dissolve in water, alcohol and ether; the reddish aqueous solution assumes a deep carmine colour by traces of alkalies or of alkaline earths.

**Brassic Acid**=Erucic Acid.

**Breidin, Brein** and **Bryoidin.** *See* Arbolabrea Resin.

**Brindonia Tallow,** from the seeds of Garcinia Indica. Almost white, fuses at 44°, contains olein and stearin.

**Brucin**$=C_{46} H_{26} N_2 O_8 + 8HO$. Distribution and preparation—*See* under Strychnin. The mother-ley obtained after the first crystallisation of the strychnin, containing the whole of the brucin and only a little strychnin, has to be mixed with as much bi-oxalate of potash as to constitute $1/50$ of the weight of the seeds of Strychnos Nux vomica employed, and is then evaporated to dryness. Grind the dry mass, treat with absolute alcohol for two days, if possible at 0°; filter; wash out with absolute alcohol; dissolve the remainder in water; remove from the solution the last traces of alcohol; digest with hydrate of magnesia for some days; filter; extract the remnant without previous drying with alcohol of 90%, and leave to evaporate. The Brucin crystallises by slow evaporation in colourless quadrangular prisms, often with a yellowish tinge and efflorescent at the air, fusible by heat under loss of water and decomposed in higher temperatures. It is of an intensely bitter taste, dissolves in 800 parts water, most readily in alcohol, not in ether, easily in chloroform, in 70 parts of glycerin, very little in volatile, still less in fixed oils, without colour in concentrated sulphuric acid; with rose-red colour in chlorine-water, and changing to a dirty-yellow by ammonia without turbidity. Nitric acid yields a vividly red solution, which becomes yellow on warming, and is coloured purple-violet and precipitated by subchloride of tin.

**Bryonin** $= C_{96}$ $H_{80}$ $O_{38}$. The bitter substance of the root of Bryonia alba. Treat the alcoholic extract of the root with cold water, precipitate the solution by acetate of lead, filter, remove from the filtrate the lead by sulphuret of hydrogen, neutralise with carbonate of soda, precipitate with tannic acid, dissolve the precipitate in alcohol, and mix intimately with hydrate of lime, digest, filter, decolourise with animal charcoal and evaporate. Colourless, very bitter substance, friable to a white powder, dissolves readily in water and alcohol, not in ether; yields with diluted acids sugar and other products.

**Buphthalmum Stearopten.** Obtained by distilling the flowers of Buphthalmum salicifolium (of a pleasant roselike odour) with water and cooling the distillate to 0°, at which temperature the stearopten separates. Yellow, silky, pointed crystals, melting by the warmth of the hand to a yellowish oil of a faint but pleasant odour, of faintly acid reaction, readily soluble in alcohol.

**Butyric Acid** $= C_8$ $H_7$ $O_3$ + HO. As yet only found in the pulp of the fruits of a few trees—Ceratonia Siliqua, Sapindus Saponaria, Gingko biloba, Tamarindus Indica—but undoubtedly widely diffused throughout the vegetable kingdom. It is obtained by distilling the respective substances with water containing a little sulphuric or phosphoric acid. The distillate, in which the presence of butyric acid may by the odour of rancid butter be easily detected, has to be mixed with carbonate of baryta ; concentrate; filter when of neutral reaction, and bring to dryness. The remnant (butyrate of baryta), when dried at 100°, is anhydrous, and contains 50·77% acid. In order to isolate the acid, the salt is dissolved in three parts of cold water, mixed with one-third of its weight of concentrated sulphuric acid, well stirred, filtered off from the sulphate of baryta and rectified. In order to remove traces of water, the acid has to stand for some days over chloride of calcium, and is then decanted and rectified.

Colourless liquid of a penetrating smell of rancid butter and acetic acid and of strong and pungent sour, afterwards sweetish taste, has a density of 0·96 to 0·98, boils at 164°, is inflammable, mixes in every proportion with water, alcohol and ether. Its compounds are all soluble in water, and have, when moist, the odour of fresh butter. Butyric acid is not unfrequently contaminated by acetic acid, and may be recognised by the formation of acetic ether, when heated with sulphuric acid and alcohol, but the exact separation of the two acids from each other is very difficult and has as yet not been carried out to satisfaction.

**Buxin**=Bebirin.

**Cacao Fat.** Obtained by pressing the prepared beans of Theobroma Cacao. White or yellowish-white, as hard as mutton tallow, of 0·90 specific gravity, of a faint cacao-like odour and mild taste, fuses at 30°, contains olein, stearin and a little palmitin.

**Caffeic Acid** or **Coffea Tannic Acid**=$C_{14} H_8 O_7$. In the seeds and leaves of Coffea Arabica, in the root of Chioccoca racemosa, in the leaves of Ilex Paraguayensis. Exhaust the pulverised coffee-seeds with ether, boil the remaining powder with alcohol of 40%, mix the alcoholic filtrate with twice its volume of water, remove the precipitated fatty flocks, boil, add solution of acetate of lead and boil for a few moments, in order to make the precipitate less voluminous and more easy to collect. Wash the precipitate with diluted alcohol, decompose under water with sulphuret of hydrogen and evaporate the filtrate on the water bath. Brittle substance, friable to a yellowish powder, of faintly acid and somewhat astringent taste, dissolves readily in water and in alcohol of any strength, little in ether, imparts a green colour to chloride of iron. Its compounds with lime and baryta are yellow and turn green at the atmosphere (by the formation of a new acid called Viridic acid=$C_{14} H_6 O_7$).

**Caffein**=$C_{16} H_{10} N_4 O_4 + 2 HO$. In the fruit and leaves of Coffea Arabica, in the leaves of Thea Chinensis, of Ilex Paraguayensis, in the fruit of Paullinia sorbilis, of Lunanea Bichi. Its preparation from the coffee-seeds is done in the following way. Mix five parts seeds, ground as finely as possible, with one part hydrate of lime, digest with 25 parts alcohol of 80%, filter and mix with alcohol, distil off the alcohol and leave the remnant to become cold, remove the oil, evaporate and purify the crystals by recrystallising. It forms long, flexible, white, silky needles without odour and of faintly bitter taste, loses the water of crystallisation at 100°, fuses at 177°, sublimates at 384° undecomposed (according to more recent observations: fusing at 224° to 228° and sublimating already at 177°); dissolves in about 60 parts water of 20°, in 9·5 parts boiling water, in 21 parts alcohol of 0.825, in 545 parts ether, in 9 parts chloroform; the solutions have a neutral reaction. It is also dissolved by volatile

oils but not by fixed oils. Tannic acid precipitates the aqueous solution. When evaporated with hydrochloric acid and chlorate of potash (or with chlorine-water), it leaves a remnant which reddens the skin like alloxan and imparts to it a peculiar smell; the solution of the said residue, when mixed with alkalies and salts of sub-oxyd of iron, acquires an indigo-blue colour, and ammonia imparts to the residue a purple-red (murexid) colour which does not turn violet by alkalies (distinction from uric acid).
[J. Williams gives the following method for preparing Caffein. Mix finely-pulverised guarana with one-third hydrate of lime and moisten with water. After an hour or two exsiccate at a gentle heat, and exhaust with boiling benzol; filter and evaporate, but not to dryness. Treat with boiling water and digest on the water-bath to the expulsion of all traces of benzol. Filter through wet paper, and evaporate to a small bulk, from which the C. will separate after twenty-four hours pure and colourless.— According to H. M. Smith, Caffein is contained to the amount of $0.133\%$ in the leaves of Ilex Cassine L.—Thomson avers that in roasting coffee a great amount of Caffein is lost, which may be regained by adapting to the burner a tube of about three feet length wherein the vapours of C. are allowed to condense. One pound of coffee yields on an average 75 grains of Caffein. C. is insoluble in a concentrated solution of potassic carbonate. By treating an infusion with subacetate of lead, concentrating and adding carbonate of potash, the C. is precipitated and may be obtained pure by dissolving in alcohol and evaporating. By passing the gas, evolved from chlorate of potash and hydrochloric acid, into an aqueous solution of Caffein and evaporating to dryness on the water-bath, a blood-red residue is obtained. One part Caffein in 1000 parts of water may thus be detected.]

**Cailcedrin.** Bitter substance of the Cailcedra-bark (from Khaya Senegalensis). Extract with hot water, evaporate to a honey consistence, treat with alcohol of 33° Baumé, add to the solution sub-acetate of lead which precipitates a red dyeing matter, filter, remove from the filtrate the alcohol by distillation, dissolve again in a little alcohol, remove the lead by sulphuret of hydrogen, evaporate, shake the aqueous liquid with chloroform and bring the latter solution to dryness. A yellowish, gumresinous, brittle, very bitter substance, mollifies easily in hot water, dissolves readily in alcohol, also in ether, chloroform, little in water, of neutral reaction, yields a voluminous white precipitate with ether. Consists of 64·9 C, 7·6 H and 27·5 O.

**Calendulin.** A substance, similar to vegetable mucilage, in the flowers and leaves of Calendula officinalis. It is obtained by treating the alcoholic extract, after it has been freed from a green, waxlike substance, with water; a voluminous, slimy mass remains, which does not dissolve in cold and only sparingly in boiling water; it is transparent, yellow and brittle when dry, and swells up in water. It dissolves readily in alcohol, acetic acid and caustic alkalies, not in ether, oils and other diluted acids; is not precipitable by tannic acid.

**Californin.** Neutral, bitter substance of the so-called Quina California (from Cascarilla hexandra). Extract the bark with alcohol, evaporate the extract, dissolve in water, precipitate the solution by acetate of lead, filter, remove from the liquid the excess of lead by sulphuret of hydrogen, evaporate, treat with strong alcohol, agitate with charcoal, filter and add ether, which precipitates the Californin. When dry, it is gold-yellow, amorphous, dissolves abundantly in water and alcohol, not in ether; of salicin-like taste, does not become coloured with sulphuric acid; not precipitable by tannic acid, chloride of platinum, chloride of mercury, chloride of iron and acetate of lead.

**Calluna Tannic Acid** $= C_{14} H_7 O$ . In Calluna vulgaris. The alcoholic extract of the herb is distilled, the residue mixed with water, filtered, and precipitated by acetate of lead. The sediment has to be treated with diluted acetic acid, the filtered solution is precipitated boiling hot with sub-acetate of lead, and the deposit decomposed under water by sulphuret of hydrogen; the liquid is filtered and evaporated in a current of carbonic acid gas. Amber-yellow mass, dissolves in water and alcohol, greens the salts of oxyd of iron.

**Camphor** $= C_{20} H_{16} O_2$. Contained in all parts of Cinnamomum Camphora; obtained by distilling with water and sublimating the raw product. A tough, white, transparent or translucid substance of crystalline-granular structure of the octahedrous form, of a peculiar, penetrating smell and bitter, aromatic taste, of 0·988 to 0·998 specific gravity at ordinary temperature, at $0° = 1·000$, fuses at 175°, boils at 204° and sublimates undecomposed, dissolves in about 1000 parts water, most readily in alcohol, ether, wood-spirit, aceton and oils.

**Camphor Oil** (volatile) from Cinnamomum Camphora. Of sherry-colour, has a density $= 0·945$, deposits much camphor in the cold and when left to evaporate by itself. By repeated rectifications a distillate free from camphor is obtained $= C_{20} H_{16} O$, which is as clear as water, of great light-refracting power and very mobile, of the odour of camphor and oil of cajeput and of 0·91 specific gravity, and leaving a resin but no camphor, when allowed to evaporate spontaneously.

**Camphor Oil** from Borneo, obtained from Dryobalanops. See BORNEEN.

[**Carnauba Wax**, the coating of the leaves of Copernicia cerifera. It is yellow, harder than bees' wax, of 0·99907 specific gravity and 84° fusing point. Contains, according to N. Stony-Waskelyne, amongst other ingredients, Melissin.]

**Cane-Sugar** (common sugar) $= C_{12} H_{11} O_{11}$. Diffused throughout the vegetable kingdom, but especially in the stalks of Gramineæ, in succulent roots (beet-roots for instance), in the sap of the stem of several trees (maple, birch, lime, palm, walnut), in fruits (always accompanied by other sorts of sugar), in seeds and flowers. On a small scale it is obtained best by boiling the respective vegetable substance, reduced to a proper state, with alcohol of 90%, filtering and keeping cold for rather a long time, to form in crystals. It crystallises in large, pellucid klinorhombohedra, is the hardest kind of sugar, of a pure sweet taste, has a density of 1·589 to 1·630, fuses at 160° without loss of weight to a clear, pale-yellow liquid, turns brown above 180° under loss of weight, acquires a bitter taste, swells up, becomes darker and is finally reduced to coal; dissolves in one-third part cold, in any quantity of hot water, little in strong alcohol, not in ether. The aqueous solution is tolerably permanent in the cold, but is converted into grape-sugar and fruit-sugar by continued boiling, or more readily by heating with diluted (not oxydising) acids, but gives rise to coloured humus-like products when treated so for a longer time; it becomes black by concentrated sulphuric acid and yields with nitric acid in the heat much oxalic, but no mucic acid. When boiled with alkalies and alkaline earths, the cane-sugar does not become perceptibly brown; its aqueous solution, mixed with carbonate of soda and boiled with subnitrate of bismuth, does not colour the latter. Cane-sugar does not reduce the alkaline tartarate of copper when boiled with it; ferments with yeast only after being converted into grape-sugar; is not precipitated by acetate of lead unless ammonia be added.

*Quantitative estimation of Cane Sugar.*—Since its isolation in the pure state is always connected with loss, methods have to be applied which permit an estimation even in presence of other matters; these are commonly, fermentation or treatment with an alkaline solution of tartarate of copper.

1. By fermentation with yeast.—The quantity of sugar may be estimated either by the weight of the carbonic acid or of the alcohol formed by this method. One hundred parts cane-sugar were formerly supposed to yield after combining with 5·26 parts water: 51·44 parts carbonic acid and 53·82 parts alcohol; but according to Pasteur's direct estimations only 49·12 parts carbonic acid and 51·01 parts alcohol are formed, while the rest of the sugar is used up in the formation of glycerin and of succinic acid. This way of estimation can, of course, only be applied in

D

the absence of other kinds of sugar capable of fermentation, which is a rare case. When other fermentable sugars are present, they may be destroyed by heating with an alkali (hydrate of lime), afterwards the whole has to be acidified again and mixed with yeast.

2. By treating with an alkaline solution of tartarate of copper. —The mode of operation is similar to that indicated under Starch, i.e., convert the cane-sugar into grape-sugar by heating the respective substance for two hours with water containing $2\%$ sulphuric acid, saturate the acid liquid (cold) with soda-ley and determine the amount of the grape-sugar by means of the copper solution. One hundred parts grape-sugar, as found by the above method, correspond to 95 parts cane-sugar. If, besides the cane-sugar, other kinds of sugar be present which directly reduce the solution of copper, the latter have to be determined first; afterwards a new portion has to be treated with acid and the whole of the sugar estimated as above. The first quantity, when subtracted from the second, will represent the amount of cane sugar, converted into grape-sugar by the acid.

**Caoutchouc**$= C_{40} H_{32}$. It is probably contained in every milky juice of plants but has been obtained on a large scale chiefly from Hevea elastica, Castilloa elastica, C. Markhamiana, Hancornia speciosa, Urceola elastica, Vahea Madagascariensis, V. Comorensis, V. gummifera, V. Senegalensis, Landolphia Owariensis, L. Heudelotii, L. florida, Willoughbya edulis, W. Martabanica, Ficus macrophylla, Ficus elastica and some other species of these genera. The commercial india-rubber is purified by dissolving in chloroform and precipitating with alcohol. White and opaque as long as it contains water in its pores, but colourless and transparent after prolonged drying at the atmosphere, only in thicker layers of a yellowish colour; elastic; without taste and smell; fuses at $160°$ and decomposes in higher temperatures; does not dissolve in water, but swells up in hot water and becomes sticky; insoluble in acids and in alkalies; swells up in cold and more so in boiling alcohol without dissolving, dissolves only partly in sulphide of carbon and in anhydrous ether (to about two-thirds), readily and completely in chloroform.

**Capric Acid**$= C_{20} H_{19} O_3 + HO$. In the oil of the Cocoanut (*See* CAPROIC ACID). One hundred parts by weight of the Caprate of baryta, consisting of scales of fat-lustre, are decomposed by a mixture of $47.5$ parts concentrated sulphuric acid and $47.5$ parts water, and the separated fat-acid crystallised in alcohol.—White mass, consisting of fine needles, of a faint goat-like smell, similar to caproic acid, of an acid, burning and disagreeable taste; has a density of $0.910$, fuses at 30 to $46°$, boils above $100°$ and volatilises without decomposition, dissolves in 1000 parts cold and a little more copiously in hot water, most readily in alcohol and in ether. Its salts are fatty to the touch, and dissolve, with the exception of those of the alkalies, sparingly in water, the salt of baryta in 200 parts cold water and contains, when dried at 100°, 68.03% acid.

**Caproic Acid** $= C_{12} H_{11} O_3 + HO$. Discovered, besides other aromatic acids, in the oil of the fruit of Cocos nucifera, in the root of Arnica montana and in the flowers of Orchis hircina. It is obtained from the above oil by saponifying with soda-ley, distilling with sulphuric acid, neutralising the distillate, containing caproic, caprylic and capric acids, with baryta, and evaporating. The caprate of baryta crystallises first, afterwards the caprylate and last the Caproate. which are purified by recrystallising. In order to separate the Caproic acid from the salt of baryta, 100 parts of the latter have to be treated with a mixture of 29·63 parts concentrated sulphuric acid and 29·63 parts water. Decant the Caproic acid floating on the surface after 24 hours, treat the residue once more with a like mixture of acids, desiccate the whole of the decanted acids by chloride of calcium and rectify.— Limpid, thin oil of sweat-like odour and of pungently acid and afterwards more sweetish taste of nitrous ether than butyric acid ; of 0·930 spec. grav.; not solidifying at − 9°, boils at about 200°, dissolves in 96 parts water, in every proportion in alcohol and ether. Its salts are all soluble in water, least so those of the heavy metals. The Caproate of baryta dried at 100° is anhydrous and contains 58·28% acid.

**Caprylic Acid** $= C_{16} H_{15} O_3 + HO$. Occurrence—*See* CAPROIC ACID. Decompose the Caprylate of baryta by diluted sulphuric acid; the Caprylic acid rises to the surface like an oil, is poured off, washed with water and distilled.—Colourless liquid of sweat-like odour, and very acid and acrid taste, of 0·911 density, solidifies below 12°, fuses at 14° to 15°, crystallises on slowly cooling in crystalline leaflets, boils at 236°, dissolves in 400 parts water at 100°, and on cooling separates again almost completely, mixes with alcohol and ether in every proportion. The salts, except those of the alkalies, dissolve slowly. The baryta compound, when dried at 100°, is anhydrous, and contains 63·80% acid.

**Capsicin.** A substance obtained from the Chillies or Cayenne pepper (the fruit of Capsicum annuum, C. baccatum, C. fastigiatum, C. frutescens, C. longum) and representing the acridity of the latter, but as yet not prepared in the pure state. Obtained by extracting with alcohol, treating the alcoholic extract with ether, and evaporating the ethereous solution.—Yellow or red-brown, soft, of at first balsamic, afterwards extremely burning taste, dissolves little in water, readily in ether, alcohol, oil of turpentine and alkalies. Capsicum contains also a crystalline resin, which, perhaps, is Capsicin in a purer state, and does not dissolve in water and ether.

**Capsulæscic Acid**$=C_{26} H_{12} O_{16}$. (Might be called Æsculic acid.—F. v. M.). In the capsules of the ripe fruit of Æsculus Hippocastanum. Sublimates undecomposed ; isomeric with triacetyl-gallic acid, behaves like the latter towards salts of oxyd of iron, and likewise reddens the solution of caustic potash.

**Caranna.** Exudation of the stem of Bursera gummifera. Dark-brown or green-brown, transparent at the edges, viscous, or solid and brittle, smells faintly like ammoniacum, of a bitter taste, and, when warmed, of a pleasant, balsamic odour, fuses readily. Contains 96% resin, which readily dissolves in alcohol, ether and alkalies.

**Carapin.** Bitter substance of the bark of Carapa Guianensis. Is very similar to cailcedrin and tulucunin, and also prepared in a like manner. Amorphous, resinous, readily soluble in alcohol and chloroform, less so in ether and water. It differs from tulucunin by striking no decided colour with acids. It consists of 55·04C, 6·54H and 38·42O.

**Cardol**$=C_{42} H_{30} O_4$. The acrid, oily ingredient of the pericarp of Anacardium occidentale, Semecarpus Anacardium and of several other species of both genera. Free the nuts from the mild oily seeds, contuse the pericarp, exhaust with ether, distil off the latter, and free the residue from tannic acid by washing with water. Dissolve the remaining mixture, of about 90% anacardic acid, 10°/₀ Cardol, and a little of an ammonia compound in 15 to 20 parts alcohol, digest the solution with freshly-precipitated hydrated oxyd of lead, whereby ammonia is evolved and a violet compound of lead is precipitated; filter again, and distil off the alcohol. Cardol of a dark claret colour remains behind, the slightly-concentrated alcoholic solution of which is mixed with water to turbidness and afterwards with an aqueous solution of acetate of lead, boiled and decolourised by adding subacetate of lead in minute quantities, whereby a brown, sticky precipitate is formed. The lead is removed from the solution by means of sulphuric acid, the alcohol distilled off, and the remaining Cardol washed with water.—Yellow, in thicker layers reddish oil of 0·978 density at 23°, without smell in the cold, when warm of a faint, pleasant odour, neutral, irritates the skin and raises blisters; does not volatilise without decomposition; insoluble in water, readily soluble in alcohol and ether, burns with a bright but very sooty, smoking flame, dissolves in concentrated sulphuric acid with red colour, also in strong potash-ley, which solution becomes red at the air, and precipitates the salts of the earthy and the heavy metals with red or violet colour.

**Carminic Acid** $= C_{28} H_{14} O_{16}$. Occurs, according to Belhomme, in the flowers of Monarda didyma.

**Carotin** $= C_{36} H_{24} O_2$. In the root of the cultivated Daucus Carota. Press the bruised roots and precipitate the liquid with diluted sulphuric acid under addition of a little tannic acid. The precipitate, consisting of Carotin, Hydrocarotin and chiefly of albumin, is pressed, boiled six or seven times, when half dry, with five to six times its quantity of alcohol of $80°/_0$ which dissolves hydrocarotin and mannit, and is then exhausted by boiling six to eight times with sulphide of carbon. Distil off the bulk of the sulphide of carbon and mix the residue with an equal volume of absolute alcohol, leave to stand quiet and to crystallise. The crystals of Carotin are washed at first with alcohol of $80°/_0$, afterwards with absolute alcohol until the latter assumes only a slight yellow tinge.—Dark-red, microscopic, quadrangular, tabular crystals of velvet lustre and free of water, the latter combining with it below $0°$ to a colourless, very unstable hydrate, which disunites at a few degrees above $0°$; smells, especially when warmed, strongly like Iris-root, becomes of a vivid red at $100°$, similar to metallic copper, fuses at $168°$ to a thick dark-red fluid, of amorphous structure when cold; is destroyed in higher temperatures, becomes readily discoloured by light and loses completely its faculty of crystallising; does not dissolve in water or in alkalies, scarcely in alcohol, while the amorphous C. dissolves in it; slowly soluble in ether and chloroform, readily in sulphide of carbon, benzol and oils, in concentrated sulphuric acid with beautiful purplish-blue colour and precipitable from it by water in dark-green flocks as amorphous Carotin.

**Carthamin** or **Carthamic Acid** $= C_{28} H_{16} O_{14}$. The red pigment of the flowers of Carthamus tinctorius. Exhaust the flowers first with cold water, press out, mix the remaining mass with water containing $15°/_0$ crystallised carbonate of soda, allow to stand for a few hours, press, neutralise the red alkaline liquid almost completely with acetic acid, throw down the C. on cotton which is left in the liquid for twenty-four hours, wash the cotton with pure water and withdraw the C. from it by macerating in water containing $5°/_0$ crystallised soda for half an hour. The solution, when mixed with citric acid, throws down the C. in flocks, which must be collected and dissolved in strong alcohol; on evaporating the latter solution, the C. subsides.—Dark brown-red, amorphous powder of a green metallic lustre, not volatile, dissolves very little in water with a faint red colour, in alcohol with beautiful purple colour, not in ether and oils, readily in solutions of the hydrates and carbonates of alkalies with deep yellow-red colour, precipitable by acids; in concentrated sulphuric acid blood-red.

**Carthamus-Yellow.** Is withdrawn from the safflower by water. Precipitate the aqueous liquid, acidulated by acetic acid, with acetate of lead; remove the white deposit, neutralise the filtered liquid with ammonia, decompose the dirty orange-yellow precipitate by diluted sulphuric acid, remove from the yellow filtrate an excess of sulphuric acid by means of acetate of baryta, evaporate the filtrate in a retort to a syrup, and withdraw the C. by absolute alcohol. This solution is to be evaporated under exclusion of the air to the consistence of syrup and mixed with water, whereby decomposed C. subsides and the unaltered C. remains dissolved. The solution is of a dark brown-yellow colour and of acid reaction, of a bitter and salty taste, decomposes readily, when allowed to stand or warmed in contact with the air, and throws down brown substances.

**Carven** $= C_{20} H_{16}$. In the oil of caraway, besides carvol; is thin, limpid, of a faint pleasant taste and smell, has a density of $0.861$, boils at $173°$.

**Carvol** $= C_{20} H_{14} O_2$, contained, besides carven, in the oil of caraway. Isolate from the crude oil by oft-repeated fractioned distillation, the part which distils at $225$ to $230°$. Limpid, thin, of the odour of carven, and of $0.953$ density; boils above $250°$.

**Caryophyllic Acid** $= C_{20} H_{11} O_3 + HO$. In the oil of cloves, besides a hydrocarbon isomeric with oil of turpentine, likewise in the oils of pimento and of cinnamon-leaves, in the oils of Canella alba and Dicypellium caryophyllatum. It is usually obtained from the oil of cloves by heating with ley of potash, whereby the hydrocarbon is volatilised; adding sulphuric or phosphoric acids to the remaining liquid and distilling the caryophyllic acid. It is a colourless oil of the smell and taste of cloves and of $1.068$ density, boils at $242°$ to $251°$, has an acid reaction, dissolves little in water, readily in alcohol, ether and acetic acid. Its compounds are mostly crystallisable and, except the salt of baryta, readily decomposed by water and alcohol.

**Caryophyllin** $= C_{20} H_{16} O_2$. In the cloves (from Eugenia caryophyllata). Crystallises in the alcoholic tincture in white, silky, spherically united needles, is without taste or smell and of neutral reaction, begins to evaporate at $280°$, without melting, and sublimates completely; not soluble in water and sparingly in cold alcohol, but dissolves in boiling alcohol and readily in ether, little in diluted acids and in alkalies.

**Cascarillin.** Bitter ingredient of the bark of Croton
Eleuteria and Croton Sloani. Precipitate the aqueous extract
with acetate of lead, remove the lead from the filtered liquid by
means of sulphuret of hydrogen, evaporate, digest with animal
charcoal and concentrate until a pellicle forms on the surface,
wash the subsiding mass with alcohol and recrystallise in boiling
alcohol.—White, fine needles and tabular crystals, neutral, bitter,
fusible, non-nitrogenised, not volatile, very slowly soluble in
water, more readily in alcohol and ether, with purple-red colour
in concentrated sulphuric acid.

**Castin.** Bitter substance of the seeds of Vitex Agnus castus
(to be looked for in the numerous other species of the genus).
Separates in crystals from the alcoholic tincture, dissolves little in
water; soluble in alcohol and ether, precipitable by alkalies.

**Castor Oil.** Obtained by pressing the seeds of Ricinus communis. Almost colourless, with a slight green-yellow tinge, of
thick fluidity, of indifferent smell and mild afterwards a little acrid
taste, has a density of 0·960, does not congeal at − 15°, but yields in
the cold a little granular matter; dissolves in any quantity of
absolute alcohol and ether, thickens at the air and becomes dry at
last; begins to boil at 265°, yielding various products. It
consists in the main of the glycerid of the ricinoleic acid. As
regards the solid fat there are different opinions; Bouis asserts
that it is the glycerid of a peculiar acid called by him Isocetic
acid, on account of its having the same composition as the cetic
acid of spermaceti.

**Catechin or Catechuic Acid** = $C_{12} H_6 O_5 + 2 HO$. In the
Catechu—the aqueous extract of the wood of Acacia Catechu, of
the nuts of Areca Catechu, and of the leaves of Nauclea Gambir;
also in the kino—the hardened juice of the stem of Pterocarpus
Marsupium. To prepare it, exhaust finely pulverised catechu with
cold water first and boil it afterwards with eight times its quantity
of water in a tubulated retort, while adding subacetate of lead,
until the solution is of a wine-yellow colour; filter the boiling
liquid and allow to cool under exclusion of the air. The white
granular mass which will have formed after 24 hours has to
be washed with cold water until every trace of lead is removed; it
is then pressed and dried in the vacuum.—A white, granular substance, consisting of fine needles, and after trituration representing
a white loose powder, but of a partly yellow-brown colour when
the air in drying was not completely excluded; without smell and
of a sweetish taste; dissolves in 16,000 parts of cold and in four
parts boiling water, readily in alcohol, in 120 parts cold and in
seven parts boiling ether; all these solutions have an acid reaction.
It imparts a dark-green colour to salts of oxyd of iron, yet pro-

duces no turbidity in a solution of glue, unless it be previously treated with nitric acid, whereby a brown substance is precipitated.

**Catechu Tannic Acid.** In the Catechu (*see* Catechin) [also in the bark of the Prickly Banksia of Western Australia.] Not known in the pure state; is a product of decomposition of catechuic acid (not the substance from which the latter is formed, as formerly stated). Macerate the catechu with ether, agitate the solution with water, decant the ethereous liquid and evaporate to dryness; dissolve in water and allow the catechuic acid to crystallise. The mother-ley contains nearly pure Catechu tannic acid. It precipitates the salts of oxyd of iron with a dirty-green colour, also glue.

**Cedrin.** The bitter ingredient of the fruit of Simaba Cedron. Remove a fatty substance by treating with ether, and dissolve the bitter substance by means of alcohol. White, silky needles, still more bitter than strychnin, little soluble in cold, readily in boiling water, of neutral reaction.

**Cellulose.** *See* Fibrin.

**Centaurin** = Cnicin.

**Ceradia Resin** = $C_{40} H_{28} O_4$. From Othonna furcata. Amber-yellow, smells like elemi, has an acid reaction when dissolved in alcohol.

**Cerasin.** *See* Cherry Gum.

**Ceratophyllin.\*** In Parmelia physoides [and probably also in other Parmelias]. Pour lime-water on the lichen washed before in cold water, let stand no longer than 15 hours, and precipitate the slightly yellow solution with hydrochloric acid. Wash the deposit with cold water and dry, exhaust with boiling alcohol of $75°/_0$, and boil the residue that has not been dissolved, with a concentrated solution of carbonate of soda. The C. is said to form in crystals when the solution has cooled down.—Thin white prisms of at first faintly afterwards strongly rancid and burning taste; fuses at 147° and sublimates in laminæ; dissolves more readily in hot than in cold water; also in cold concentrated sulphuric acid unaltered, readily in alcohol, ether and alkalies, and precipitable from the latter solution by acids; becomes purple-violet by choride of iron.

**Cerealin.** Ingredient of the bran of the grains of cereals, possessing in a high degree the faculty of converting starch into dextrin—therefore a kind of diastase. To obtain it, treat the bran with several changes of diluted alcohol in order to remove foreign matters; afterwards exhaust with cold water, filter and evaporate

---

\* This term is apt to lead to a very different plant and might advantageously be changed to Parmelin.—F. v. M.

at 40° to dryness. The remaining Cerealin is similar to albumin, amorphous, nitrogenised, soluble in water, insoluble in alcohol, ether and oils. The aqueous solution curdles at 75°, and at the same time the C. loses its activity.

**Ceropinic Acid** $= C_{36} H_{34} O_5$. In the bark of Pinus sylvestris. Obtained in the same manner as indicated under Pinocorretin, and purified by recrystallising in alcohol with aid of animal charcoal. White powder, consisting of microscopic crystals, fuses at 100° and congeals like wax.

**Cerosin** $= C_{43} H_{60} O_2$. Wax-like substance which forms on the surface of the stalks of the sugar-cane (Saccharum officinarum) and is easily collected by scraping. To purify it, digest first with cold alcohol, dissolve afterwards in boiling alcohol and allow to crystallise. It forms pale-yellow light laminæ of mother-of-pearl lustre, without odour and of 0·961 sp. gr., is hard, easily friable, fuses at 82°; insoluble in water and cold alcohol, readily soluble in boiling alcohol and congealing, when cold, to an opodeldoc-like mass; insoluble in cold, slowly soluble in hot ether, combines slowly with alkalies.

**Ceroxylin** or **Ceroxylon Resin** $= C_{40} H_{32} O_2$. It is contained in the Palm-wax, obtained by scraping the stem of Ceroxylon Andicola and boiling the substance with water. It is obtained pure by boiling the Palm-wax with alcohol, filtering while hot and allowing the liquid to cool. The wax is then removed and the mother-ley evaporated to form crystals. The crystallised resin appears in white, fine needles, melts above 100°, dissolves little in cold, readily in hot alcohol, also in ether and volatile oils.

**Cetraric Acid** $= C_{36} H_{16} O_{16}$. In the Iceland-moss (Cetraria Islandica). Boil with alcohol under addition of carbonate of potash, strain, precipitate the decoction with diluted hydrochloric acid and water, and remove from the deposit foreign matters as lichenostearic acid, thallochlor, &c., by successively treating with boiling alcohol of 42% and ether, mixed with oil of rosemary or camphor. From the remaining grey-white mixture of Cetraric acid and an indifferent white compound, the former is dissolved by a cold aqueous solution of bicarbonate of potash, and has to be precipitated with hydrochloric acid and recrystallised in the least possible quantity of boiling alcohol.—Snow-white, loose tissue of shining, hair-shaped needles, very bitter, not volatile, loses at 100° nothing of its weight, turns brown at 125° and decomposes; does not dissolve in water, but imparts to it a faint, bitter taste when boiled; is slowly soluble in cold, readily in boiling alcohol, little in ether, not in oils; most readily in the hydrates and carbonates of alkalies; the bright-yellow solutions have a very bitter taste, and are precipitable by acids.

**Chelerythrin** $= C_{38}$ $H_{17}$ $NO_8$. In the root of Glaucium luteum and allied species, also in all parts of Chelidonium majus, Eschscholtzia Californica and Sanguinaria Canadensis (all plants of the order of Papaveraceæ). Production from the root of Chelidodium majus: macerate with water containing sulphuric acid, precipitate the liquid with ammonia, edulcorate, press and dissolve the precipitate in alcohol acidulated with sulphuric acid, add water and distil off the alcohol, precipitate the remaining liquid with ammonia, wash the deposit and dry between blotting-paper at a gentle heat as quick as possible, reduce to powder and treat with ether, which dissolves principally the Chelerythrin. After evaporating the ether, a viscid, turpentine-like mass remains, which must be dissolved in the least possible amount of water, containing hydrochloric acid, in order to remove resin. The solution is evaporated to dryness and rinsed with ether, which leaves behind chloride of Ch.; dissolve the latter in little cold water which leaves undissolved in the main chloride of Chelidonin, evaporate and redissolve, &c., as long as chloride of Chelidonin is formed. From the last aqueous solution the Ch. is isolated by ammonia, and has, after rinsing and drying, to be purified by dissolving in ether and evaporating.—White grains of pearl lustre; remains after the ethereous solution has been evaporated, as a turpentine-like mass which slowly solidifies to a shining, friable substance; without taste by itself, but the alcoholic solution tastes burning, acrid and bitter; its dust induces vehement sneezing; it softens resin-like at 65°, is decomposed in higher temperatures, is of alkaline reaction; assumes slowly a yellow-white colour when exposed to the air and becomes red by traces of acid vapours; does not dissolve in water, but most readily in alcohol, ether and oils. With acids, which impart to it a beautiful orange-red tint, it forms crimson-red, neutral, partly crystallisable salts of a burning, acrid taste, readily soluble in water and precipitable by alkalies and by tannic acid.

**Chelidonic Acid** $= C_{14}$ $H_2$ $O_{10} + 3$ $HO + 2$ $Aq$. As yet only found in Chelidonium majus, and present in all parts. Is obtained when the aqueous extract is precipitated with nitrate of lead, the deposit washed and treated first with very diluted nitric acid (30 parts water and one part acid of 1·22 density), in order to remove other compounds of lead, and afterwards decomposed boiling hot by sulphide of sodium. The excess of the latter is destroyed by means of an acid (acetic acid), the filtered liquid is evaporated, and the acid which will have separated is recrystallised. It forms small, colourless needles, dissolves in 166 parts cold and 36 parts boiling water, in 709 parts alcohol of 75%, is without smell, of a strongly acid taste, is carbonised by heat, and precipitated white by salts of lead, nitrate of silver and the nitrates of mercury; most of its salts dissolve in water and are easily crystallisable.

**Chelidonin** $= C_{38} H_{17} N_3 O_6 + 2 HO$. In Chelidonium majus, principally in the root. After the extract of the root, prepared with water and sulphuric acid and obtained for the production of chelerythrin, has been precipitated by ammonia and the chelerythrin withdrawn from the precipitate by ether—the residue has to be dissolved in the least possible quantity of water, containing some sulphuric acid and the solution mixed with double its volume of concentrated hydrochloric acid, which throws down chloride of Ch. This is decomposed by water containing ammonia; is purified by repeatedly dissolving in a little acidulated water, precipitating with hydrochloric acid and decomposing with ammonia, and is recrystallised in boiling alcohol.—Colourless, glassy, tabular crystals, of bitter taste similar to sulphate of quinin (according to others : acrid not bitter); loses the water at 100°, fuses at 130° and is decomposed in higher temperatures; volatilises with the vapours of water, is insoluble in water, slowly soluble in alcohol and ether. Its salts are colourless, crystallisable and of acid reaction, dissolve in water, have a strong and pure, bitter taste, and are precipitable by alkalies and by tannic acid.

**Chelidoxanthin.** In the root, herb and flowers of Chelidonium majus. Withdraw the chelidonin and chelerythrin from the root by water acidulated with sulphuric acid, exhaust the residuum with hot water, mix the liquid with acetate of lead and give access to sulphuret of hydrogen. The sulphide of lead, washed with cold water, yields to boiling water Ch., which is obtained on evaporating as a yellow, granular mass. This has to be digested successively with water, containing ammonia, and with ether to remove foreign matters; exhaust the remnant with absolute alcohol, filter, evaporate and rinse the remaining Ch. with cold diluted sulphuric acid, ammoniacal water and ether. Dissolve it in hot water and allow slowly to crystallise.—Yellow, friable mass or yellow, short needles; tastes very bitter; dissolves very slowly in cold, better in hot water, imparting to it a strong yellow colour, slowly in alcohol, not in ether, in concentrated sulphuric acid with yellow-brown colour ; in the alcoholic solution it is precipitable by tannic acid.

**Chenopodin** $= C_{12} H_{13} NO_8$. Peculiar, nitrogenised body, closely related to the alkaloïds, from Chenopodium album and other species of that genus. To obtain it, bruise the fresh plant gathered before flowering, press, heat the juice to 80°, strain off from the coagulated albumin, evaporate to honey consistence, mix with alcohol of 90%, boil for a few moments, pour off the liquid, boil the residue with several changes of alcohol, leave the alcoholic liquids to stand in the cold, strain off from the crystals of saltpetre, distil the alcohol and allow the residue, concentrated to a syrup-like consistence, to stand cold for a few days. The Ch., which will then have separated in granular crystals, must be collected, agitated with ether and recrystallised in boiling alcohol.—White, lustreless, permanent powder, under the microscope, of the appearance of concentrically united needles, without taste or smell; it loses nothing of its weight at 100°, begins to sublimate at 200° in snow-white crystalline flocks, sublimates completely at 225°, becomes liquid, begins to boil and evolves a very penetrating nauseous odour. It dissolves in 11 parts cold and in 3 to 4 parts boiling water, in 202 parts cold and in 77 parts boiling alcohol of 90%, is of neutral reaction; dissolves readily in diluted acids; the chloride crystallises in cubes and is precipitated with a light-yellow colour by chloride of platinum.

**Cherry Gum.** Exudation of the stems and branches of Prunus Amygdalus, P. Persica, P. Cerasus, P. domestica and P. Armeniaca. It is a mixture of about equal parts of arabin and cerasin (compound of the metagummic acid $= C_{12} H_{11} O_{11}$ with lime); and leaves the cerasin when treated with cold water as a colourless, pellucid easily friable mass, which swells up in cold water without dissolving, and is insoluble in alcohol.

**Chica Red** $= C_{16} H_8 O_6$. In the Chica or Carajuru, a pigment obtained from the leaves of Bignonia Chica in South America. Digest the chica with alcohol acidulated with sulphuric acid, neutralise the extract with carbonate of ammonia, wash the precipitate with boiling water and dry at 100°. It turns brown at the light, is not soluble in water, readily so in alcohol with ruby-red colour, little in ether with yellow, in alkalies with dirty-red colour and precipitable by acids.

**Chimaphilin.** Yellow, crystalline compound in Pyrola (Chimaphila) umbellata. Agitate with chloroform the tincture of the herb prepared with diluted alcohol, let subside, throw away the upper stratum and allow the liquid beneath to evaporate; the substance, which will separate, has to be recrystallised in alcohol. In the purest state it is to be obtained by distilling the herb with water. The stalks yield more of it than the leaves.—Beautiful, gold-yellow, long needles, without smell or taste, fusible, sublimate unaltered; almost entirely insoluble in water, soluble in alcohol, ether, chloroform, in fixed and in volatile oils; is carbonised by concentrated sulphuric acid, not altered by concentrated nitric or hydrochloric acids. The alcoholic solution is not affected by tannic acid or chloride of mercury.

**Chinamin** = QUINAMIN.

[**Chiratin** $= C_{52} H_{48} O_{30}$. Bitter substance, found by Hoehn in all parts of Ophelia Chirata. Contained in the extract, prepared with alcohol of 60°/$_o$. It forms a light-yellow, very hygroscopic powder of a strong and lasting bitter taste, dissolves sparingly in cold, better in hot water, easily in alcohol and in ether, has a neutral reaction, does not act on alkaline copper solution, forms a copious, white, flaky precipitate with tannic acid. With acids the Ch. separates into Ophelic acid and Chiratogenin $= C_{26} H_{24} O_6$.]

### Chlorogenic Acid (Payen) $=$ Caffeic Acid.

**Chlorogenin.** A substance contained in the root of Rubia tinctorum, and distinguished by the green colour it assumes when boiled with acids, but as yet not obtained in the pure state. According to Kraus it appears to be identic with Rochleder's Rubichloric acid and with Runge's Rubiaceic acid. Remains, besides sugar and mineral substances, in the liquid obtained for the preparation of Rubian, after the precipitate produced by acids has been filtered off. Precipitate the aqueous decoction of madder with oxalic acid, filter and neutralise the liquid with chalk, filter again and evaporate the liquid on the water-bath until a dark-brown thick syrup is obtained. The latter dissolves in water and leaves behind brown products of decomposition originated in the course of the evaporation (while the solution has an acid reaction on account of phosphoric acid, and assumes a green colour when boiled with acids). Precipitate the solution with subacetate of lead, remove the excess of lead in the filtered liquid by sulphuret of hydrogen, filter again and evaporate over sulphuric acid; a brownish-yellow, honey-like substance is obtained, which does not dry up again. This is Chlorogenin, mixed with a little sugar, which exists readily formed in the madder, and with the acetates of potash, lime and magnesia. It has a nauseous, sweetish and bitter taste, throws down a brown powder when evaporated in the aqueous solution, dissolves in alcohol, not in ether; emits an offensive smell when boiled with diluted hydrochloric or sulphuric acid, becomes dark-green and throws down a dark-green powder.

**Chlorophyll** is according to Fremy a mixture of a yellow and a blue pigment. By shaking an alcoholic tincture of chlorophyll with a mixture of two parts of ether and one part of moderately diluted hydrochloric acid, the ether dissolves the yellow matter, while the acid beneath assumes a beautiful blue colour. By mixing the two liquids by means of alcohol the green colour is restored. The yellow pigment, isolated in the above manner, is called by Fremy Phylloxanthin, the blue Phyllocyanin. Closer investigations of these two compounds are wanting.

[According to the researches made by Kraus, Sorby and others on the green colouring matters of leaves, spectrum analysis has revealed the existence of at least ten different colouring substances more or less separable by means of sulphide of carbon, alcohol and water, used in varying combinations. Amongst these substances are a blue-green and a yellow-green, called respectively blue and yellow chlorophyll, and five yellow substances: orange xanthophyll, xanthophyll, yellow xanthophyll, orange lichnoxanthin and lichnoxanthin. As these colouring matters are rapidly decomposed by acids, Fremy's blue and yellow Chlorophylls are but products of decomposition. Of the above colouring matters, orange xanthophyll occurs in considerable quantity in Oscillatoriæ, blue chlorophyll in olive Algæ, xanthophyll in Porphyra vulgaris (Algæ), yellow xanthophyll in various pale-yellow flowers as Chrysanthemums, &c., and lichnoxanthin in Clavaria fuciformis. Xanthophyll and yellow xanthophyll give absorption-bands, not lichnoxanthin. On addition of a little hydrochloric acid to the alcoholic solutions, the first and the third of the three last-named matters fade slowly without altering their colour, while the second is changed into another yellow substance with two absorption-bands and then into deep blue. These three yellow substances are all soluble in sulphide of carbon, they exist in green leaves and also generally in yellow flowers though mixed in different proportions.]

[Solutions of Chlorophyll are decomposed by small quantities of hydrochloric acid, and yield on filtration a solid black residue and a yellowish-brown liquid. The latter, on addition of more hydrochloric acid, assumes a deep-green colour and yields by filtration a yellow residue and a pure blue liquid. The above black matter shows, according to Filhol, a crystalline structure of radiated needles under the microscope in the case of Monocotyledons, but is amorphous with Dicotyledons. It dissolves very sparingly in cold alcohol of 85%, abundantly on boiling, also in ether and benzol with yellowish-brown, in sulphide of carbon with yellow, in chloroform with violet, and in acetic acid with blue-violet colour. Concentrated hydrochloric and sulphuric acids dissolve it slowly with green colour, probably under decomposition. The solution of the crystalline black matter in acetic acid assumes a splendid green colour when mixed with traces of the acetates of copper or zinc and heated to boiling.]

**Cholestearin** $= C_{18} H_{24} O + HO$. Exists, according to recent investigations, also in the vegetable kingdom and especially in the seeds and young parts of various plants (in malt, &c.) From peas it is obtained in the following manner. Digest with warm alcohol of 94%, evaporate to honey-consistence, dissolve in water, add oxyd of lead, boil until clear, allow to cool, throw away the water, treat again with hot alcohol, remove the lead in the tinctures by means of sulphuret of hydrogen, leave to crystallise in the cold, and recrystallise the Ch. in alcohol.—It forms large, white, leaf-like crystals of mother-of-pearl lustre, without smell or taste, lighter than water, loses the water when heated, fuses afterwards (at 137°) and distils by careful heating undecomposed. In higher temperatures it separates into various volatile products. It does not dissolve in water, little in cold alcohol and oil of turpentine, readily in hot alcohol, oils and ether; is not affected by caustic alkalies.

**Chrysophanic Acid** = $C_{20}$ $H_8$ $O_6$ (called also according to its origin Lapathin, Parietin or Parietic acid, Parmel-yellow, Rhaponticin, Rheic acid, Rumicin). In the roots of the genus Rheum and Rumex, in Parmelia parietina and Squamaria elegans, and probably many other Lichenes. Digest the Parmelia parietina or the root of rhubarb with weak alcohol, containing caustic potash, give to the strained liquid access of carbonic acid gas, dissolve the deposit in alcohol of 50% mixed with a little caustic potash, filter and precipitate with acetic acid. Dissolve the precipitate, after filtering, in boiling alcohol, filter hot and add water, whereby yellow flocks of Chr. are separated, which have to be purified by recrystallisation in alcohol.—Delicate orange-yellow, felted needles, similar to iodide of lead, of gold-lustre, almost without taste, fusing at 162°; sublimate partly unaltered by careful heating; scarcely soluble in cold, a little more in boiling water with bright yellow colour, in 1125 parts alcohol of 85% at + 30°, in 224 parts boiling alcohol, in ether, glacial acetic acid, amylalcohol, remarkably well in benzol and oil of coal-tar, in concentrated sulphuric acid with beautiful red colour and precipitable without alteration by water with yellow colour, readily soluble in alkalies with red colour.

**Chrysopicrin** = Vulpulin.

**Chrysorhamnin** = $C_{46}$ $H_{22}$ $O_{22}$. According to Kane in the unripe so-called Yellow Berries (Grains of Avignon, from Rhamnus tinctorius, R. infectorius, R. saxatilis, R. amygdalinus, R. oleoides); yet Gellatly did not succeed in obtaining it either from the ripe or the unripe berries. Obtained by extracting with ether, it represents beautiful gold-coloured, silky, concentrically united needles, dissolves scarcely in cold water, readily in alcohol and in ether; by boiling with water, xanthorhamnin is produced; it is not altered by acids and dissolves, apparently under decomposition, in alkalies.

**Cicutin.** Volatile alkaloïd, occurring in all parts of Cicuta virosa, but as yet only obtained in the aqueous solution, and is very imperfectly investigated.

[**Cicuten** = $C_{20}$ $H_{16}$. A hydrocarbon contained in the volatile oil of the root of Water-hemlock or Cowbane (Cicuta virosa). It boils at 166°, rotates to the right and is of 0·87038 specific gravity at 18°. It dissolves in every proportion in alcohol, ether, chloroform, benzol and sulphide of carbon, dissolves iodine, sulphur and phosphorus, and forms with water a crystalline hydrate. It absorbs chlorine in large quantities, producing a thick camphor-like liquid of the formula $C_{20} H_{12} Cl_4$.—Van Ankum].

[**Cinchona Alkaloids.** F. E. de Vrij gives the following method for the separation and quantitative estimation of the five Cinchona alkaloïds—Quinin, Cinchonidin, Cinchonin, Quinidin and the amorphous base, soluble in ether.—Mix at least five grammes of the pulverised crude alkaloïd with fifty grammes ether, shake repeatedly and allow to rest till the next day. Filter and evaporate the ethereous solution.

A. *Portion soluble in ether.* Dissolve the residue in 10 parts proof spirit, mixed with 5°/$_0$ sulphuric acid. Add cautiously of tincture of iodine as long as a precipitate ensues. The precipitate contains the whole of the *Quinin* as herapathit. Collect on a filter, wash with proof spirit and dry at 100°. One part by weight contains 0·565 parts *Quinin.*

The filtered liquid contains the *amorphous base.* Add sulphurous acid in alcohol until colourless, neutralise with caustic soda, drive away the alcohol on the water-bath and precipitate by slight excess of soda-ley.

B. *Portion insoluble in ether.* Add 40 parts hot water and diluted acid, until only slightly alkaline and a sufficient quantity of tartarate of potassa et soda, stir well and leave to rest till the following day, when *Cinchonidin,* if present, will have separated under the form of crystals of tartarate of Cinchonidin. Collect on a filter, wash with a little water and dry at 100°. One part contains 0·804 Cinchonidin.

The liquid after separation is mixed with dissolved iodide of potassium and stirred. *Quinidin* falls down as a coarse crystalline powder of hydriodide of Quinidin. Collect, wash with little water and dry at 100°. One part contains 0·718 parts Quinidin.

The filtered liquid contains the *Cinchonin,* which is obtained by precipitating with soda-ley, washing with water and drying.]

[**Cinchona Oily Alkaloid,** Howard's = $C_{20} H_{12} NO_2$. Found in the mother-ley of Quinin. Obtained by treating the mother-ley with ether, evaporating, dissolving in the smallest possible quantity of oxalic acid, crystallising and recrystallising under addition of animal charcoal. The alkaloïd is separated by carbonate of potash or soda.—A yellowish oil of bitter taste, easily decomposable by heat, readily soluble in alcohol and ether. Exhibits with chlorine water and ammonia the reaction of Quinin and Quinidin. Nitric acid produces a green-yellow colour. Of the salts the oxalate crystallises best, the other compounds are exceedingly soluble in water and partly of an oily nature.—Found by Howard in the bark of Cinchona succirubra.]

**Cinchonidin**=C₂₀ H₁₂ NO. In the Quina rubra granatensis (called originally Quina pseudoregia), from Cinchona succirubra, C. Calisaya, C. officinalis, C. lancifolia. Exhaust the bark reduced to powder, with water acidulated by hydrochloric acid, treat the solution with milk of lime, collect the deposit, dry, pulverise, draw out with alcohol, evaporate the fluid, let crystallise and purify by recrystallising in alcohol.—Forms colourless, long, shining prisms; fuses at 167 to 170° and decomposes in higher temperatures; without smell, of bitter taste; becomes electric by friction; dissolves in 3287 parts cold, and in 596 parts boiling water, in 88 parts cold and in 19 parts boiling alcohol of 0.833, in 398 parts cold ether of 0·740; the alcoholic solution has an alkaline reaction; dissolves readily in diluted acids, also in concentrated sulphuric acid without change; dissolves colourless in chlorine-water, and ammonia produces in the solution a grey-white precipitate, soluble in an excess of ammonia with pale sherry colour. It saturates the acids completely, and forms well crystallisable salts, which are permanently precipitated by caustic alkalies and their carbonates, phosphate of soda, chloride of mercury, chloride of platinum, chloride of gold, iodide of potassium, sulphocyanide of potassium, nitrate of palladium and by tannic acid.

**Cinchonidin**=PASTEUR'S QUINIDIN.

**Cinchonin**=C₂₀ H₁₂ NO. Prevails in the grey and brown and especially in the Huanoko Quina-barks, from Cinchona micrantha, C. nitida, C. Peruviana, C. Pahudiana, C. scrobiculata, C. cordifolia. Digest with water acidulated with hydrochloric acid, press, concentrate the extracts, filter, precipitate with ley of soda, wash the deposit and dissolve it in acetic acid, precipitate again, dry the deposit, boil with alcohol and allow to crystallise.—Forms white pellucid needles of alkaline reaction, of slightly bitter taste, fuses only when decomposition is setting in, and evaporates partly undecomposed (according to others: it fuses at 150 to 160° and sublimates partly in white needles); dissolves in 3670 parts cold, and in 2500 parts boiling water, in 126 parts absolute, and in 115 parts alcohol of 90% at 15°, in 40 parts boiling alcohol, in 600 parts cold and in 470 parts boiling ether, in 23 parts chloroform, little in oils; dissolves colourless in chlorine-water, and is precipitated white by ammonia; dissolves colourless in concentrated sulphuric acid and is not coloured on heating; insoluble in alkalies. Neutralises the acids and yields mostly crystallisable salts, the solutions of which turn dark red-brown in the sunlight.

E

**Cinchovatin** = Aricin.

**Cinnamein** = Oil of Balsam of Peru.

**Cinnamen** = Styrol.

**Cinnamic Acid** $= C_{18} H_7 O_3 + H O$. In the balsam of Peru (from Myroxylon Pereiræ), balsam of Tolu (Myroxylon toluiferum), liquid storax (from Liquidambar orientalis), perhaps also in other odoriferous resins.    May be prepared like benzoic acid.    Crystallises in colourless klinorhombic prisms and needles, smells and tastes faintly like cinnamon, liquifies at 129°, boils at 300° and sublimates in pungent, cough-producing vapours; dissolves little in cold, readily in boiling water, in alcohol and ether; is of acid reaction; the aqueous solution evolves, when boiled with peroxyd of lead, or with chromate of potash and sulphuric acid, oil of bitter almonds and the remaining liquid contains benzoic acid. Its salts are soluble in water and crystallisable, very similar to the benzoates; the least soluble is the silver-salt; the lead-salt is also little soluble.

**Cissampelin** = Bebirin.

**Cisso-Tannic Acid** $= C_{20} H_{12} O_{16}$.    The pigment of the red autumnal leaves of Vitus hederacea, occurs also in the ripe fruits of Fragaria vesca and its congeners.    Treat the leaves with alcohol of 80%, add to the tincture one-fifth of water and distil the alcohol, evaporate the remaining liquid to honey consistence and treat with cold water, which yields a dark-red solution, while a kermes-red powder (see below) remains undissolved.    The red solution has an adstringent and bitter taste, is coloured dark-green by chloride of iron, and is precipitated by glue, tartarated antimony and acetate of lead (olive-green by the latter).    The solution becomes turbid by age and throws down the same kind of powder as forms on dissolving the extract.    This red powder, *insoluble or changed Cisso-Tannic Acid* $= C_{52} H_{28} O_{25}$, is, when dry, dense, dark-brown, shining, brittle, of bitter and acerb taste, carbonises quietly when heated, is insoluble in water, readily soluble in alcohol with blood-red colour, in ammonia with dark brown-yellow colour; the alcoholic solution behaves towards chloride of iron, glue, tartarated antimony and acetate of lead like the original acid.

**Citric Acid** $= C_{12} H_5 O_{11} + 3 HO + 2 Aq$.    Widely diffused, free and mixed with more or less malic acid in fruits, or bound to bases in various roots, herbs, barks, &c.    It is precipitated from the extracts by acetate of lead; the deposit, when decomposed by sulphuret of hydrogen and evaporated, yields the acid in crystals. Rectorhombic, translucent, colourless crystals, without odour and of a pure and strong acid taste, lose their lustre at the air, give up at 100° 2 eq. water, fuse at 150° and decompose in higher temperatures ; dissolve readily in water, alcohol and ether. Citric acid is distinguished from the similar tartaric acid by the following tests. It evolves on heating no smell of burning sugar, but a different empyreumatic odour ; it forms no scantily soluble acid salt with potash, and is not turbidified by lime-water in the cold, but yields with it when heated a precipitate, which disappears again on cooling. The salts of the alkalies are readily soluble in water, those of the other bases slowly or not in water. The salt obtained by acetate of lead and dried at $120°$ is $=3$ PbO $+$ $\overline{Ci}$ + HO, and contains $32·68\%$ acid.

**Cnicin** $= C_{42} H_{28} O_{15}$. In the leaves of Carbenia benedicta, Centaurea Calcitrapa and other Cynarocephalous plants of Compositæ. Is prepared like Salicin. Crystallises in white, silky needles, has a remarkably bitter taste; permanent at the air, neutral, fusible, not volatilisable without decomposition ; dissolves sparingly in cold, much better in boiling water, readily in alcohol, wood-spirit, scarcely in ether, not in oil of turpentine and in fixed oils, in concentrated sulphuric acid with beautiful red colour, turning black when heated.

**Cocain** $= C_{32} H_{19} NO_8$. In the leaves of Erythroxylon Coca. Exhaust with water of $60°$ to $80°$, precipitate the liquid with acetate of lead, precipitate the filtered solution with a saturated solution of sulphate of soda, filter, concentrate, render slightly alkaline by means of carbonate of soda, and shake repeatedly with ether. Distil off the bulk of the ether from the ethereous solutions, leave the rest of the solvent to evaporate in the cold, and withdraw a part of the pigment from the remaining impure C. by distributing it in cold water. Now pour a thin layer of the impure chloride of C. into Graham's dialysator made of parchment-paper, and leave to diffuse in three changes of water for three days, whereby most of the chloride of C. will be diffused into the water, while much of the pigment remains in the dialysator. Throw down the C. in the solution with carbonate of soda, dissolve in alcohol, add of acetic acid enough to render it acid, and allow to evaporate spontaneously over sulphuric acid. The C. remains free of acetic acid, and is withdrawn from the remnant by ether, while the foreign matters are left undissolved and combined with acetic acid under the form of oily drops.—It crystallises in large, colourless klinorhombic prisms of at first somewhat bitter, afterwards cooling taste; of alkaline reaction, fuses at $98°$, may be sublimated for a small part with careful heating, dissolves in 704 parts cold and a little more copiously in hot water, readily in alcohol, and better still in ether; readily in diluted acids, in concentrated sulphuric acid without colour, but is carbonised by heat; separates when heated with concentrated hydrochloric acid into benzoic acid and a new crystalline base of sweetish bitter taste (Ecgonin). Forms with acids crystallisable salts; liquor of ammonia, carbonate of ammonia, and caustic potash produce in the chloride of C. solution white precipitates, soluble in excess of the reagent; carbonate of soda gives a permanent precipitate which becomes crystalline on keeping; the bicarbonates of alkalies and phosphate of soda do not affect the solution; bi-iodiole of potassium causes a kermes-brown, picric acid a yellow precipitate, tannic acid (only when mixed with some hydrochloric acid) white flocks; Coca-tannic acid precipitates nothing.

**Coca-Tannic Acid.** In the leaves of Erythroxylon Coca. It is obtained besides cocain, when the leaves are exhausted with alcohol acidulated with sulphuric acid; the tincture digested with hydrate of lime, filtered, neutralised with sulphuric acid and freed from the alcohol by distillation; in the remaining liquid the cocain is precipitated by carbonate of soda, and can be removed by shaking with ether, while the Coca Tannic acid, remains dissolved in the aqueous solution. Evaporate from the latter any traces of ether; neutralise with nitric acid, remove the sulphuric acid by means of nitrate of baryta, the excess of the latter with carbonate of ammonia, neutralise and precipitate with acetate of lead; the washed deposit has to be decomposed under water by sulphuret of hydrogen, and the liquid is then evaporated.—Brown-red, amorphous, hygroscopic substance of slightly acerb taste, fusible by heat, colours' chloride of iron brown-green, precipitates tartarated antimony, but not glue.

**Coccognic Acid.** In the seeds of Daphne Gnidium. Digest with alcohol, treat the alcoholic extract with water, and evaporate the aqueous solution.—Colourless, quadrangular prisms of a peculiar acidulous taste, the solution of which is not turbidified by lime-water, chloride of baryum, acetate of lead or subsulphate of iron.

**Cocoa Oil,** obtained from the fruit-seeds of Cocos nucifera by boiling with water and skimming; is white, of butter-consistence, of a rather disagreeable, somewhat cheese-like odour and mild taste, fuses at $20^\circ$ to $22^\circ$, turns easily rancid, dissolves readily in alcohol, and contains, combined with glyceryl, much laurostearic, also oleic, a little palmitic, myristic, caproic, caprylic and capric acids.

**[Coccognin.** Indifferent substance, discovered by A. Casselmann in the fruits of Daphne Mezereum. The fruits are pulverised and freed from oil by pressing and treating with sulphide of carbon. The residue is then extracted three times with alcohol of 95% and the solvent distilled off. The extract is freed from resin by alcohol of 70% and the residue dissolved in boiling alcohol of 95°/$_\circ$. The C. deposited on cooling is purified by recrystallisations.—The C. appears as a white powder, under the microscope as needle-shaped, sericeous crystals; of poignantly acrid taste and without odour. It is insoluble in cold, very sparingly in hot water, readily soluble in alcohol, insoluble in ether; of neutral reaction. Alkalies dissolve it with yellow, afterwards red colour. Subacetate of lead forms with it on heating a yellow precipitate. It may be fused and sublimed by careful heating without any residue, while emitting the odour of cumarin.]

**Codein** = $C_{36}$ $H_{21}$ $NO_6$ + 2HO. In the opium. Exhaust, according to Robertson-Gregory's method for obtaining the most important ingredients, the opium with water of 38°, evaporate the liquid under addition of pulverised chalk to a syrup, add chloride of calcium in excess, and boil for a few minutes. Dilute the fluid cold with a moderate quantity of water, whereby resinous flocks, meconate of lime and dyeing substances are thrown down, filter, evaporate under addition of a piece of chalk to the point of crystallisation, and separate the liquid from the sediment. The crystals, a mixture or double-salt of the chloride of morphin and Codein, obtained on cooling and concentrating, are freed from the black mother-ley by pressing and purified by recrystallising. Dissolve these in water and add ammonia, which precipitates the morphin. In concentrating the filtered liquid chloride of Codein crystallises first predominantly, and has to be freed from the bulk of chloride of ammonium by recrystallisation; it is then dissolved in hot water and decomposed by concentrated ley of potash, whereby a part of the Codein separates in the form of an oil, afterwards congealing, and another part crystallises on cooling. Wash with water, dissolve in ether, and evaporate under addition of water.— Forms small, white, silky scales, of little taste (its salts are bitter), fuses at 100° under loss of water, and decomposes afterwards; dissolves in 80 parts cold and in 17 parts boiling water, more readily in alcohol and in ether, in liquor of ammonia about as much as in water, in concentrated sulphuric acid without colour and turning brown by heat, in concentrated nitric acid under explosion with red colour, is not affected by chloride of iron and iodic acid. Combines with acids to mostly crystallisable salts which are almost insoluble in ether and not precipitated by ammonia or carbonate of soda, but are so incompletely by caustic potash.

**Colchicin** = $C_{34}$ $H_{19}$ $NO_{10}$. In all parts of Colchicum autumnale. Exhaust the seeds with alcohol of 90%, evaporate to a syrup, and mix hot with twenty times its quantity of water. Allow to rest, remove the fat-oil from the surface and precipitate the filtered liquid with subacetate of lead, remove from the filtered liquid the lead with phosphate of soda, and precipitate the Colchicin with tannic acid. Collect and wash the deposit and mix intimately in the moist state with oxyd of lead, dry the mixture by heat and exhaust with hot alcohol, evaporate the tincture to a syrup-like consistence and allow to dry over sulphuric acid in the vacuum. To get it completely pure, it must be thrown down repeatedly with tannic acid and treated with oxyd of lead as above.— Light-yellow, brittle, adhering like a resin when triturated, of a faintly aromatic hay-like odour; dissolves readily in water and alcohol, not in ether, has even in a very diluted solution a strong and lastingly bitter taste, is of neutral reaction, fuses at 140° and is decomposed by higher temperatures, yields an intensely yellow precipitate with mineral acids and alkalies, is thrown down by tannic acid, the chlorides of mercury and of gold; dissolves in concentrated sulphuric acid with dark-green colour which changes soon into yellow and is altered by nitric acid successively into blue, violet, and brown, and lastly yellow again, which becomes of a dark-red colour with ammonia. The Colchicin possesses no basic properties, and is even changed into a weak acid when treated with a mineral acid, and assumes thereby a crystalline appearance, *but without altering its composition.* This slightly acid compound has been called Colchicein; it forms white warty masses and needles, tastes much less bitter than Colchicin, and is of a slightly acid reaction, but possesses the same poisonous properties as the latter.

**Colocynthin**$=C_{56} H_{42} O_{23}$. In the pulp of the fruit, less in the seeds, of Cucumis Colocynthis. Treat the aqueous extract after drying with cold water, precipitate the filtered liquid with acetate of lead, filter again and precipitate with sub-acetate of lead, strain and remove the excess of lead by sulphuret of hydrogen, precipitate with tannic acid, collect and wash the deposit, mix it with hydrated oxyd of lead, warm and exhaust with alcohol, decolourise the tincture with animal charcoal, filter, dry and shake with ether (anhydrous) which leaves the C. in the pure state.—Amorphous, yellow, crystallises in white-yellow tufts when the alcoholic solution is slowly evaporated, has an intensely bitter taste, burns when heated, dissolves in 8 parts cold and in 6 parts boiling water, in 6 parts aqueous and in 10 parts absolute alcohol, not in ether, in concentrated sulphuric acid with crimson-red colour; secedes with diluted acids into sugar and another product.

**Colombic Acid**$=C_{42} H_{22} O_{12} + HO$. Besides colombin and berberin in the Colombo-root (from Jateorrhiza palmata). Digest hot the dried alcoholic extract of the root with milk of lime, saturate the filtered liquid with hydrochloric acid, which causes a yellow amorphous precipitate, remove from the latter the berberin by washing with water, and the colombin by boiling with ether, dissolve the rest in ley of potash and adduce carbonic acid gas, oversaturate the fitered liquid with hydrochloric acid, wash the white flocks, which will separate, with water and dry.— Amorphous, pale straw-yellow powder, of strongly acid reaction, tastes less bitter than colombin, is almost insoluble in water, dissolves with yellow colour in alcohol, very little in ether, better in acetid acid, with brown-yellow colour in potash-ley.

**Colombin**$=C_{42} H_{22} O_{14}$. The indifferent bitter ingredient of the Colombo-root (from Jateorrhiza palmata). Exhaust with alcohol of 75°/$_o$, distil off the alcohol from the tincture, evaporate to dryness on the water-bath, dissolve in water, shake with ether, evaporate the ethereous solution and purify the crystallised C. by pressing and recrystallising in ether.—Forms white or transparent prisms and needles, without smell, very bitter, neutral; fuses like wax, decomposes in higher temperatures; dissolves little in cold water, alcohol and ether, in 30 to 40 parts boiling alcohol, in acetic acid, a little in volatile oils, in alkalies and re-precipitable by acids, in concentrated sulphuric acid with orange, afterwards dark-red colour; not precipitable by metallic salts.

**Colophony.** *See* Abietic Acid.

**Conessin**$=C_{25} H_{22} NO$. In the seeds and bark of Wrightia antidysenterica. Exhaust the bark with water containing a small admixture of hydrochloric acid, precipitate the extract with ammonia, treat the deposit with alcohol, evaporate the tincture to a syrup's thickness, bring to dryness with acetate of lead and with a little ammonia and treat with ether. The C. remains after evaporating.—White, amorphous powder of very bitter, acrid and rancid taste, is destroyed by heat, dissolves in alcohol, ether and chloroform; according to other observers it dissolves little in alcohol, ether and sulphide of carbon.

**Conhydrin**$=C_{16} H_{17} NO_2$. Alkaloïd of Conium maculatum, existing besides coniin in the flowers and seeds. Exhaust with water containing a little sulphuric acid, saturate the moderately concentrated extract with an excess of lime or caustic potash, distil and saturate the alkaline distillate with sulphuric acid, evaporate to a syrup's consistence, shake with alkohol and evaporate the alcoholic solution, add potash-ley, shake with ether and let the ether evaporate from the ethereous solution. The residue when subjected to distillation yields at first oil-like coniin with a little ether and water and afterwards a dry crystalline substance. This when purified by pressing, &c., is pure Conhydrin.—It represents colourless, iridescent laminæ of mother-of-pearl lustre, smells faintly like coniin, liquifies with a gentle heat and sublimates already below 100°, dissolves moderately in water, readily in alcohol and in ether, is of strongly alkaline reaction and dislodges the ammonia from its compounds. It yields with hydrochloric acid an uncrystallisable salt (whereas chloride of coniin crystallises and is permanent at the air); is converted into coniin when heated with anhydrous phosphoric acid.

[**Coniferin** $= C_{48} H_{32} O_{24} + 6 HO$. Glucosid, detected by Th. Hartig in the cambium liquid of coniferous trees, viz.:—Pinus Abies, P. picea, P. Strobus, P. Cembra, and P. Larix. It is obtained from the juice by boiling, filtering, and inspissating to one-fifth of its volume, when the C. after cooling separates in crystals, which have to be pressed and recrystallised from water or alcohol under addition of animal charcoal.—It forms white, sericeous, delicate needles or warty masses consisting of concentrically grouped spearlets. They lose their water of crystallisation at 100°, fuse at 185°, and congeal after cooling to a glassy mass. In higher temperatures, the C. turns brown and becomes carbonised under emission of caramel odour and a peculiar aromatic smell. It dissolves little in cold, copiously in hot water, sparingly in strong alcohol, not in ether. The taste of the aqueous solution is slightly bitter. It turns the polarised light to the left. Sulphuric acid colours C. dark violet. The solution in hydrochloric acid, on heating and evaporating, throws down an intensely-blue body. Treated with chromate of potash and sulphuric acid, C. is converted into vanillic acid, which might serve as a substitute for vanilla.]

**Coniin** $= C_{16} H_{15} N$. In all parts of Conium maculatum, most copiously in the seeds. Distil the latter with water and an admixture of lime, saturate the distillate with sulphuric acid, evaporate almost to dryness, agitate the granular remnant with a mixture of two parts absolute alcohol and one part ether, evaporate the solution, distil the remaining mass with potash ley and desiccate with chloride of calcium.—Limpid, oily fluid of penetrating nauseous hemlock-odour, and of acrid, nauseous, tobacco-like taste, of 0·89 density, boils at 170°, has a strongly alkaline reaction, dissolves in 100 parts water; the solution becomes cloudy when warmed, and clears again on cooling; dissolves readily in alcohol, ether and oils, and neutralises the acids completely. The aqueous solution is precipitable by tannic acid.

**Convallamarin** $= C_{46} H_{44} O_{24}$. Precipitate the aqueous decoction of the root of Convallaria majalis or the decoction of the whole herb, as obtained for the preparation of convallarin, with subacetate of lead, remove the lead from the filtered liquid by means of carbonate of soda, precipitate with tannic acid, wash and dry the deposit, exhaust with alcohol, remove the tannic acid by digesting the tincture with slaked lime, distil off the alcohol from the filtrate, free the remaining liquid from the rest of the lime by adducing carbonic acid gas and evaporate to dryness.

The C. thus obtained, contains resin and mineral substances; the former may be removed by means of ether; the latter by dissolving repeatedly in water, precipitating with tannic acid and treating as above.—White powder, intermixed with small crystals of a peculiar, lasting, sweet-bitter taste; is softened by heat and afterwards destroyed; dissolves readily in water and in alcohol, not in ether, is soluble unaltered in liquor of ammonia, assumes a beautiful violet colour with concentrated sulphuric acid; is resolved by diluted acids into sugar and a resin (Convallamaretin).

**Convallarin** $= C_{34} H_{31} O_{11}$. In Convallaria majalis. Boil with water the whole herb, collected with the roots while blossoming or shortly after the time of inflorescence, employ the decoction for the preparation of Convallamarin (*see* this), exhaust the remnant with alcohol of 0·84, precipitate the tinctures with subacetate of lead, remove the lead from the filtrate by sulphuret of hydrogen, distil off the alcohol and let crystallise. The crystals, mixed with resin and chlorophyll, must be pressed and washed with ether.—Forms rectangular prisms, has a rancid taste when dissolved in water or alcohol, fuses at 100° and is afterwards destroyed, dissolves very little in water, readily in alcohol, not in ether, divides under the action of diluted acids into sugar and a resin (Convallaretin).

**Convolvulin** $= C_{62} H_{50} O_{32}$. The main ingredient of the resin of Ipomaca Purga. Boil the root first with water, afterwards with alcohol, add water to the tincture until it shows traces of turbidity, boil with animal charcoal, filter and evaporate. The yellowish, brittle resin is reduced to powder, exhausted with ether, dissolved in absolute alcohol and precipitated by ether.—Colourless substance, at 100° brittle and reducible to a white powder, without taste or smell, fuses at 150° without loss and decomposes afterwards, has a slightly acid reaction when dissolved in alcohol, dissolves very little in water, in any quantity of alcohol, not in ether or in oils, both fixed and volatile ones, readily in acetic acid, in concentrated sulphuric acid with carmine-red colour, in alkalies and and alkaline earths under formation of an acid and not reprecipitable by acids, breaks up with diluted acids into sugar and an oily product.

**Copaiva Balsam.** Exudation of the stems of various kinds of Copaifera. Is pale-yellow, of honey consistence, of 0·94 to 0·95 density, has a peculiar, unpleasant balsamic smell and an aromatic, bitter taste, of acid reaction, dissolves copiously in alcohol of 90%, mixes with ether and oils in every proportion, dissolves mostly clear in liquor of ammonia and in potash-ley, hardens with 1-16th caustic lime or magnesia to a solid paste. Consists of volatile oil (up to 80°/₀), of two resins, the one of which is prevailing and hard (copaivic acid), the other soft, and of metacopaivic acid.

**Copaivic Acid** $= C_{40} H_{29} O_3 + HO$. In the balsam of Copaiva as hard resin besides a volatile oil and a soft resin. To prepare it, mix nine parts balsam with two or more parts liquor of ammonia of 0·95, and let stand cold in a closed vessel whereby at 10° only after several weeks, at − 12° after twenty-four hours, white, shining needle-like crystals will be formed. Drain and shake the crystals with a little ether, which dissolves the balsam and leaves the crystals little altered, press and recrystallise in absolute alcohol.—Long, translucid or transparent, colourless crystals of faint smell and bitter taste, of acid reaction, insoluble in water, readily soluble in absolute alcohol, less so in aqueous alcohol and in ether, in permanent and volatile oils, in concentrated sulphuric acid with red-brown colour; also in potash-ley but reprecipitated by an excess of the solvent.

**Copal.** Resinous exudation of several trees, viz., Valeria Indica, Hymenæa Courbaril, Trachylobium Gaertnerianum, Cynometra Spruceana, Vouapea phaseolocarpa, Bursera Copal, and other species of these genera. Shows, according to its origin, some physical and chemical distinctions, but is in general pale or brownish-yellow, translucid or transparent, hard and brittle, without taste or smell, heavier than water, fusible without noticeable smell, but with a change of properties, yields in higher temperatures volatile oil and water. It dissolves little in absolute alcohol, more readily when exposed to the air for a long time, more readily still after melting or with an admixture of camphor; swells up in ether and dissolves completely, also in caustic alkalies, copiously in chloroform, slowly in benzol and castor oil, partly in sulphide of carbon and in volatile oils. Five mostly acid resins have been obtained from Copal through the agency of different solvents.

**Copalchin.** Alkaloïd of the Copalchi-bark (from Croton niveus), of which is only known that it is distinctly crystalline, tastes similar to quinin, dissolves in ether, is precipitated white by alkalies from its solution in acids, and behaves towards chlorine and ammonia like quinin (turns dark-green).

**Coriamyrtin** $= C_{40} H_{24} O_{14}$. Bitter ingredient of the leaves and fruits of Coriaria myrtifolia. Precipitate the juice of the fresh or the aqueous extract of the dried plant with subacetate of lead, remove the excess of lead from the filtered liquid by means of sulphuret of hydrogen and evaporate to the consistence of syrup. The residue yields up to ether the C. which is obtained by evaporation, and purified by recrystallising in alcohol.—Forms white, four to six sided klinorhombic prisms, very bitter; loses nothing of its weight at 200°, fuses at 220°, is decomposed by higher temperatures; of neutral reaction; dissolves at 22° in 70 parts cold and a little more in boiling water; in 50 parts cold, very copiously in boiling alcohol, also in ether, benzol, chloroform, scarcely in sulphide of carbon; is converted by diluted acids into sugar and a resin.

**Corksubstance = Suberin.**

**Cornin or Cornic Acid.** Crystalline bitter substance of the root and bark of Cornus florida, and probably also in that of other species. The aqueous extract is repeatedly shaken with freshly precipitated oxyd of lead; filtered; evaporated to honey-consistence; treated with absolute alcohol, and the tincture evaporated.—Fine silky needle-like crystals of very bitter taste, readily soluble in water and alcohol, slowly in ether, of slightly acid reaction, not precipitable by alkalies, tannic acid, salts of iron and baryta, or neutral salts of lead; but precipitated by subacetate of lead and nitrate of silver.

**Cortici-pino Tannic Acid = $C_{16}$ $H_7$ $O_7$.** In the bark of aged stems of Pinus sylvestris. Distribute the lead-salt obtained in the preparation of Pinicorretin, in water, decompose with sulphuret of hydrogen, remove the sulphide of lead by filtration, and evaporate in a current of carbonic acid gas.—Bright red-powder, the aqueous solution of which turns green with chloride of iron.

**Corydalin = $C_{46}$ $H_{29}$ $NO_7$.** Alkaloïd of the tubers of Corydalis tuberosa and C. fabacca. Exhaust with water acidulated with hydrochloric acid, press, throw down the liquid with carbonate of soda, collect the deposit, wash, dry, and exhaust with water and an admixture of hydrochloric acid, precipitate the filtered liquid with carbonate of soda, collect, wash and dry the deposit, shake with oil of turpentine, pour off the oil containing the C., shake it with water acidulated with hydrochloric acid, precipitate the aqueous solution with carbonate of soda, collect, wash, and dry the deposit.—White voluminous strongly adhering powder, without smell or taste, very bitter when dissolved, assumes a citron-yellow colour when exposed to the sunlight, fuses at 70° without loss of weight, and decomposes in higher temperatures; insoluble in water, dissolves in nine parts of alcohol of 90% at ordinary temperature, and in equal parts when boiling hot, in three parts of ether, not in solutions of alkalies; in concentrated nitric acid with yellow colour, and with the same colour in concentrated sulphuric acid, but changing slowly into violet; is precipitable by tannic acid, by the hydrates and carbonates of alkalies, the chlorides of mercury, of gold, and of platinum. Yields amorphous resinous salts.

**Crataegin.** Crystalline, bitter substance of the bark of the branchlets of Crataegus Oxyacantha. Obtained by boiling with water, treating the decoction with caustic lime, &c. Crystallises in warty masses similar to grape-sugar, is grey-white, tastes very bitter, dissolves readily in water, less so in alcohol, not in ether, is neutral, and contains no nitrogen.

**Crocin** $= C_{58} H_{42} O_{30}$. The yellow pigment of the stigmata of Crocus sativus and of the fruits of Gardenia grandiflora, also occurring in Fabiana imbricata. Remove the fat substances of saffron by ether, boil with water, precipitate the decoction with sub-acetate of lead, wash the deposit, and decompose under water by sulphuret of hydrogen, collect the sulphide of lead which retains the C., treat with boiling alcohol, and evaporate the solution.— Bright-red powder, without odour, decomposes above 100°, dissolves in water with orange colour, readily in alcohol, with great difficulty in ether, readily in diluted alkalies, turns blue with concentrated sulphuric, green with nitric, black-brown with hydrochloric acids; changes when heated with diluted acids into sugar, and a deep-red powder, which is sparingly soluble in water (Crocetin).

**Croton Oil.** Obtained by pressing the seeds of Croton Tiglium. Brown, of thick fluidity and of very rancid odour, irritates strongly the skin, and acts as a drastic; becomes turbid at a moderate cold, and slowly changes at the air to a thick, viscous mass. Dissolves in twenty-three parts alcohol of $85\%$, is not solidified by hyponitric acid. It contains Crotonol (substance irritating the skin), a product of decomposition of the latter, causing the smell, and as glycerids: stearic, palmitic, myristic and lauric acids; two or more liquid acids belonging to the same series as oleic acid, but differing from the latter; also, angelic [Tiglinic acid, according to Genther and Froehlich] and crotonic acids. A mixture of the two latter acids appears to constitute the Iatrophic acid of Pelletier and Caventon, and the Crotonic acid of the earlier chemists.

**Crotonol** $= C_{18} H_{14} O_4$. In the fat-oil of the seeds of Croton Tiglium, forming $4\%$ of the oil; is the ingredient which irritates the skin, but not the drastic principle. Shake the oil with an alcoholic solution of caustic soda sufficient to form an emulsion; warm gently for a few hours, add water or a solution of common salt, and remove the oil, which will form on the surface, by repeatedly filtering through wet paper. The aqueous liquid, when mixed with water and hydrochloric acid, separates another oil which is to be dissolved in cold alcohol, digested with hydrated oxyd of lead until the acid reaction has disappeared (whereby a flocky, afterwards conglutinating precipitate is formed), and mixed with a little caustic soda and much water. A milky liquid is produced, which becomes clear after the oil has subsided. Wash the latter, first by itself, and afterwards dissolved in ether, with water, and evaporate the ethereous solution in a vacuum. The Crotonol remains pure. Colourless or slightly wine-yellow, viscid, turpentine-like mass of a faint, peculiar odour; not distillable, is changed when boiled with potash-ley into a brown resin, which does not rubify the skin.

**Cubebin**$=C_{20} H_{16} O_6$. Must not be confounded with cubeb-stearopten. The cubebs, freed from the volatile oil by distillation with water and pressing, are to be dried and exhausted with boiling alcohol. Evaporate the tinctures to honey-consistence, and remove foreign matters by treating with potash-ley; wash the remaining C. with water, and purify by repeatedly recrystallising in alcohol.—Forms small, white needle-like crystals or silky laminæ, neutral, without taste or smell; fuses at 120°, loses at 200° nothing of its weight, and decomposes afterwards; dissolves scarcely in cold, very little in hot water, at 20° in 76 parts absolute, and in 140 parts alcohol of 0·850, in ten parts boiling alcohol, in 26 parts ether, little in chloroform; but in oils, also in acetic acid; is not changed by alkalies, becomes blood-red through concentrated sulphuric acid, and is destroyed by heat; is not changed by hot hydrochloric acid.

**Cumarin**$=C_{18} H_6 O_4$. Found as yet in the following vegetables: Aeranthus fragrans, Nigritella alpina, Orchis purpurea, Anthoxanthum odoratum, Hierochloa borealis and other species of that genus; Asperula odorata, Dipterix odorata, Melilotus officinalis and congeneric plants; fruit of Myroxylon toluiferum; Herniaria glabra, Liatris odoratissima; bark of Prunus Mahaleb. To prepare it, exhaust with alcohol, distil off the alcohol from the tinctures, let the remnant crystallise and recrystallise in water or alcohol.—Forms colourless silky leaflets, of very aromatic odour, of aromatic and pungent taste; fuses at 67°, boils at 270° and sublimates unaltered; is of neutral reaction; dissolves scarcely in cold, readily in boiling water, most readily in alcohol, ether, volatile and fixed oils; yields by fusion with caustic potash salicylic acid.

**Cuminol**$=C_{20} H_{12} O_2$. In the volatile oil of the seeds of Cuminum Cyminum and of Cicuta virosa besides Cymen$=C_{20} H_{14}$. To be looked for in many other Umbelliferæ. The crude volatile oil is submitted to distillation; at 200° the whole of the cymen has distilled besides much Cuminol, the remaining oil is pure Cuminol which has to be distilled in a current of carbonic acid gas.—Colourless or yellowish oil of strong odour of cumin and of acrid, burning taste, of 0·972 density, boils at 220°.

**Curarin**$=C_{20} H_{15} N$. Alkaloïd of the vegetable arrow-poison Curare of South America, obtained from Strychnos Guianensis. To prepare it, boil the curare under addition of a little carbonate of soda with absolute alcohol, distil off the alcohol from the filtered tincture and dissolve the remnant in water; the aqueous solution is thrown down with chloride of mercury, the deposit washed and decomposed with sulphuret of hydrogen and the solution containing impure chloride of C. purified by repeatedly precipitating with chloride of mercury.—It crystallises in colourless, quadrangular prisms of lasting bitter taste, becomes moist at the air, dissolves in water and alcohol in every proportion, less readily in chloroform and amylalcohol; not soluble in anhydrous ether, benzol, oil of turpentine and sulphide of carbon; has a slightly alkaline reaction, saturates the acids completely and yields crystallisable salts; assumes with concentrated sulphuric acid a beautiful blue, with chromate of potash and sulphuric acid (like strychnin) a violet and with concentrated nitric acid a purple-red colour.

**Curcumin.** The yellow resinous pigment of Curcuma longa. Treat the root, exhausted by water, with alcohol, evaporate to dryness, treat with ether, which leaves a brown extractive substance, and evaporate the ethereous solution.—Red-brown, of a beautiful yellow colour when finely divided or dissolved, of acrid, pepper-like taste, fuses above 50°, dissolves little in boiling water, readily in alcohol, ether and oils, becomes brown-red by alkalies, yellow-red by boric acid, carmine-red by diluted sulphuric acid.

**Cusconin**=ARICIN.

**Cusparin**=ANGUSTURIN.

**Cyanin**=ANTHOCYAN.

**Cyclamin**=$C_{28} H_{16} O_{12}$. The poisonous ingredient of the tubers of Cyclamen Europæum; also, but in less quantity, in the roots of Primula veris, Anagallis arvensis, Limosella aquatica, and doubtless in many other Primulaceæ, Scrophularinæ and allied plants. Treat the bruised root with cold water, evaporate the liquid to honey-consistence, exhaust with alcohol and allow to evaporate.—Forms small, white crystals or white, amorphous, lustreless, friable mass, becomes brown in the light, swells up in damp air, has a very acrid, rancid taste and neutral reaction; is not volatile; dissolves in 500 parts water, becomes transparent in contact with cold water, afterwards viscid and dissolves readily; the solution yields a froth like soap-water, becomes cloudy at 60° to 70° from the separation of coagulated C. and resumes its clearness when cold and kept for several days; dissolves in alcohol, wood spirit, acetic acid and glycerin, not in sulphide of carbon, chloroform, ether and oils, is completely precipitated by tannic acid, but not by metallic salts; dissolves in alkalies, in diluted acids more readily than in water; is decomposed by synaptase and yields sugar.

**Cylicodaphne Fat,** obtained from the fruit of Tetranthera calophylla. Fuses at 45°, contains 14% olein and 85% laurostearin.

**Cymen** or **Cymol** $= C_{20} H_{14}$. In the volatile oils of Cuminum Cyminum, Cicuta virosa, Carum Ajowan, also in the oil of Thymus vulgaris and expected to exist in many other plants of Umbelliferæ and Labiatæ. That portion of the oil of cumin which distils below 200°, has to be rectified over melting caustic potash, which retains the whole of the cuminol as cumate of potash.— Colourless oil of great light-refracting power, smells pleasantly like lemons; of 0·86 density; boils at 171°.

**Cynapin.** Alkaloïd said to occur in Æthusa Cynapium, to crystallise in prisms, to dissolve in water and alcohol, but not in ether, to possess an alkaline reaction and to yield a crystallisable salt with sulphuric acid.

**Cyperus Oil** $=$ Oil of Cyperus esculentus.

**Cytisin.** Alkaloïd of the ripe seeds of Cytisus Laburnum. Throw down the aqueous extract with subacetate of lead, filter, precipitate with tannic acid, mix the deposit with oxyd of lead, dry, exhaust with alcohol, evaporate the tincture, saturate with nitric acid, mix the well-crystallising salt with oxyd of lead, dry and exhaust with absolute alcohol. The pure C. crystallises in the latter as a colourless, radiated crystalline, deliquescent mass which sublimates undecomposed by careful heating; has a strongly alkaline reaction and saturates the acids completely.

[**Dambonit** $= C_8 H_8 O_6$ . Variety of sugar, found by A. Girard in the Gaboon caoutchouc, which is the exudation of Landolphia Owariensis and other species of that genus. It is obtained from the recently-inspissated juice by treating with aqueous alcohol.— It is white, of a sweet taste, easily soluble in water and aqueous alcohol, sparingly in absolute alcohol. It crystallises from alcohol in six-sided prisms, from water in oblique prisms with $1\frac{1}{2}$ HO. It is not fermentable and does not reduce the alkaline copper solution. D. becomes carbonised by hot sulphuric acid, and is oxydised to oxalic and formic acids by hot nitric acid. Concentrated alkalies are without any influence on D. even at 100° C., though they lessen its solubility in water. Lime-water, barytawater, and acetate of lead produce no precipitates. Iodide of potassium forms with D. beautiful crystals of a double salt. With fuming hydroiodic and hydrochloric acids, D. separates into methyl compounds and a substance polymeric with glucose (Dambose $= C_6 H_6 O_6$ ). Dambose is not fermentable and does not reduce the alkaline copper solution.]

**Dammara Resin.** (a) Indian, from Dammara orientalis, is paleyellow, clear and transparent, sometimes with a white powder on the surface; it breaks easily with a concheous, shining fracture; almost without smell and taste, fuses at 73° (according to other experiments at 150°), dissolves in alcohol and ether only partially, completely in volatile and fixed oils, not in acetic and hydrochloric acids, liquor of ammonia and potash-ley. Contains dammaryl, hydrate of dammaryl and dammarylic acid, described below. (*b*) New Zealand D., from Dammara australis, is of a pale amber colour, transparent inside, translucid in the outer layers and covered with an opaque crust, evolves on friction a turpentine odour, fuses in boiling water and does not appear to be chemically different from the Indian dammara. (The Dammaras of Australia and the South Sea Islands produce similar resins.—F. v. M.)

**Dammaryl** $= C_{40} H_{32}$. In the Indian dammara-resin, which yields $13^\circ/_\circ$ of it, probably also in that of New Zealand. Exhaust with warm alcohol of $82^\circ/_\circ$, which removes soluble resins, treat the rest with ether, concentrate the ethereous solution and throw the remnant into boiling water for a few seconds, collect and dry as quickly as possible in a vacuum.—White, amorphous, magnesia-like powder of high lustre, not electric by friction, mollifies at $145^\circ$, fuses at $190^\circ$ and becomes decomposed in higher temperatures; does not dissolve in water or in alcohol, readily in cold ether and warm oils, not in alkalies; is easily oxydised by the air.

**Dammaryl Hydrate** $= 2 C_{40} H_{32} + HO$. Remains after treating the dammara-resin with alcohol and ether to the extent of $8^\circ/_\circ$, as a grey, dough-like mass which retains the ether tenaciously, and becomes after drying shining, brittle and easily friable; mollifies at $205^\circ$, fuses at $215^\circ$; is insoluble in alcohol, ether, acetic acid, and alkalies; dissolves completely in hot oil of turpentine, better in petroleum, little in fixed oils.

**Dammarylic Acid** $= C_{45} H_{36} O_3$. Dammara-resin when treated with weak alcohol, yields to it $36^\circ/_\circ$. By evaporating the solution, boiling the residue with water, to remove traces of alcohol, and drying, the hydrate of the acid $= C_{45} H_{36} O_3 + HO$ is obtained as a white, highly electric powder, fusing at $56^\circ$ and of acid reaction when dissolved.—By treating the resin with absolute alcohol, $46\%$ are dissolved; the dissolved part is the anhydrous acid, similar to the above, but whiter, fusing at $60^\circ$ and of more acid reaction.

**Daphnin** $= C_{62} H_{38} O_{38} + 8 HO$. Bitter glucosid of the bark of Daphne alpina and Daphne Mezereum and others, also obtained from the flowers of these plants. Prepare an alcoholic extract of the fresh bark and exhaust the former with ether, boil the undissolved portion in water, allow to rest for twenty-four hours in order to remove the resin, throw down with acetate of lead, remove the deposit and precipitate the D. in the liquid with sub-acetate of lead. Decompose the deposit under water with sulphuret of hydrogen, evaporate the filtered liquid to the consistence of syrup, dilute, filter off from the resin, concentrate again, &c.,

and lastly free the syrup-like liquid from any resin by frequently shaking with ether. The aqueous liquid, separated from the ether, consolidates soon afterwards to crystals of D. which have to be washed and recrystallised in hot water.—Colourless, rectangular prisms or needles of moderately bitter, afterwards acerb taste, neutral; lose the water at 100° and become opaque; fuse at 200° and decompose in higher temperatures; dissolve little in cold, readily in hot water, a little more readily in cold alcohol than in water, most readily in boiling alcohol, not in ether, in the oxyds and carbonates of alkalies with gold-yellow colour; are decomposed when boiled with diluted acids, also by emulsin and by yeast, into sugar and another crystalline product (Daphnetin $= C_{38} H_{14} O_{18}$).

**Datiscin** $= C_{42} H_{22} O_{24}$. Bitter glucosid in the herb and root of Datisca cannabina. Exhaust the root with wood-spirit, evaporate to syrup-consistence, throw down resinous substances with hot water, evaporate the decanted liquid and let crystallise. The impure D. is pressed, dissolved in alcohol, mixed with water, filtered off from the resin and obtained pure by evaporation.— Colourless needle-like or laminate silky crystals, soft and transparent like grape-sugar, neutral; fuses at 180° and is destroyed afterwards while evolving an odour of burnt sugar, dissolves little in cold, better in hot water, readily in alcohol, less so in ether; the solutions have a strongly bitter taste and neutral reaction; dissolves in alkalies with yellow colour, and is in these solutions precipitated and discoloured by acids; decomposes on heating with diluted acids into sugar and Datiscetin $= C_{30} H_{10} O_{12}$. Is precipitable by tannic acid, acetate and subacetate of lead, by the salts of oxyds of iron, of copper and zinc.

**Daturin** $=$ ATROPIN.

**Delphinin** $= C_{48} H_{35} NO_4$. Alkaloid of the seeds of Delphinium Staphisagria. Mix the bruised seeds with water so as to form a thin pulp, warm over the water-bath for several hours, strain, remove the bulk of the fat by pressing, digest the remaining mass with strong alcohol, press again, distil off the alcohol from the tincture, treat the remaining oleo-resinous mass with water to which some hydrochloric acid is admixed, precipitate the filtered liquid with ammonia, wash and dry the deposit and exhaust with ether, which leaves impure staphisagrin as a brownish resin and dissolves the D. without colour. The D. remains after evaporating as a transparent, resinous, slightly-coloured mass. To purify it completely, dissolve again in water and hydrochloric acid, precipitate with ammonia, dissolve the deposit in ether, and leave to evaporate by itself. The pure precipitated D. is entirely white; when obtained by evaporation from the ethereous solution, it has an amorphous resinous appearance. It has a lasting acrid taste and a strongly alkaline reaction, is almost insoluble in water, readily soluble in alcohol, ether, and acids.

**Deutocatechuic Acid**=CATECHUIC ACID.

**Dextrin** = $C_{12} H_{10} O_{10}$. With this name a gum-like product, obtained from starch under the agency of diastase or acids, is designated, which differs from common gum by turning the polarised light strongly to the right, while gum arabic turns it to the left, and by yielding no mucic acid with nitric acid. Neither are its solutions clouded by subacetate of lead, chloride of iron or silicate of potash. Consequently it ought not to be coloured (red) by iodine, a fact which must here be particularly stated, because such a property is sometimes ascribed to D., but which only takes place when it is contaminated with a kind of modified starch, as indeed often occurs with different commercial kinds of D. As is the case with gum, D. seems not to be wanting in any plant; yet it is often confounded with the former on account of its great similarity, and perhaps because it is constantly associated with it.—Its isolation and quantitative estimation correspond with that of gum. It will be easy from the above to distinguish it; but not when it is associated with gum, because it is distinguished from the latter more by negative than by positive qualities, except in the behaviour towards polarised light, which is entirely opposite to that of gum and becomes modified, when the two substances are mixed together.

**Dextroglucose**=GRAPE SUGAR.

**Diastase.** Nitrogenised ingredient of the germinated barley (Hordeum vulgare) which possesses in a high degree the property of converting starch into dextrin and sugar. It is obtained when the contused malt is macerated with cold water, pressed, the liquid mixed with a little alcohol, the filtered liquor completely precipitated with alcohol, the deposit washed, redissolved in water, precipitated with alcohol, and lastly dried on a glass-plate at a very gentle heat. After triturating, it forms a white, amorphous powder without taste or smell; dissolves in weak, not in strong alcohol; is not precipitable by subacetate of lead. The aqueous solution when heated to the boiling point becomes cloudy and inactive; it even becomes decomposed on keeping and turns acid.

**Dica Fat** obtained from the dica-bread (the seeds of Irwingia Barteri). Similar to Cacao fat; fuses at 30° to 33°, contains laurin and myristin.

**Digitaletin** = $C_{44} H_{38} O_{18}$. (According to Waltz, the Digitalin of Homolle). In Digitalis purpurea and D. lutea, and doubtless several congeners; ingredient of the crude digitalin. The leaves are reduced to powder and freed from chlorophyll by ether. They are then exhausted with alcohol, and the tincture is precipitated with alcoholic acetate of lead: the filtered liquid is freed from the surplus of lead by sulphuret of hydrogen, decolourised with animal charcoal and allowed to evaporate spontaneously. —White, warty mass; has a very bitter taste when dissolved in cold water; fuses at 175°, and decomposes in higher temperatures; dissolves in 848 parts cold, in 500 parts at 45°, and in 222 parts boiling water, in $3\frac{1}{2}$ parts cold alcohol of 0·85, in $2\frac{1}{5}$ to $2\frac{1}{2}$ parts at the boiling heat; in 1960 parts cold, and in 1470 parts boiling ether; changes into sugar, and a resin when boiled with diluted acids.

**Digitalic Acid, Non-Volatile.** In Digitalis purpurea. Evaporate the aqueous infusion of the leaves to syrup-consistence, add alcohol of $92°/_{\circ}$ as long as a precipitate ensues, filter after a few days, distil off the alcohol and evaporate to the consistence of thick honey. Exhaust the latter with several changes of hot ether until it has nearly lost the bitter taste whereby digitalin or an allied substance and Digitalic acid are dissolved. Add to the ethereous tinctures by degrees, caustic baryta until alkaline, collect the yellow deposit, wash with ether until it has lost the bitter taste, afterwards with alcohol until the latter passes off colourless and decompose (the deposit) under water with (best an insufficient quantity of) diluted sulphuric acid. Evaporate the reddish, very acid, filtered liquid as much as possible under exclusion of the air, allow to stand cold, decant from brown flocks that will have subsided, precipitate any traces of Digitalate of baryta with alcohol of 95%, and evaporate the filtered liquid in a vacuum. The crystals, obtained from the brown mother-ley, are recrystallised in alcohol, if possible under exclusion of the air.—White needle-like crystals of not unpleasant acid taste, and of faint peculiar smell; fuses by heat, and is decomposed afterwards; changes also readily under the influence of light and air, especially in the presence of alkalies; dissolves most readily in water and alcohol, less in ether. The salts of the alkalies, alkaline earths and of zinc oxyd dissolve in water.

**Digitalic Acid, Volatile** $= C_{10}\,H_9\,O_3 + HO$ (of the same atomic composition as valerianic acid). In Digitalis purpurea. Distil the herb with water, saturate the distilled liquid with carbonate of soda, evaporate to dryness, dissolve in a little water, add oxalic acid and distil. The acid is floating on the distillate as a colourless oil of strongly acid reaction, smells like the bruised herb, faintly resembling valerianic acid, of disagreeable taste, forms with lead oxyd a crystallisable salt, soluble in water.

**Digitalin** $= C_{56}\,H_{48}\,O_{28}$ (according to Walta). Occurrence like Digitaletin. Exhaust the pulverised herb with water in a displacement apparatus, digest the liquids with lead oxyd and a little subacetate of lead until the latter test ceases to cause any

68

cloudiness, throw down the lead in solution by sulphuric acid, neutralise with ammonia, precipitate with tannic acid, mix the washed deposit with hydrated lead oxyd, dry, exhaust with alcohol, decolourise (if necessary) with animal charcoal, and evaporate.—Yellowish, amorphous mass, very bitter, fuses at 137° and is destroyed in higher temperatures, dissolves in 125 parts cold and in 42 parts hot water, in $2\frac{1}{2}$ parts cold and in $1\frac{2}{3}$ parts hot alcohol, in 20,000 parts cold and in 10,000 parts hot ether, in 80 parts chloroform, breaks up when boiled with diluted acids into sugar and two other products.

[**Digitalin, crystallised** $= C_{50} H_{40} O_{30}$. Alkaloïd, discovered by C. A. Nativelle in the herb of Digitalis purpurea, gathered in spring. To prepare it, add to 100 parts of the coarsely pulverised herb 25 parts acetate of lead and 100 parts water, and after twelve hours exhaust with water in a displacement apparatus. The residue is exhausted with 300 parts of proof spirit, and to the liquid is added a solution of 4 parts phosphate of soda. Remove the precipitate by filtering and drive off the spirit by distillation on a water-bath. By evaporating the residue on a water-bath to about one-tenth, a pitch-like mass is obtained, which is washed with water and left on blotting-paper to dry at the air. The two or three parts of the pitch-like body, obtained in this way, are dissolved in double their quantity of alcohol of 60% and allowed to stand until yellow radiated crystals have formed, together with those of an inactive body. Separate the crystals from the mother-ley and wash with alcohol of 35% to remove the inactive principle. Dissolve the remaining crystals in boiling alcohol of 80%, boil a few minutes with animal charcoal, filter, evaporate half the alcohol and allow to crystallise. Separate the crystals from the mother-ley and dry over a water-bath. Finely pulverise the crystals and shake strongly with 20 parts chloroform pure and especially free from alcohol. After twenty-four hours filter and distil to dryness. The remaining yellow Digitalin is dissolved in alcohol of 80%, boiled with a little washed animal charcoal and crystallised. After repeating the latter process several times, the Digitalin is obtained pure.—D. is neutral, without odour, of an intense, lasting bitterness. It dissolves in chloroform in every proportion; in twelve parts cold and six parts boiling alcohol, not so well in absolute alcohol. Ether, benzol and water dissolve only traces. Sulphuric acid dissolves it with green colour which turns to a light-red through vapour of bromine and becomes green again on addition of water. Nitric acid dissolves it without colouration, but soon changing to yellow. Hydrochloric acid dissolves it with green-yellow colour, turning slowly to an emerald green. Heated above 100° it becomes very elastic; heated on platinum-foil it melts without colouration to a transparent mass, evolves copious white vapours, becomes brown and disappears completely. It acts as a strong poison.]

**Digitoleic Acid.** In Digitalis purpurea. Precipitate the extract of the leaves cold with subacetate of lead, boil the washed deposit with a solution of carbonate of soda for a quarter of an hour, saturate the brown filtrate with an excess of diluted sulphuric acid which throws down flakes that have to be collected, washed, dried and boiled with alcohol of 85°/$_{0}$. Evaporate the alcoholic solution, treat the remnant of crystalline structure repeatedly with ether, and let the liquids evaporate. A green oil remains, which soon solidifies into a granular, crystalline mass. Dissolve in aqueous bicarbonate of soda, precipitate with acetic acid, wash, dissolve in ether and evaporate.—Green, concentrically united crystalline needles, fusing at 30°, of no disagreeable odour, of bitter and acrid taste; has an acid reaction when dissolved in alcohol, leaves greasy spots on paper; little soluble in water, readily in alcohol and in ether, forms with the oxyds of metals insoluble yellow or green floccous salts.

**Diosmin.** Bitter ingredient of the Buccu-leaves (from Barosma serratifolia, B. crenulata, B. betulina.) Crystallises from the alcoholic tincture, is insoluble in water, dissolves in alcohol, ether, volatile oils and diluted acids.

[**Ditamin.** Alkaloïd, obtained by Jobst and Hesse from Dita-bark (from Alstonia scholaris). Precipitate the aqueous decoction of the bark successively with acetate and subacetate of lead, remove the deposit, treat the liquid with sulphuret of hydrogen, filter, acidulate the filtrate and precipitate with phospho-tungstic acid. Mix the precipitate with excess of solution of baryta, shake with ether and treat the ethereous solution of impure Ditamin with acetic acid. Acetate of D. is thus obtained, which is subsequently converted into the pure base.—It forms a white amorphous powder of a slightly bitter taste, easily soluble in ether, chloroform, benzol and alcohol. Concentrated sulphuric acid dissolves it with reddish colour, which becomes violet on heating. Concentrated nitric acid dissolves it yellow, on gently heating dark-green, then orange-red under evolution of orange vapour. It melts at 75° to a yellow liquid, which at 130° becomes deep brown-red. Dissolved in alcohol it has an alkaline reaction and neutralises acids, forming solutions of very bitter taste, which are precipitated, yellow by chloride of platinum; impure yellow by chloride of gold; white by chloride of mercury, iodide of potassio-mercury, iodide and sulphocyanide of potassium, tannic and phospho-tungstic acids.

By exhausting the Dita-bark with petroleum ether, the same authors obtained ECHICAUTSCHIN = $C_{25}$ $H_{20}$ $O_2$, soluble also in chloroform, ether and benzol, sparingly even in boiling alcohol; ECHICERIN = $C_{30}$ $H_{24}$ $O_2$, ECHITIN = $C_{32}$ $H_{26}$ $O_2$, ECHITEIN =

$C_{42}$ $H_{35}$ $O_2$ and ECHIRETIN $= C_{35}$ $H_{28}$ $O_2$, all of which are obtained from the dried petroleum-extract by means of boiling alcohol. Echitin is an indifferent body, crystallising in delicate white leaflets; Echitein is also indifferent and crystalline; Echiretin is resinous and closely resembles resins, obtained by Heintz from the milky juices of Brosimum (Galactodendron) utile and of Taberna-montana utilis.]

**Dragon's-blood.** Exudation of different plants, viz.:—Calamus Rotang and some other species, Dæmonorops Draco, Dracæna Draco, and Pterocarpus Draco. Dark red-brown, when pulverised carmine-red, without taste or smell, readily soluble in alcohol with red colour, also more or less completely in ether, oils, and alkalies, fuses at 210° and is destroyed in higher temperatures. The puri-fied resin is composed according to the formula $C_{40}$ $H_{21}$ $O_8$ .

**Dulcamarin** = $C_{65}$ $H_{50}$ $NO_{29}$. Alkaloïd, found besides solanin, in the twigs of the bitter-sweet (Solanum Dulcamara). To pre-pare it, exhaust with alcohol of 80%, press, mix the tincture with water, distil off the alcohol completely, filter the remaining liquid, throw down with tannic acid, mix the washed deposit with newly-precipitated hydrate of lead oxyd, dry, triturate, draw out with alcohol of 90%, and let the tincture evaporate.—Pale-yellow, translucid, resinous, brittle, permanent mass, when pulverised white with a tinge of yellow; without smell, evolving the odour of fresh bitter-sweet when moistened with alcohol or water and warmed; tastes at first very bitter, afterwards lastingly sweet. When heated to 100° the D., dried at the air, loses 8% of water, up to 120° it undergoes no further diminution of weight, but becomes a little sticky, at 165° it softens, above 200° it liquifies, but then already partially decomposed, by more heat it is reduced to coal, while emitting a very faint horny odour and evolving vapours of acid reaction. Water dissolves of the D. only $1/_{1075}$ its own weight; the solution smells and tastes slighty like bitter-sweet, is of neutral reaction, is precipitated white by tannic acid not altered by other reagents. Alcohol of 90% dissolves $1/_{10}$, when warm considerably more; the solution has a faintly alkaline re-action and is precipitable by the chlorides of platinum and of mercury. Ether dissolves $1/_{1440}$. In liquor of ammonia it swells up like a jelly without dissolving.

[According to the latest researches of Dr. Geissler, the Dulca-marin, prepared as above, is not pure. To purify it, dissolve in water, allow to stand for some time, add to the filtrate acetate of lead, filter, wash the precipitate, suspend in alcohol and decompose with sulphuret of hydrogen. Evaporate the filtered liquid, dis-solve the remnant in alcohol and evaporate to dryness. The re-maining pure Dulcamarin has the composition $C_{44}$ $H_{34}$ $O_{20}$ and behaves as a glucosid, separating into Dulcamaritin $= C_{32}$ $H_{26}$ $O_{12}$

and glucose when boiled with dilute mineral acids. Dulcamarin forms a slightly yellowish powder, permanent, without odour, of at first very bitter, afterwards lasting sweet taste. It melts by heat and decomposes at 205°. It dissolves in 30 parts cold and 25 parts hot water, in $8\frac{1}{2}$ parts cold and 5 parts hot alcohol; also in acetic ether. The solutions have a more or less red-brown colour and are neutral. The aqueous solution, on shaking, gives a dense froth. It is insoluble in ether, chloroform, benzol, sulphide of carbon.]

**Dulcit** $= C_{12} H_{14} O_{12}$. In a sweet product of unknown origin from Madagascar, dissolves in boiling water and crystallises on cooling. Also in Melampyrum nemorosum and probably in its congeners and named at first melampyrit. To obtain it from the last-named plant, the decoction is precipitated with acetate of lead, the filtered liquid boiled with lead oxyd, filtered and treated with sulphuret of hydrogen. Remove the sulphide of lead and evaporate to a thin syrup-consistence, allow to recrystallise in the cold and purify the crystals of D. by repeatedly recrystallising in water.—Colourless, very lustrous, klinorhombic prisms, slightly sweet, not rotating; fuses at 182°, behaves otherwise in the heat like mannit, dissolves readily in water, scarcely in boiling alcohol, yields with nitric acid oxalic and mucic acids, behaves toward concentrated and diluted sulphuric acid, alkalies, solution of copper, yeast and salts of lead like mannit.

**Ecbalin, Elaterid, Hydroelaterin** and **Prophetin,** bitter and resinous substances of Ecballion Elaterium and Cucumis Prophetarum, closely related to Elaterin; their individuality has to be further proved yet.

**Ecbolin.** One of the three alkaloïds (Ecbolin, Ergotin and Trimethylamin) of the ergot (the mycelium of Cordyceps purpurea) of Secale cereale. Is obtained, when the aqueous extract is precipitated with acetate of lead, the excess of lead removed from the filtered liquid by sulphuret of hydrogen, the liquid concentrated and chloride of mercury is added gradually until it ceases to produce a deposit. The latter is washed, decomposed with sulphuret of hydrogen, the filtered liquid (containing hydrochloride of Ecbolin), after driving off the sulphuret of hydrogen, mixed with phosphate of silver and, after filtering, with hydrate of lime, any excess of lime removed by carbonic acid, and the filtered liquid evaporated with a gentle heat.—Brownish, amorphous mass, slightly bitter, of alkaline reaction, dissolves in water and alcohol, little in wood spirit, not in ether and chloroform; neutralises the acids completely and forms amorphous, mostly deliquescent salts. Is precipitable by phospho-molybdic acid, the chlorides of mercury, of platinum and of gold, by tannic acid, cyanide of potassium, bi-iodide of potassium, chlorine-water and ammonia, bromine-water.

**Elaic Acid** = OLEIC ACID.

**Elain** = OLEIN.

**Elaterit.** *See* ECBOLIN.

**Elaterin** = $C_{40} H_{28} O_{10}$. In the fruit of Ecballion Elaterium. Press the fruits, evaporate the juice to honey-consistence, treat with alcohol, precipitate the solution with alcoholic acetate of lead, remove the lead from the filtrate by sulphuret of hydrogen, evaporate and edulcorate the remnant with ether, dissolve the residue in alcohol, and throw down the E. from the liquid with water.— Colourless, lustrous, tabular or needle-like crystals; tastes very bitter; fuses at 200°, decomposes in higher temperatures; has a neutral reaction; is insoluble in water, dissolves in 15 parts cold and in 2 parts hot alcohol, in 290 parts ether, little in oils, in concentrated sulphuric acid and precipitable from it by water without change, not in diluted acids and alkalies.

**Elecampane-Stereopten** = HELENIN.

**Elemi.** Exudation of the stem of Bursera Icicariba. Tough light-yellow or greenish-yellow mass, has a fennel-like smell and aromatic bitter taste, sinks in water, dissolves with difficulty but completely in alcohol. Contains 60% readily soluble resin of acid reaction = $C_{40} H_{32} O_4$; 24% slowly soluble, crystalline resin, Elemin = $C_{40} H_{33} O$; and $12\frac{1}{2}\%$ volatile oil.

**Ellagic Acid** = $C_{28} H_6 O_{16} + 6 HO$. In the Turkish nutgalls (from Quercus infectoria), and, it appears, in the root of Potentilla Tormentilla. By fermenting the pulverised nutgalls with a little water and at a moderate heat, the tannic acid changes into gallic acid, the products of decomposition of sugar and Ellagic acid. Press out, boil the press-cake with water, and press and filter the liquid, milky, from separated E. while still hot; a yellow-white powder impure E. remains on the filter, it is dissolved in diluted potash-ley, and the filtered liquid evaporated at the air, whereby green-white scales of Ellagate of potash subside, which have to be rinsed with water and decomposed with diluted hydrochloric acid, which throws down the E.—Pale-yellow, light powder, consisting of microscopic needles, without taste, of slightly acid reaction, loses at 120° four equivalents and up to 200° two more equivalents of water, sublimates afterwards, without fusing, partly undecomposed in sulphur-yellow needles, dissolves even in boiling water very little, little in alcohol, not in ether, readily in fixed alkalies, little in liquor of ammonia, in concentrated sulphuric acid unaltered in the heat and re-precipitable by water. Its salts easily decompose in the moist state, and only those of the alkalies dissolve in water.

**Emetin** $= C_{40} H_{30} N_2 O_{10}$. Alkaloïd of the different kinds of the Ipecacuanha-root, from Cephælis Ipecacuanha and possibly in roots of other Rubiaceæ of emetic property. Boil with water, evaporate the decoction to dryness, treat with alcohol, filter and distil off the alcohol, evaporate again to dryness and exhaust with water acidulated with hydrochloric acid. Precipitate the solution with chloride of mercury, wash the deposit with cold water and dissolve in alcohol, precipitate the mercury with sulphide of baryum, remove the latter with sulphuric acid, dilute with water, distil off the alcohol, throw down the E. with ammonia, wash with cold water and dry.—White powder, slightly bitter, of alkaline reaction, fuses at 50° and is destroyed in higher temperatures, dissolves slowly in cold, a little more in hot water, readily in alcohol, not in ether and oils, readily in acids, not in the hydrates and carbonates of alkalies; becomes dark olive-green with concentrated sulphuric acid. Its salts are not crystallisable.

[Glénard prepares Emetin by mixing the pulverised root or the extract with lime, exhausting with ether, evaporating the ethereous solution to dryness, dissolving the residue in dilute hydrochloric acid and precipitating the E. with ammonia. The composition of the E., dried at 110°, was found to be $C_{30} H_{22} NO_4$. The chloride forms tufty, united needles, and is almost colourless.]

**Emodin** $= C_{40} H_{15} O_{13}$. In the root of Rheum Emodi, probably also in other species of this genus of Polygoneæ. Macerate the root with water, dry again and exhaust with benzol, distil off most of the benzol, press the remnant converted into a crystalline pulp after cooling, treat again with benzol, which dissolves chrysophanic acid and leaves intact a reddish, sparingly soluble substance; dissolve the latter in hot benzol, leave to cool and purify by recrystallising in hot and highly concentrated acetic acid and in hot alcohol.—Lustrous, deep orange-red, brittle, klinorhombic prisms; does not fuse below 250° and sublimates to a slight extent in yellow fumes. It dissolves, similar to chrysophanic acid, in liquor of ammonia with violet-red colour, more readily in alcohol and amylalcohol, not so readily in benzol.

**Emulsin.** The peculiar albuminous matter of almonds and of the seeds of some trees allied to Prunus Amygdalus; exists dissolved in the emulsions of these seeds, and causes their milky appearance by suspending the fat oil. To the E. belongs the remarkable property of decomposing amygdalin into hydrocyanic acid and other products. It is obtained, when an emulsion of sweet almonds is agitated with four times its volume of ether, and allowed to rest for a few days; the clear aqueous solution is separated from the supernatant ether, and from insoluble particles; precipitated with alcohol and dried in a vacuum.—Horny substance, or when triturated white powder without smell or taste, soluble in cold water;

but the solution coagulates in the heat like common albumin, and the coagulated E. has lost its power of decomposing amygdalin. *See* also Synaptase.

**Emulsin of Mustard**=Myrosin.

**Equisetic Acid**=$C_4 HO_3 + HO$.  As yet only found in some species of the genus Equisetum; it has also been named Maleic acid, because it is formed during the destructive distillation of malic acid (at 200°).  To separate and obtain it, boil the comminuted herb with water, neutralise the decoction with carbonate of soda, remove sulphuric and phosphoric acids with acetate of baryta, precipitate the Equisetic acid with acetate of lead, wash the deposit, decompose with sulphuret of hydrogen and evaporate the filtered liquid.—It crystallises in large, klinorhombic prisms, tastes very acid, fuses by heat (according to Lassaigne at 47·5°, according to Pelouze at 130°), and sublimes almost undecomposed, dissolves readily in water, alcohol and ether.  All its salts are soluble in water, except the neutral lead-salt and those of silver and copper.  Its quantitative estimation is best effected through the neutral lead-salt, which is anhydrous when dried at 100°, and contains 30·44% acid.—According to Baup, Equisetic acid is identical with Aconitic acid, consequently tribasic, and having the formula $C_{12} H_3 O_9 + 3 HO$; nor is it sublimable.  *See* Aconitic Acid.

**Ergotic Acid.**  Peculiar volatile acid of ergot.  To obtain it, distil the extract prepared by cold water with sulphuric acid, and remove any formic acid by gently warming the distillate.  An acid liquid remains, which, with lead, silver and baryum, yields salts, insoluble in water, but soluble in acids.

**Ergotin.**  One of the three alkaloïds (Ecbolin, Ergotin, and Trimethylamin) of ergot.  Precipitate the aqueous extract of the drug with acetate of lead, free the filtrate from lead by sulphuret of hydrogen, concentrate the liquid and add pulverised chloride of mercury as long as a precipitate ensues; filter and precipitate the filtrate with phospho-molybdic acid, digest the precipitate, while still moist, with water and carbonate of baryta; filter and evaporate the liquid to dryness.  The Ergotin closely resembles ecbolin, but differs from it by not being precipitated by the chlorides of mercury and platinum (unless under addition of ether-alcohol) and by cyanide of potassium.

The name of Ergotin has been formerly applied to another different substance of ergot, and prepared by boiling the pulverised drug, previously freed from fat by means of ether, with alcohol, evaporating the tincture, adding water to the residue, and collecting the separated flocky mass.  Fine reddish-brown powder; smells specifically, especially when hot; tastes acrid, bitter, and aromatic; decomposes by heat without fusing; dissolves only by traces in water and in ether, with red-brown colour in alcohol, also in alkalies, not in diluted acids.

[**Ergotinin.** An alkaloid, found by Touret in ergot. Treat the coarsely pulverised ergot with boiling alcohol of 86°/$_0$, and remove the alcohol from the tinctures by distillation on a water-bath. The residue, when cold, separates into three layers, a solid fat, an aqueous solution and resin. Wash the resin with ether and dissolve the fat in the latter, filter, shake the solution repeatedly with dilute sulphuric acid (1:15), then the separated acid liquid with excess of potash-ley, finally with chloroform, and evaporate the latter solution under exclusion of air. The aqueous solution, obtained together with the fat, likewise yields E. It is filtered and subjected to distillation on an oil-bath, and in a current of hydrogen. When all the alcohol is expelled, the residue is rendered alkaline by carbonate of potash and distilled again. Impure trimethylamin passes over. The remaining syrupy liquid is acidulated, washed with ether, supersaturated with carbonate of potash, and treated as before with chloroform. —The E. is solid, soluble in ether, alcohol and chloroform, of strongly alkaline reaction, is precipitated by iodide of potassio-mercury, biiodide of potassium, phosphomolybdic and tannic acids, the chlorides of gold and platinum, and by aqueous bromine. Concentrated sulphuric acid colours it first orange, then deep violet-blue. It changes easily at the air, its salts turning first pale, then darker red. Alkalies likewise decompose it, especially when warm, under formation of trimethylamin. The rapid spoiling of pulverised ergot seems to be attributable to the decomposition of the Ergotinin.]

**Ericinol** $= C_{20} H_{16} O_2$. It is a product of decomposition of ericolin, but exists ready-formed in the volatile oils of the Ericaceæ, mentioned under Ericolin. The volatile oil of Ledum palustre, for instance, is a mixture, containing valeric and other volatile acids, amongst which an oil-like acid $= C_{16} H_{10} O_8$; further, a hydrocarbon isomeric with oil of turpentine, and boiling at 160°; and Ericinol. After the oil of Ledum has been repeatedly shaken with concentrated potash-ley and thereby freed from acids, it is washed and desiccated, a mixture of the hydrocarbon with some resin and Ericinol remaining. By distilling the mixed oils and reserving the portion which passes over at 236° to 250°, Ericinol is obtained almost pure. This Ericinol boils at 240° to 242°, is blue-green, of disagreeable smell and burning, nauseous, bitter taste. After rectifying it with a little caustic potash, it is partly decolourised, of 0·874 density at 20°, and has the composition $= C_{20} H_{16} O_2$. On boiling with excess of lime-hydrate, a hydrocarbon $= C_{20} H_{16}$ is obtained.

**Ericolin**$=C_{68}$ $H_{56}$ $O_{42}$. Bitter glucosid of Arctostaphylos Uva ursi, Calluna vulgaris, Erica herbacea, Ledum palustre, Rhododendron ferrugineum, consequently a widely distributed constituent of the Ericaceæ, and to be sought for in many other co-ordinal plants. Boil (preferably Ledum palustre) with water, strain, precipitate with subacetate of lead, evaporate the filtered liquid to one-third in a retort, remove by filtering the lead compound which will have separated; throw down the rest of the lead from the liquid with sulphuret of hydrogen, evaporate to honey-consistence and extract the Ericolin by means of dehydrated ether-alcohol. The residue left after the evaporation of the latter is dissolved several times with ether-alcohol and evaporated until it dissolves without any residue.—Brown-yellow powder, conglutinating at 100°; of very bitter taste, breaks up, when heated with diluted sulphuric acid, into sugar and Ericinol.

**Erucic Acid**$=C_{44}$ $H_{41}$ $O_3$ + HO. In the fixed oils of white and black mustard and of rape. Saponify the oil with soda-ley, decompose with hydrochloric acid, dissolve the fat acids in hot alcohol, keep in a cold place, press the separated mass and recrystallise.—White, glossy, thin crystallised needles, without smell or taste, fusing at 32° to 33°; decomposed in higher temperatures; of acid reaction; dissolve most readily in alcohol and in ether.

**Erucin.** Peculiar crystalline substance of the seeds of Brassica alba. Mix the seeds, ground to a powder, with a little water, allow to rest (in order to develop the acridness), treat with ether, evaporate the ethereous solution to honey-consistence and leave it to stand at the open air. After some time, hard crystalline bodies will form, which are insoluble in water and alkalies, dissolve slowly in alcohol, readily in ether and oils, do not redden the salts of iron oxyd and contain no sulphur.

**Erythric Acid**$=C_{40}$ $H_{22}$ $O_{20}$. Exists in the lichen Lecanora tartarea, more abundantly in Roccella Montagnei. Boil the lichen with water, strain, rinse the white flocks and crystals which will have separated after cooling with cold water, re-dissolve in hot water, filter off from the little black or brown deposit and leave to stand in a cold place.—It crystallises in colourless, fine needles, has neither taste nor smell, fuses a little above 100°, is destroyed in higher temperatures, dissolves in 240 parts of boiling water, better in alcohol and in ether, becomes red through ammonia under access of the air, yields orsellic ether on boiling with alcohol. The alcoholic solution assumes a purple-red colour with chloride of iron and turns yellow without cloudiness on addition of ammonia; nitrate of silver effects no alteration in the alcoholic solution; ammonia, when added to the mixture, yields a white precipitate that turns black on boiling, while it covers the glass with a bright film of silver.

**Erythrophyll.** *See* CISSOTANNIC ACID.

**Erythroretin.** *See* APORETIN.

**Erythrozym.** Peculiar nitrogenised matter of the root of Rubia tinctorum. Treat one part madder on a linen cloth with nine parts water of 38°, mix the extract with an equal volume of alcohol, collect the subsiding dark red-brown flocks and exhaust with boiling alcohol, wash the remaining impure E. with cold water until the latter is no longer precipitated by acetate of lead, and dry on a water-bath.—Black, hard mass, difficult to pulverise, burns with a horny odour, yields with water a thickish red-brown fluid, but does not properly dissolve.

**Eschscholtzia Alkaloids.** From the root and herb of Eschscholtzia Californica Waltz obtained, besides a little chelery-thrin, an acrid alkaloïd and a bitter one. (*a*) Acrid alkaloïd. Exhaust with water and an admixture of acetic acid, throw down chelerythrin and the acrid alkaloïd with ammonia (the bitter alkaloïd remaining in solution), wash the deposit, dry and dissolve in ether, evaporate, dissolve again in water acidulated as before, precipitate with ammonia, &c.; or remove the dyeing matters by digesting with animal charcoal.—White powder, almost tasteless by itself, but very bitter when dissolved in alcohol or ether; of alkaline reaction, insoluble in water, readily soluble in alcohol and ether. Forms neutral, colourless salts, precipitable by the hydrates and carbonates of alkalies and by tannic acid. Becomes *not* violet with sulphuric acid. (*b*) Bitter alkaloïd. After the chelerythrin and the acrid base have been removed by ammonia from the extract of the herb as above, the liquid is neutralised with acetic acid, precipitated with tannic acid and otherwise treated like Porphyroxin (from Sanguinaria). — Crystalline, easily fusible mass of nauseous bitter taste and alkaline reaction, imparts to concentrated sulphuric acid a beautiful violet colour, which is observable even with a solution containing 1% of the alkaloïd by allowing one drop of the solution to float on the sulphuric acid.

**Eserin** = PHYSOSTIGMIN.

**Essential (ethereal or volatile) Oils** are compounds of carbon and hydrogen with or without oxygen, nitrogen, or sulphur. Their distribution throughout the vegetable kingdom is so general that it is difficult to single out a plant which does not, by its peculiar odour, betray the presence of at least traces of volatile oils contained either in the flowers, seeds, pericarps, leaves and barks, or, though less frequently, in woods, and which oils often show a marked difference in physical and chemical properties when obtained from different parts of the same plant. The taste of these oils is generally hot and aromatic and their odour analogous to the material employed for their preparation, but intensified to the highest degree. Besides these, another class of oils occurs, which like oil of bitter almonds, mustard, &c., do not exist ready formed in the living organism, but owe their formation to a kind of fermentation or breaking up of an inodorous substance into different products, one of which is the respective volatile oil.

Three processes are commonly employed for obtaining the isolated oils, viz.:—*Pressing, Extraction* by means of solvents, and *Distillation.*

The *first* of these methods is only practicable with materials which, like orange-peels, are very rich in oils. As it is wasteful and generally yields an oil contaminated with fat, wax, resin, &c., this operation is restricted to a few materials only.

The method of *Extraction* (commonly by means of ether) is adopted either with small quantities of the raw material, or with substances very poor in oil, or with oils, like those of turnsole, jasmin, reseda, violets, suffering decomposition by heat. By this process all the other constituents, soluble in ether, of the raw material are of course likewise extracted, and accordingly the remnant, after the spontaneous evaporation of the solvent, contains generally also fixed oils, wax, chlorophyll, resins, &c., although the larger portion of these impurities, with the exception of the fixed oils, separate during evaporation from the volatile oil. It is therefore not possible to obtain a pure product by either pressing or extracting in the usual way.

*Distillation* by means of water is by far the most common method of obtaining ethereal oils. Usually the properly comminuted substance is mixed with six or eight times its weight of water, and the mixture left to rest for a day. The whole is then submitted to distillation (with proper means for refrigeration), until the distilled water ceases to separate any oil, or until it is nearly devoid of odour. To prevent any contamination of the oil, by partial charring of the contents of the still with empyreumatic products, the distillation is effected more properly by forcing steam through the substance operated on.

The essential oils obtained by distillation, and when freshly prepared, are as a rule colourless or pale-yellowish, seldom, like the oils of a few of the Compositæ and Myrtaceæ, of a blue or green colour. They are also commonly of thin fluidity (at mean temperature) and of great light refracting and dispersing power; sometimes they are of thickish consistence, partly or completely solid and in the latter case distinguished by the name of camphor or stearopten, while the liquid part is called in contradistinction eleopten and which may often be obtained solid or in crystals by refrigerating to 0° or below. The specific gravities of volatile oils range from 0·800 to 1·100; their boiling points are in all instances higher than 100°, seldom below 150°, but often as high as 300° and more. The hydrocarbon-oils are, as a rule, more volatile than those containing oxygen, and the volatility seems to diminish with the increasing percentage of the latter.

The quantitative estimation of volatile oils is very difficult and can only be done approximately. When the oil has been obtained by distillation and spontaneous separation from the aqueous distillate, that portion of the oil which remains dissolved in the water, has also to be accounted for [irrespective of the fact, that the proportionate quantity of volatile oils is subject in the same species to much fluctuation according to the soil, climate and local influences.—F. v. M.], and which is calculated as constituting one-tenth per cent. of the weight of the whole aqueous distillate, when the oil was a hydrocarbon; and one-fifth per cent. when an oxygenised oil.

A thorough analysis of volatile oils requires not only a vast quantity of material but also much time and circumspection. After distillation the oil is first freed from any water by agitating occasionally and for several days with fused chloride of calcium. It is then poured off and examined in regard to reaction with litmus-paper, specific gravity, boiling point, and its behaviour in low temperature.

The determination of the boiling-point is carried on in a retort, not more than half filled with the oil, and through the tubulus of which a thermometer, indicating at least 300° above zero, is passed so as to touch with its bulb the surface of the oil, while the neck of the retort is adapted to a Goebel's or Liebig's condensor. As crude volatile oils are commonly mixtures of different oils, in most instances the boiling point gradually rises during the operation. A difference of only a few degrees is not regarded as of much consequence, but as soon as the boiling-point rises 10° or 20° and remains constant there, the receiver is changed and the first distillate poured into a glass-phial with a well closing glass stopper. Meanwhile, the distillation is carried on, but not quite to dryness, to prevent contamination with empyreumatic products, and each portion of the distillate differing by ten or more degrees, is collected separately. These different oils are then examined separately in regard to their physical properties (specific gravity, colour, smell, taste, fluidity, vapour-density, rotatory power, &c.), and chemical constitution. In cases, where material enough is at hand, the same process of fractional distillation is advantageously repeated with each separate portion, to secure greater precision of results.

Some volatile oils are combinations of an acid and a neutral body, and often separate into these different parts during distillation, the neutral portion (mostly a hydro-carbon or an ether) being more volatile and passing over first, while the acid oil (always oxygenised) distils much later.

But to effect a complete separation, it is better to previously mix the oil with potash, or soda-ley, which combines with the acid, allowing the neutral or basic oil to distil first. By adding to the residue in the retort afterwards sulphuric or better phosphoric acid and distilling again, the acid oil is obtained pure.

[*Estimation* of *Essential Oils* according to Osse.—Treat five grammes of the finely pulverised substance cold with twenty-five cubic centimeters of petroleum ether (of a boiling-point not above 40°), shake for a few hours, let subside, evaporate one cubic centimeter of the clear liquid in a current of air, weigh the remaining mixture of wax or fat and oil, evaporate the volatile oil at 110° and weigh again. The latter weight deducted from the first gives the amount of the essential oil.]

[**Eucalyptus Constituents.** The leaves of Eucalyptus globulus, according to H. Weber, contain besides the volatile oil the following substances:—1. A white body, crystallising in needles, mixed with a resinous, amorphous mass, both of acid reaction. 2. An acid, yellow resin of bitter taste. 3. Eucalyptic acid, precipitable by lead salts. 4. Eucalyptin, a neutral, crystallisable, bitter substance, soluble in ether and alcohol, partly in water. Used as a remedy against intermittent fevers.]

**Eucalin.** *See* MELITOSE. ·

**Eugenic Acid** = CARYOPHYLLIC ACID.

**Eugenin** = $C_{20} H_{12} O_4$ (isomeric with cuminic acid and caryophyllic acid). Separates slowly from the turbid aqueous distillate of cloves overcharged with volatile oil.—Delicate white, pellucid, pearly laminæ, turning yellow by keeping; of a faint odour of cloves; dissolves readily in alcohol and ether; assumes, like caryophyllic acid, a blood-red colour with cold nitric acid.

**Eupatorin.** Bitter ingredient of the leaves and flowers of Eupatorium cannabinum (probably obtainable from numerous species of this genus.—F. v. M.) It is obtained by precipitating the aqueous extract with slaked lime, digesting the precipitate, after it has absorbed carbonic acid from the air, with alcohol and evaporating the tincture to dryness.—White powder of a bitter and pungent taste, insoluble in water, soluble in absolute alcohol and in ether, said to yield a crystalline salt with sulphuric acid.

**Euphorbium.** The hardened milky juice of several species of the genus Euphorbia. Dirty-yellow, white inside, brittle, porous lumps without smell and of excessively acrid taste. Contains about 60% resin (Euphorbin = $C_{20} H_{16} O$), 14% wax, 5% caoutchouc and various salts.

**Euphrasia Tannic Acid**$=C_{32} H_{20} O_{17}$. Peculiar tannic acid of Euphrasia officinalis, greening the iron salts, and only obtained in combination with lead; precipitable by glue and by tartarated antimony.

**Evernic Acid**$=C_{34} H_{16} O_{14}$. Besides Usnic acid in Evernia prunastri. Precipitate the extract of the lichen, prepared with diluted milk of lime, 'with hydrochloric acid and boil the deposit with weak alcohol; or dry and exhaust with ether. The solution of the E. is purified by recrystallisation under aid of animal charcoal.—Colourless crystalline needles, inodorous and tasteless, of acid reaction; lose at $100°$ nothing of their weight; fuse at about $164°$; are afterwards decomposed and yield beside other products a sublimate of orcin; are insoluble in cold and dissolve slightly in boiling water, also in cold and very abundantly in hot alcohol; readily in ether, also in alkalies, the solution in ammonia turning slowly to a deep-red by exposure to the air.

**Extractive Substances.** By extracting vegetable matters with water or alcohol, a more or less yellow or brown liquid is obtained which becomes darker on evaporating under the influence of the air, and remains as a stiff pasty mass difficult to dry and quickly absorbing moisture when exposed to the air. This product has the general name of extract, and represents a mixture of the most different matters and salts of the plant employed. The general characteristics of extracts are as follows: solubility in water and alcohol, insolubility in ether, amorphous condition, non-volatility, hygroscopicity, and finally inclination to decompose when in solution under concurrence of heat and air. Among the substances common to all extracts are gum, sugar, pigments, salts, &c., because they are soluble in water and alcohol, and are accordingly carried into the extracts. In cases where one or the other of the aforesaid substances predominates and in default of sufficiently accurate analysis, or when no mode of isolation is known, the remaining mixture, called extract, is spoken of and designated according to its predominating character, as bitter, sweet, astringent, gummous, colouring, resinous, &c., extractive substance.

It follows therefore that the name "extractive substance" does not mean a well-defined body, but comprises a mixture of substances, which advancing knowledge will teach to separate.

**Fats.** Non-volatile compounds of carbon with hydrogen and comparatively little oxygen; widely distributed throughout the vegetable empire, and most abundantly occurring in seeds; at a mean temperature of all consistencies from that of an oil to a solid; colourless or yellowish, with a greenish or brownish shade; of imperceptible or faint odour and taste; leave a permanent spot on paper, are lighter than water (spec. grav. between 0·88 and 0·95); the solid fats fuse almost without exception below $100°$; not inflammable by themselves, but burning by means of a wick with a bright, smoky flame; insoluble in water, slightly soluble in alcohol, mostly abundantly soluble in ether and volatile oils, also in chloroform and sulphide of carbon. They form with strong bases, under formation of glycerin, soaps, which dissolve in water and alcohol when formed by the real alkalies, but are insoluble in those two liquids when composed of any other base. From parts rich in oil (for instance, from the seeds) most of the oil may be obtained by proper comminution and strong pressure at a gentle heat; but as this method gives very inaccurate results, it is necessary for a quantitative analysis to extract the oil by means of ether (benzol, ether, petroleum ether) as shown in No. II., Division III., Part II.

**Fiber**$= C_{12} H_{10} O_{10}$. After the whole or part of a plant has been exhausted by means of ether, alcohol, water, diluted acids and alkalies, a substance remains which is commonly called Fiber, Cellulose or Skeleton. Treated as above, it is generally of a brownish colour on account of impurities, and contains besides a variable amount of mineral matters, which remain after incineration as ash, the weight of which has to be deducted from the weight of the Fiber, previously dried thoroughly at 110°.

To obtain Fiber in a purer state the colouring matters have to be destroyed by a proper agent which has no influence on the Fiber itself. Best adapted for this purpose is so-called chloride of lime. Reduce one part of the latter with water to a soft pulp (in a glass or porcelain mortar), dilute with more water, until the whole represents ten parts, let settle in a high cylindrical glass vessel covered with a glass-plate; filter, and mix the Fiber, treated as above, in a similar vessel with enough of the solution to be covered completely; cover with a glass-plate and let stand cold for twenty-four hours in a dark place. After the bleaching has been completed, the solution is poured off, and the Fiber washed with a few changes of pure water, and afterwards with water strongly acidulated by hydro-chloric acid, in order to dissolve the small quantity of lime, precipitated on the Fiber. It is then collected on filtering-paper, washed and dried. Should the Fiber by the one day's exposure not be completely decolourised, reaction will become more powerful by adding to the above mixture of chloride of lime solution and Fiber, enough of hydro-chloric acid to render it strongly acid; leave to stand covered in a dark place for another day, bring on a filter, wash and dry.

There are certain dark pigments which are not destroyed even by this second treatment, but which may be bleached to white-yellow by means of warm nitric acid of 1·20. To use a stronger acid, would not be advisable on account of its destructive action on the Fiber itself.

Cellulose, thus purified, is white, of loose texture, without taste, becomes carbonised by heat without fusing, and burns completely away (leaving nothing or only traces of ash), is not affected by ether, alcohol, water, diluted acids, alkalies or solution of chloride of lime; but dissolves even at ordinary temperature in the blue ammoniacal solution of sulphate of copper, and acquires, when kept in a cold concentrated solution of chloride of zinc, the property of turning blue with iodine, also to dissolve in it, when warmed, under conversion into sugar, which latter, however, is destroyed by the continued application of heat.

**Filicic Acid** $= C_{26} H_{14} O_8 + HO$. Found in the rhizoma of Aspidium Filix mas. Exhaust with ether (free from alcohol), evaporate to an oily consistence, leave to stand cold for a few days, collect on a filter the acid, which will have separated in yellow crusts, wash with small quantities of a mixture of equal volumes of absolute alcohol and ether, afterwards with a mixture of two volumes of alcohol and one volume of ether, until the remnant becomes of a pale lemon-yellow, and recrystallise in hot ether.— Small, green-yellow, rhombic laminæ or loose, light-yellow, crystalline powder of faint odour, and slightly nauseous taste; becomes electric by rubbing; fuses at 161°, and decomposes afterwards; has an acid reaction when dissolved in ether; is insoluble in water, slightly soluble in aqueous, better in boiling absolute alcohol, little in cold, better in hot ether, readily in sulphide of carbon, fixed and volatile oils. Yields with alkalies salts, which dissolve in water, with other bases insoluble compounds.

**Filixoleic Acid.** In the rhizoma of Aspidium Filix mas; is scarcely different from oleic acid.

**Flavequisetin.** Yellow pigment, found in Equisetum fluviatile, is only known in the impure state.

**Formic Acid** $= C_2 HO_3 + HO$. As yet it has been found only in a few plants, viz., in the leaves, bark and wood of the Coniferæ, in the fruits of Sapindus Saponaria, Tamarindus Indica, the leaves of Urtica urens, Sempervivum tectorum; but exists, in all likelihood, in many other plants. As it is a volatile acid (it has an acrid, pungent odour, similar to acetic acid), it must be first and best looked for in the water which has been distilled from the respective plants. Saturate this distillate with a little excess of carbonate of soda, evaporate to a small bulk, neutralise carefully with nitric acid, add nitrate of silver and heat; if the liquid becomes grey from reduced silver Formic acid was present.

The whole of the Formic acid is obtained by distilling the respective substance with water containing phosphoric acid. (*See* No. IX., Division III., Part II.)

To estimate the amount of Formic acid, over-saturate a weighed or measured quantity of the distillate with carbonate of soda, evaporate to dryness, expose the triturated mass to a temperature of 110° until thoroughly dry; weigh; extract with alcohol of 95%, and note down the weight of the residue, consisting of carbonate of soda, chloride of sodium, &c. The difference of the two weights represents the Formate of soda, which contains 54·38% Formic acid.

On the other hand, should, as frequently happens, such organic acids, as likewise yield soda-compounds soluble in strong alcohol, be present in the aqueous distillate, then the alcoholic solution of the soda-salts is evaporated to dryness, and the residue mixed with nitrate of silver in the proportion of 1 part residue to $2\frac{1}{2}$ parts nitrate of silver; the whole is then gently heated, until the reduction of the resulting Formate of silver to the metallic state is completed; 100 parts of this reduced silver, washed and dried on a filter, indicate 34·26 parts Formic acid.

The Formate of silver, like all compounds of that acid, is freely soluble in water, and only separates from very concentrated solutions as a white, curdled mass, which, on addition of water, dissolves again. As the silver salts of acids, belonging to the same series as Formic (like acetic, propionic, valerianic, &c.), do not undergo reduction under the same conditions, their presence does not interfere with the above result.

If the reduced silver, obtained by this process, should be found to contain chloride of silver, it is necessary to treat it with nitric acid, which leaves the chloride behind. The latter is collected, weighed, and serves to correct the estimate by deducting for every 100 parts of chloride, 25·8 parts Formic acid, or the amount of the Formic acid is calculated from the reduced silver, minus the chloride of silver.

**Frangulin** = RHAMNOXANTHIN.

**Fraxin** = $C_{54} H_{30} O_{34} + HO$. Glucosid of the barks of Fraxinus excelsior, Aesculus Hippocastanum, A. Pavia, and of allied species of those genera. Precipitate the decoction of ash-bark, collected in spring during the floral season of the tree, with acetate of lead, the filtered liquid with subacetate of lead; press the latter deposit, decompose under water with sulphuret of hydrogen, filter, evaporate the filtrate to syrup consistence, collect any crystals formed, wash the latter with water, afterwards with a little spirit of wine, and purify by recrystallisation.—Tuftily united, white, lustrous needles, inodorous, slightly bitter and acerb, losing the water at 110°, fusing at 320° under decomposition; dissolve in 1000 parts cold, readily in hot water, little in cold, readily in hot alcohol, not in ether; the largely-diluted alcoholic or aqueous solution shows, especially in the presence of traces of alkalies, a blue or bluish-green fluorescence which disappears through acids ; it separates on access of diluted acids into sugar and Fraxetin $= C_{30} H_{12} O_{16}$, a crystalline body slightly acerb to the taste.

**Fraxinin** obtained, according to Keller, from the bark of the Ash-tree, is mannit.

**Fruit Sugar** (LÆVO-GLUCOSE) $= C_{12} H_{12} O_{12}$. Cane-sugar in solution by keeping, or more quickly by heating with diluted acids, is converted into inverted sugar, and instead of rotating to the right, acquires the property of rotating to the left; which change amounts for every 100° of the original dextro-rotation, to 38° lævo-rotation at 14°, provided the inversion has been complete. During inversion cane-sugar assimilates $5 \cdot 26°/_o$ water, forming inverted sugar, *i.e.*, a mixture of equal parts of dextro-glucose (grape-sugar) and of lævo-glucose : $2 C_{12} H_{11} O_{11} + 2 HO = C_{12} H_{12} O_{12}$.

Fruit-sugar occurs as inverted sugar in honey, in many succulent fruits, and other saccharine vegetable parts. Its isolated occurrence has as yet not been proved with certainty, yet some kinds of pears and apples contain more fruit than grape-sugar. To obtain Fruit-sugar pure, mix ten parts inverted sugar intimately with six parts lime-hydrate and 100 parts water, whereby the at first fluid mass congeals after some shaking, and on hard pressure allows grape-sugar lime to run off, while the solid residue, decomposed after washing by oxalic acid, yields Fruit-sugar. In like manner the inverted sugar, occurring in fruits, may be decomposed. Pure Fruit-sugar is, after heating to 100°, a colourless, uncrystallisable syrup, or an amorphous, hard mass, as sweet as cane-sugar, dissolves more readily in alcohol than grape-sugar, undergoes directly with yeast the vinous fermentation, and behaves towards alkaline solutions of tartarate of copper like grape-sugar.

**Fruittannic Acid.** Occurs in unripe apples, pears, and some other fruits, and disappears on ripening, while the amount of sugar increases proportionately. By mixing the filtered juice of the above fruits with starch paste and adding iodine, iodine-starch is formed, but not before the whole of the tannic acid has been converted into an iodine compound. To obtain the latter, add to the juice of unripe fruits tincture of iodine in small quantities until the colour of the iodine no longer disappears ; a brown precipitate is formed after a few moments, which has to be washed with water. Yellow, amorphous powder, insoluble in water and alcohol, is decomposed by boiling with diluted acids, yielding grape-sugar.

**Fumaric Acid** $= C_8 H_2 O_6 + 2 HO$, also called boletic, lichenic, and paramaleic acid, the latter on account of its occurrence, together with maleic (equisetic) acid, among the products of the destructive distillation of malic acid ; has been found hitherto in Fumariaceae, Lichenes (for instance, Cetraria Islandica), Fungi (species of Boletus, &c.), and Glaucium luteum. To obtain it, precipitate the decoction with acetate of lead, separate the deposit by filtration, wash cold and boil with water, filter hot and allow the liquid to stand cold. The crystallised fumarate of lead is reduced to a fine powder, decomposed under water with sulphuret of hydrogen, and the liquid evaporated.—The Fumaric acid crystallises in needles, scales, or warty granules; is inodorous, of a strongly acid taste and reaction, fuses with heat and sublimates, dissolves only in two to three hundred parts cold water, abundantly in hot water, readily in alcohol and ether. The fumarates of the alkalies dissolve readily in water, those of most of the other bases slowly, the fumarate of silver not at all; the latter becomes anhydrous at $100°$, and contains $29.70\%$ acid.

**Fumarin.** Alleged alkaloïd of Fumaria officinalis. It is said to be obtained in the following way. Treat the bruised herb with diluted acetic acid, evaporate to a syrup, exhaust with alcohol, decolourise the tincture with animal charcoal, and evaporate. It crystallises in fine needles, as acetate of Fumarin. Dissolve the salt in water, and add an alkali which throws down the Fumarin, as a cloddy mass which may be obtained in crystals from hot alcohol. The compounds of Fumarin have a lasting bitter taste.

**Fungic Acid.** *See* Malic Acid.

**Galambutter.** From Bassia longifolia, B. latifolia, and B. butyracea, obtained by boiling the fruits with water. Dirty red and white, transparent, of lard consistence, has a faint smell and pronounced cocoa-like taste. Oil of Illipe seems identical with this.

**Galbanum.** Gum-resinous exudation of Peucedanum galbaniferum and Polylophium Galbanum. More or less brown-yellow; at ordinary temperature tough, brittle when cold, of disagreeable smell and acrid, bitter taste; contains as main ingredients: volatile oil, resin and gum.

**Galbanum Resin** $= C_{52} H_{38} O_{10}$. Free the galbanum from the volatile oil by distillation with water, separate the remaining resin from the supernatant turbid liquid, boil it with milk of lime, throw down the dark-yellow solution with hydrochloric acid, wash the white-yellow flocks, dissolve in ether and evaporate. Amorphous, white flocks, after the evaporation of the ether honeyyellow mass, tasteless, insoluble in water, readily soluble in alcohol, also in common ether, not completely in absolute ether, difficultly in potash-ley, yields no sugar with diluted acids; forms a blue oil and other products when submitted to destructive distillation.

**Galipein** = Augusturin.

**Galitannic Acid** = $C_{14} H_8 O_{10}$. Found in the herb of Galium verum and G. Aparine; (occurring, doubtless, in numerous other species.—F. v. M). Precipitate the aqueous decoction with acetate of lead, filter, precipitate with subacetate of lead, decompose this second deposit with sulphuret of hydrogen, filter off from the sulphuret of lead, precipitate the liquor with acetate of lead in order to remove the rest of citric acid; throw down the filtrate with subacetate of lead, and decompose the precipitate with sulphuret of hydrogen; a solution is obtained of slightly bitter, styptic taste, imparting a beautiful green colour to chloride of iron.

**Gallic Acid** $= C_{14} H_3 O_7 + 3 HO + 2 Aq$. This acid is said to exist in many vegetables, especially in those containing tannin; but as Gallic acid has not yet been obtained with certainty as decomposition-product from any iron-greening tannic acid, but only from two iron-blueing tannic acids, *i.e.*, from ordinary and Chinese nutgalls, and from sumach, it seems at least improbable, that Gallic acid should occur ready formed in other plants than those named already (and in species closely allied to them.—F. v. M.).

In order to obtain the Gallic acid which may exist ready formed in a vegetable extract, the always co-existing tannic acid is first removed by means of glue, either dissolved in water, or better, in the form of animal membrane (isinglass) soaked in cold water. The liquid, after it has been separated from the tannate of glue, is evaporated to an extract, exhausted with strong alcohol, the solution evaporated again, and treated with ether. During the evaporation of the ether, the gallic acid forms in crystals.—It crystallises in white, silky needles, is inodorous, of an acidulous styptic taste, decomposes at 210° under partial decomposition and formation of a sublimate of pyrogallic acid $= C_{12} H_6 O_6$ ; fuses at 226°, dissolves in 100 parts cold and in 3 parts boiling water, readily in alcohol, less readily in ether; the aqueous solution throws down neither glue nor alkaloïds, but precipitates the salts of oxyd of iron with a dark-blue colour, like that produced by gallotannic acid; the gallate of iron (the above deposit) differs from the tannate of iron by its great solubility in acetic acid and in the hydrates and carbonates of alkalies (incl. ammonia), while of the tannate of iron only traces are dissolved by acetic acid, and the iron tannate is decomposed completely by the hydrates and carbonates of potash and soda, and partly by ammonia, in the former case under formation of oxyd of iron. Only the gallates of alkalies and of the alkaline earths are soluble in water.

**Gallotannic Acid** $= C_{54} H_{22} O_{34}$. In certain excrescences on the branches of a few plants, produced by the puncture of insects, viz., in Turkish nutgalls (from Quercus infectoria), in Chinese nutgalls (from Rhus semialata), and in the bark of the sumach tree (Rhus coriaria), and some other species. It cannot be decided at present, if it is still more widely distributed (*see* Kinotannic acid). To obtain it, extract pulverised nutgalls with common ether of 0·740 to 0·750; evaporate the tincture to dryness, and remove resin, fat, and gallic acid with anhydrous ether.—White, or slightly yellowish, loose, amorphous, resinous, easily friable mass of faint, mostly a little ethereous odour, of highly adstringent taste, and of strongly acid reaction; fuses incompletely with heat, swells up, and becomes carbonised, yielding a sublimate of pyrogallic acid ($C_{12} H_6 O_6$). Dissolves most readily in water, more readily in hydrated than in anhydrous alcohol, slightly in anhydrous ether, readily in hydrated ether, which solution soon separates into two portions, the lower of which contains most of the acid with water and a little ether, and the upper stratum a little of the acid only, water, and most of the ether. It secedes on boiling with diluted sulphuric acid into gallic acid and glucose, likewise with alkalies; but then the sugar is instantly changed into humus-like products. It throws down the glue in grey elastic flocks, which re-dissolve in an excess of the glue. It does not alter the salts of suboxyd of iron, but precipitates the solutions of the oxyd blue or blue-black, even when very much diluted; this precipitate is in appearance quite similar to the one obtained by gallic acid, but differs from it by its behaviour towards alkalies and acetic acid (see Gallic acid). Only the tannates of the alkalies are soluble in water, but decompose easily when exposed to the air. Even the aqueous solution of the pure acid becomes soon decomposed, turns mouldy, and deposes gallic acid.

**Gamboge.** Exudation of the stem of Garcinia Cochinchinensis, (G. pictoria and G. Morella), more or less brown yellow, yields a bright yellow powder, has a faint odour and an acrid, rancid, afterwards sweetish taste. Contains about 80% resin and 18% gum. The resin, also called Gambogic acid, obtained by extracting with ether, is cherry-red, almost opaque, yields a beautiful yellow powder, is inodorous and tasteless, dissolves readily in alcohol and ether, in alkalies, in concentrated sulphuric acid with red colour, and precipitable by water without change, has an acid reaction, and is composed according to the formula $C_{40} H_{24} O_8$.

**Gardenia Tannic Acids.** In the Chinese wongshy (the fruits of Gardenia grandiflora).

First Acid $= C_{48} H_{28} O_{26} + 8$ HO. Exhaust the fruits with alcohol, evaporate the extracts in a current of carbonic acid gas, remove by means of a moistened filter the oil, which separates on cooling and on addition of water, and precipitate the red-yellow filtrate with acetate of lead, which throws down pigment and the First acid (the filtrate serves for the preparation of chlorrubin). Decompose the precipitate under water with sulphuret of hydrogen, filter, again precipitate the filtrate with acetate of lead, decompose again with sulphuret of hydrogen, whereby the whole of the pigment remains with the sulphide of lead, filter, and evaporate in a current of carbonic acid gas. Amorphous, brown-yellow mass, the solution of which greens chloride of iron.

Second Acid $= C_{46} H_{28} O_{26} + HO$. Boil the fruits, after exhausting with alcohol, with water, concentrate the decoctions, throw down the jelly with alcohol, strain, precipitate with acetate of lead, treat the deposit with a little acetic acid, filter, remove the lead from the solution by means of sulphuret of hydrogen, and evaporate to dryness.

**Gaultierilen** $= C_{20} H_{16}$. Contained in the oil of Gaultiera procumbens, and probably other species of the genus, up to about one-tenth of its weight. Distil the oil with potash-ley, wash the distillate, consisting of wood spirit and Gaultierilen, first with a very weak solution of caustic potash, afterwards with pure water, desiccate the undissolved oil with chloride of calcium, and rectify over potassium. Colourless, thin, smells rather pleasantly pepper-like. Is lighter than water. Boils at 160°.

[**Gelsemic Acid.** Is, according to Professor Sonnenschein, identical with Aesculin.]

[**Gelsemin.** Alkaloïd discovered by Wormley in the herb of Gelsemium nitidum. Prepared from the aqueous liquid which has served for the production of gelsemic acid. After the g. acid has been extracted by chloroform, the aqueous liquid is rendered slightly alkaline with potassa, and repeatedly shaken with chloroform. On evaporating the latter, a hard gummous body remains, which has to be treated with water acidulated by hydrochloric acid in order to dissolve the alkaloid. Filter, concentrate to one-sixteenth of the original fluid extract, add caustic potash in slight excess, collect the precipitate, wash with water and dry. Dissolve again in acidulated water, precipitate with potash and extract with ether. After evaporation the G. remains as a hard shining mass, yielding a colourless powder. The G. is a strong poison of an intensely bitter taste, and forms salts with acids. It is sparingly soluble in water, easily in ether (25 parts), and in chloroform. Concentrated sulphuric acid colours it red-brown, the solution becoming purple on careful heating; nitric acid dissolves it with green, hydrochloric acid with yellow colour. A little below 100° C. it fuses to a viscid mass which congeals to a translucid body; in higher temperatures it sublimes unaltered. The Gelsemin compounds are precipitated by caustic alkalies; an excess of the latter redissolves the precipitates. Bichromate of potash produces a copious yellow, amorphous precipitate, sparingly soluble in acetic acid; picric acid yellow; biiodide of potassium brown even with diluted ( $1/10000$ ) solutions; bromine in hydrobromic acid yellow; the chlorides of gold and of platinum likewise; sulphocyanide of potassium dirty white in moderately diluted solution;

cyanide of potassium blue-green; chloride of mercury white. Concentrated solutions are also precipitated by iodide and by ferro-cyanide of potassium.]

**Gentian Bitter**$= C_{40} H_{30} O_{24} + 2 HO$. The bitter ingredient of the root of Gentiana lutea. Digest the fresh root (the dried root gives no crystalline bitter) with alcohol of 70%, evaporate to honey-consistence, dissolve in three times its weight of water, shake the solution twice with animal charcoal, rinse the latter in cold water, dry and boil with alcohol of 80%. This tincture, when freed from the whole of the alcohol by heating, separates on cooling, or better by diluting with water, a resin which has to be removed, and the liquid digested with oxyd of lead. Filter hot, remove the lead by sulphuret of hydrogen, evaporate to a syrup consistence and shake with ether. After being left in contact for one day the mixture solidifies to a crystalline mass, which has to be pressed and recrystallised in a little hot water with aid of a little animal charcoal.—Colourless, radially united needles, efflorescent at the air, of a strong and pure bitter taste; neutral; loses its water at 100°; fuses at 120° to 125°, becomes carbonised in higher temperatures while emitting an odour of burnt sugar; dissolves readily in water and in hydrated alcohol, less readily in absolute alcohol, not in ether; in concentrated sulphuric acid colourless, changing to a beautiful carmine red on heating; decomposes with diluted sulphuric acid into sugar and another product; is soluble in alkalies with yellow colour.

**Gentianin,** or **Gentianic Acid**$= C_{28} H_{10} O_{10}$. In the root of Gentiana lutea, but is not the bitter ingredient. Withdraw from the root most of the bitter ingredient by means of cold water, dry the root and exhaust with strong alcohol, evaporate to a syrup, pour water on it, and remove thereby the substances soluble in water, leaving a sediment which contains Gentianin, fat, resin, and the bitter substance. Remove the fat with ether, and recrystallise in alcohol.—Light, long, pale, yellow, silky needle-like crystals, inodorous and tasteless, neutral, unchangeable at 250°; becomes brown at 300°, sublimates with careful heating, dissolves in 5000 parts cold and in 3850 parts hot water, in 500 parts of cold and in 90 parts boiling alcohol, in 2000 parts ether; is not changed by diluted acids; dissolves readily in alkalies with yellow colour, forming salts.

**Gentisin**$=$ GENTIANIN.

**Geraniin.** After the roots of species of Geranium are extracted with alcohol, part of the alcohol distilled off from the tincture, the tannic acid removed from the remaining liquid by adding hydrate of lime, the liquid filtered and evaporated, the resin that separates removed, and the liquid brought to dryness;.

there remains a honey-yellow, transparent, very hygroscopic mass of very bitter taste, readily soluble in water and aqueous alcohol, not in ether and absolute alcohol, not precipitable by metallic salts.

**Getah Lahoe.** The milky juice of Ficus subracemosa and F. variegata, hardened at the air, has in appearance some similarity with crude gutta-percha, is outside of a blackish-grey, inside of a delicate pink colour, very porous, brittle, becomes by continued rubbing soft and plastic like wax, is easily lighted and burns with a white, smoking flame, becomes sticky at 35°, liquid at 75°, floats on water, is not soluble in cold alcohol, dissolves in hot alcohol, leaving undissolved a brownish, very viscid mass; the brown solution, on cooling, throws down most of the dissolved substances under the form of a white granular powder. Ether, chloroform, benzol, and oil of turpentine dissolve the G. at ordinary temperature, with the exception of the said brownish mass. Caustic alkalies affect it only after continued boiling, while dissolving the brownish mass and leaving the G. white. The G. may consequently be regarded as a kind of wax.

**Gingkoic Acid** $= C_{48} H_{47} O_3 + HO$. Peculiar, solid fat-acid of the fruit of Gingko biloba, is obtained by extracting with ether, evaporating the solution and cooling the remaining fat down to 0°, when the acid will crystallise in concentrically radiated needles of yellow colour. It dissolves readily in alcohol and in ether, is of strongly acid reaction, fuses at 35°. The gingkoate of lead is viscid, that of baryta sparingly soluble in water, readily soluble in alcohol; the gingkoate of silver is insoluble in alcohol.

**Glaucin.** Alkaloïd, contained in the herb of Glaucium luteum, not in the root. Bruise the herb with acetic acid, press, boil the liquid, strain, add a little nitric acid and precipitate warm with nitrate of lead. After cooling, filter off the fumarate of lead deposit, remove from the liquid excess of lead by sulphuret of hydrogen, neutralise the filtrate and precipitate with tannic acid. Mix the washed and pressed deposit moist intimately with hydrate of lime, exhaust the mixture with alcohol, impregnate the filtered liquid with carbonic acid, evaporate, wash the remnant with a little cold water, in order to remove dyeing matter, and crystallise the remaining Glaucin in hot water.—White crusts consisting of minute scales of pearly lustre, separates from ether as a turpentine-like, at first almost oily mass, which hardens by keeping, fuses under water to an oil, is destroyed in higher temperatures, tastes bitter and very acrid, has an alkaline reaction, reddens at the air and more rapidly in the sunlight, dissolves in water, better when hot, most readily in alcohol and ether, assumes, when heated with concentrated sulphuric acid—until the acid begins to evaporate, a beautiful blue-violet colour, which passes to a dark peach-flower red on addition of water, and yields an indigo-blue precipitate with ammonia. It neutralises the acids, and forms white salts of a burning, acrid taste, precipitable by ammonia with white colour.

**Glaucopicrin.** Alkaloïd of the root of Glaucium luteum, besides Chelerythrin. Exhaust with water containing acetic acid, precipitate the chelerythrin by means of ammonia, saturate the filtered liquid with acetic acid, precipitate with tannic acid, triturate the washed deposit with hydrate of lime and alcohol, warm, subject the filtrate to carbonic acid, distil off the alcohol, filter, evaporate the liquor, and exhaust the remainder with ether. The residue left after the evaporation of the ether, yields by washing with water and a little ether, a comparatively pure product, which has to be crystallised in hot water, while impure Gl. is dissolved by the ether. Dissolve the latter after the evaporation of the ether, in water and acetic acid mixed with subacetate of lead, and adduce sulphuret of hydrogen. Filter off the sulphide of lead, withdraw from it by repeatedly boiling with water and acetic acid the Gl. mixed up with it, saturate these solutions and the liquid filtered off from the sulphide of lead with sulphate of soda, and precipitate with ammonia. The precipitate may be obtained in the pure state, but with difficulty, by dissolving in ether, and evaporating the latter. Another alkaloid, precipitable by tannin together with Gl., remains with the almost pure Gl. in small quantity; it forms a salt which crystallises in needles, but soon effloresces.—Snow-white, permanent grains, bitter; dissolves readily in water, especially hot, and covers the concentrated solution with a pellicle on cooling, which soon becomes crystalline and sinks to the bottom; dissolves in alcohol, not so well in ether, yields neutral white salts of a bitter and nauseous taste.

**Gliadin.** *See* GLUTEN.

**Globularia Resin** = $C_{40} H_{36} O_{16}$. The fragrant resin of the leaves of Globularia Alypum. From the alcoholic extract of the leaves the bitter substance (Globularin) is extracted by water, while Globularia Resin remains. Dissolve the latter in alcohol, digest with animal charcoal, and precipitate the filtered solution with water. Olive-green, translucid, plastic mass, of the smell of the leaves.

**Globularin** = $C_{60} H_{44} O_{28}$. The bitter ingredient of the leaves of Globularia Alypum. The alcoholic extract is to be distilled, the remnant dissolved in water, digested hot with oxyd of lead for a considerable time, filtered, the liquid evaporated on the water-bath, the remnant freed from a yellow pigment by ether, dissolved in water, precipitated with tannin, the deposit dissolved in alcohol, the solution digested with oxyd of lead, filtered, and the liquid brought to dryness. White, bitter powder, soluble in water and alcohol, insoluble in ether, converted by digestion with diluted sulphuric acid into sugar and other products.

**Globularia Tannic Acid** $= C_{16} H_{12} O_{14}$. In the leaves of Globularia Alypum. Is precipitated from the aqueous solution of the alcoholic extract by digestion with oxyd of lead. (*See* Globularin.) Divide the precipitate in alcohol, decompose with diluted sulphuric acid, shake the green-brown liquor with a little carbonate of lead and precipitate with alcoholic acetate of lead. Isolated by sulphuret of hydrogen, it is only known in solution which imparts a dark-green colour to chloride of iron.

**Glucose = Grape Sugar.**

[**Glucosids** are organic compounds of a high atomic weight and great chemical complexity. They are resolved by heating with dilute acids, or sometimes even when left in contact with certain protein-substances, under assimilation of water, into sugar (fermentable or not), and one or more other products, generally an essential oil or a resinous body. They also in most cases exhibit the following properties:—They are crystalline or amorphous, neutral, or slightly acid ; without odour, of a bitter or acrid taste; fusible, but without sublimation, and carbonised by prolonged heating ; they dissolve easily in alcohol, less so in water, sparingly or not at all in ether, under decomposition in alkalies with yellow, and in concentrated sulphuric acid with blue, violet, or red colour. —F. v. M.]

**Gluten.** A substance which has been formerly considered as a peculiar proximate constituent of the vegetable kingdom, but which is now proved to be only a mixture of different protein substances. It occurs especially in the seeds of the cereals and of leguminous plants. It is obtained best by kneading wheat-flour under cold water until the water passes from it clear and without a milky appearance (by taking up starch). When fresh, it is of a greyish-white, very viscid, glutinous, elastic, tasteless, of insipid odour; when dried, it appears grey-yellow, horny, brittle, mollifies in cold water and dissolves a little ; the solution curdles at 62°; by boiling with water it becomes hard and insoluble. Hot alcohol withdraws from it so-called gliadin, which is itself a mixture of gluten, mucin and gum, and leaves undissolved the so-called vegetable fibrin as a brownish-grey matter, containing sulphur and nitrogen. (Some authors believe the latter to be coagulated albumin).

**Glutin**. Peculiar protein substance, ingredient of gluten. By boiling gluten with alcohol, a solution is obtained which yields on evaporating the so-called gliadin, *i.e.*, a mixture of Glutin and mucin and gum. By distilling the solution with water, most of the Glutin separates together with mucin, while a portion only remains dissolved in water by means of the gum. From this aqueous solution the gum is thrown down with alcohol and the liquid evaporated. Glutin is pale, yellow, tough, more or less glutinous, tasteless, of a faint peculiar odour, dries to a yellow transparent mass. It only swells up in water and soon becomes putrid; it dissolves in hot alcohol, in cold alcohol only partially, also partially in acetic acid, not in ether, readily in potash-ley.

**Glycerin** $= C_6 H_5 O_3 + 3 HO$. Almost in every fat combined with fat-acids, and separated from the latter by saponification. Saponify any fat with soda-ley, decompose the soap with diluted sulphuric acid, filter, saturate the liquid with carbonate of soda, evaporate, shake the pulpy remnant with strong alcohol, and evaporate the solution. To remove all water, keep for a rather long time at a temperature of 120°. Colourless syrup of a pure sweet taste, uncrystallisable, inodorous, neutral, of 1·27 density; distils partly when boiled with water; decomposes when heated by itself, while evolving vapours (acrolein) of a highly offensive and acrid odour; burns on the open fire like an oil; mixes with water and alcohol in every proportion; is insoluble in ether.

[Glycerin on a large scale, and perfectly colourless and pure, is now obtained as a bye-product from soap and candleworks, when the fats are distilled by high-pressure steam, and are thereby decomposed into fat-acids and glycerin.—F. v. M.]

**Glycyrrhizin** $= C_{48} H_{36} O_{18}$. In the root of Glycyrrhiza glabra and G. echinata and in Monesia bark (from Lucuma glycyphlaea); it is not decided yet if the substances of similar properties from the leaves of Abrus precatorius, from the root of Polypodium vulgare and from Sarcocolla are identical with Glycyrrhizin. Macerate in water, boil the solution, filter, concentrate and precipitate with diluted sulphuric acid. The at first pale yellow flocks unite slowly to a dark-brown viscid mass, which has to be washed repeatedly with water until free from sulphuric acid (on testing with soluble salts of baryta); it is then dissolved in alcohol of 80%, and the moderately-concentrated solution mixed with small quantities of ether; a brown resin, which separates after a short time, is removed, the ether-alcoholic solution evaporated on the water-bath, redissolved in alcohol, mixed with ether, filtered and evaporated.—Amorphous, yellow-white powder, smells in the alcoholic solution similar to an infusion of liquorice; has an intensely bitter-sweet taste, is of a strongly acid reaction when dissolved in water, fuses at 200°, and is destroyed in a higher temperature; dissolves slowly in cold, readily in hot water, with yellow colour, and a small portion separates on cooling under the form of resinous drops; dissolves abundantly in alcohol, less readily (according to others not at all) in ether; is not fermentable; is precipitated in flocks by mineral and vegetable acids. The flocks are a compound of G with acids (according to others pure G.); combines with bases; alkalies increase its solubility in water, metallic salts throw the G. down.

**Gœmin.** Proximate constituent of carrageen or gœmon (Chondrus crispus), and bearer of the mucilaginous properties of the drug. Contains much nitrogen and sulphur. To prepare it boil the alga with water for a few hours, strain, precipitate the liquid with alcohol, rinse with alcohol, redissolve in water and evaporate the solution.—Thin, transparent, elastic laminæ similar to isinglass, without smell and taste, swells up in cold water and dissolves when warmed to a mucilage of neutral reaction; dissolves also in hot hydrochloric and in nitric acid under formation of oxalic acid, in the ordinary mixture of nitric and hydrochloric acid under formation of sulphuric acid; soluble also in alkalies. Consists of 21·80 C, 4·87 H, 21·36 N, 2·51 S, 49·46 O.

**Grape Sugar,** or **Dextro-glucose,** called also from its origin honey sugar, starch sugar, or simply glucose $= C_{12} H_{12} O_{12} +$ 2 HO. Abundantly in sweet fruits, often associated with cane-sugar, always associated with so much lævo-glucose, that the mixture may be regarded as inverted sugar. (*See* FRUIT SUGAR.) It is produced from the juice of grapes by boiling, neutralising the free acid with chalk, evaporating to half the volume, leaving to subside, clarifying with albumen if necessary, and evaporating to the consistency of a syrup. The Grape-sugar forms into crystals on keeping, and is recrystallised in water.—White, opaque, semi-spherical, or cauliflower-like mass, consisting of microscopic, sex-angular tabular crystals, tastes less sweet than cane-sugar and at the same time mealy when in the solid state; softens at 60°, deliquesces at 90° to 100° under loss of all its water, and solidifies to a colourless amorphous mass; loses at 170° two more equivalents of water, and carbonises afterwards with the odour of burnt sugar. The fused Grape-sugar deliquesces at the air first under absorption of water, and solidifies again as soon as the quantity of water absorbed is sufficient for the formation of crystals to a crystalline granular mass. The Grape-sugar which has been desiccated with-out fusing (in a current of air of 55° to 60°), and is obtained as a white powder, absorbs no water from the air. Grape-sugar dissolves in 1½ parts cold, in every proportion in boiling water, in 50 parts cold, and in 4 parts boiling alcohol of 0·837, scarcely in ether; but yields with nitric acid, oxalic, no mucic acid; dissolves in cold concentrated sulphuric acid without colouration, but is charred by heat; turns brown on continued boiling with dilute sulphuric acid; is readily decomposed and turns dark brown when heated with alkalies and alkaline earths, colours the subnitrate of bismuth grey brown when boiled with it under addition of carbonate of soda; reduces the alkaline tartrate of copper (1 equivalent Grape-sugar forms 10 equivalents copper suboxyd $= Cu_2O$); ferments directly with yeast, is not precipitable by acetate and sub-acetate of lead, but is so by ammoniacal acetate of lead, and this precipitate assumes by keeping at the air, or more quickly when warmed, a red colour.

The quantitative estimation of Grape-sugar can, like that of cane-sugar, be accomplished either by fermentation or by means of a solution of copper. But it must be borne in mind that in the first case the water which cane-sugar assimilates, before it becomes converted into carbonic acid and alcohol, is already present in Grape-sugar. Consequently, if 100 parts cane-sugar yield 49·12 parts carbonic acid, and 51·01 parts alcohol, 105·26 parts anhydrous Grape-sugar will effect the same transformation; or, 46·66 parts carbonic aid and 48·46 parts alcohol are obtainable from 100 parts anhydrous Grape-sugar.

In the second case, of course, the saccharine liquid wants no preliminary treatment except, when acid, to be oversaturated with caustic potash or soda, before it is titrated with the solution of copper.

**Gratiolin** $= C_{40}H_{34}O_{14}$. The bitter ingredient of Gratiola officinalis [and its congeners. — F. v. M.]. Precipitate the aqueous decoction of the herb with subacetate of lead; mix the filtered liquid with carbonate of soda, but not in excess; filter, precipitate with tannic acid, collect the deposit and mix it with hydrated lead oxyd; treat the mixture with alcohol, filter and decolourise the tincture with animal charcoal; dry after filtering, exhaust the residue successively with absolute ether and with cold water, dry and recrystallise in boiling alcohol or in boiling water. The ether dissolves mainly gratiolacrin, the cold water gratiosolin.—White powder, crystallising from alcohol in warty masses, from water in fine silky needles; tastes at first very slightly, afterwards strongly bitter; has a faint smell; fuses at 200° without change, but is destroyed in higher temperatures; mollifies on beating with water and rises to the surface like an oil; dissolves in 893 parts cold and in 476 parts boiling water, most readily in alcohol, in 1000 parts cold, and in 666 parts boiling ether, in concentrated sulphuric acid with dark-red colour, and precipitable from it by water; decomposes on heating with diluted sulphuric acid into sugar and other products.

**Gratiosolin** $= C_{46}H_{42}O_{25}$. The aqueous gold-coloured solution obtained during the process of preparing gratiolin is digested with animal charcoal, filtered, dried on the water-bath, and traces of gratiolacrin removed by anhydrous ether.—Amorphous, bright-red mass, friable to a yellow powder, has a peculiar smell and a nauseous, bitter taste, is permanent at the air, fuses at 125°, is destroyed in higher temperatures, dissolves in 7 parts cold and in 5 parts boiling water, in 3 parts cold and 2 parts boiling alcohol, in 1700 parts cold and in 1100 parts boiling ether, readily in liquor of ammonia, in concentrated sulphuric acid with brown-red colour; secedes on heating with diluted acids or alkalies into sugar and some other product.

**Green Acid.** Occurring in the roots of many Dipsaceæ, Compositæ and Umbelliferæ, forming with ammonia a yellow compound which becomes blue-green at the air, resembles caffe-tannic, rubichloric and valeria-tannic acids. Exhaust (say, the root of Scabiosa succisa) with alcohol, precipitate the tincture with ether, collect and wash the white precipitate with ether, and throw down its aqueous solution with acetate of lead. After this precipitate is decomposed under water with sulphuret of hydrogen and the filtered liquid evaporated, the green acid remains as an amorphous, yellow, brittle, acid mass.

**Guacin.** Resinous bitter substance of the leaves of Mikania Guaco (and perhaps other Guaco-plants). Draw out with ether, treat the ethereous extract with alcohol, the alcoholic extract with boiling water, the aqueous extract again with alcohol, and evaporate the latter solution to dryness.—Yellowish-brown, resinous, very bitter, fuses at 100°, dissolves very slightly in cold, abundantly in boiling water, most readily in alcohol and ether.

**Guaiacic Acid** $=C_{12} H_8 O_6$. In the wood and resin of Guaiacum officinale. Dissolve the resin in alcohol, concentrate the solution to one-quarter its volume, separate after cooling the acid, yellowish liquid from the subsiding resin by filtering, evaporate the former to a syrup-consistence, exhaust with ether, evaporate the solution and sublimate the acid from the obtained warty mass containing resin over a carefully-conducted fire.—White, shining, needle-like crystals, much more soluble in water than benzoic or cinnamic acids, also soluble in alcohol and ether. Requires further comparisons with benzoic acid.

**Guaiaconic Acid** $=C_{38} H_{20} O_{10}$. Constitutes about 70% of guaiacum resin. An alcoholic solution of guaiacum resin is mixed with an alcoholic solution of caustic potash; the liquid is separated from a solid product, consisting of acid guaiacum resin and potash, and evaporated at a temperature of 30°. A thick syrup-like fluid remains, which mixes with absolute alcohol under separation of some acid guaiacum resin combined with potash. Remove the mass, pervade the liquid with carbonic acid, filter, mix the filtrate with water and a little hydrochloric acid, evaporate the alcohol, and wash the resin thus obtained with warm water. It appears, after drying, as a brittle, brown mass. This resin, by treating with ether, is decomposed into Guaiaconic acid, which dissolves, and into guaiacum beta resin remaining undissolved. Separate the acid in the ethereous solution with potash-ley, throw away the supernatant ether; dilute the alkaline solution with water, and throw down with acetate of lead. The grey-green precipitate is decomposed under water with sulphuret of hydrogen; the deposit is dried and extracted by alcohol, which dissolves the Guaiaconic acid and leaves it on evaporating.—Light-brown, brittle mass of concheous fracture, friable to a paler, inodorous, and tasteless powder: neutral; fuses at 95° to 100°, and decomposes afterwards; is insoluble in water, readily soluble in alcohol, ether, acetic acid, chloroform, scarcely in benzol and sulphide of carbon, in concentrated sulphuric acid with beautiful cherry-red colour, and precipitable from it by water in violet flocks, containing sulphur.

**Guaiacum Beta Resin** $= C_{40} H_{20} O_{12}$. Constitutes 10% of guaiacum-resin. Is obtained during the preparation of guaiaconic acid, and forms the residue insoluble in ether; withdraw from it all guaiaconic acid by treating the substance mixed with sand with ether, dissolve in alcohol, decolourise with animal charcoal and precipitate by pouring the solution into ether. The brown flocks are purified by dissolving anew and precipitating, and lastly dissolved in alcohol and again separated with water.—Red-brown powder, neutral, fusing at 200° to a black mass, is insoluble in water, readily soluble in alcohol and alkalies, not in ether, sulphide of carbon, chloroform, and benzol.

**Guaiacum Yellow.** Yellow pigment of guaiacum resin. Boil the pulverised resin with caustic lime, filter, evaporate the liquid, in order to remove most of the lime as carbonate, and with it most of the acid guaiacum resin; extract the remnant with water, oversaturate the solution with acetic acid, filter again, and allow to stand for a rather long time. Small, pale-brown, tabular crystals are formed, which dissolve slowly in much water, while leaving a resin; readily soluble in ether and alcohol, and crystallising in the latter on evaporating at ordinary temperature.—Pale-yellow, quadratic, octahedrous, or tabular crystals, inodorous, bitter, fusible by heat and afterwards decomposing, sparingly soluble in water; soluble in alkalies and alkaline earths, in alcohol, ether, sulphide of carbon; difficultly in chloroform, benzol, oil of turpentine; scarcely a little in acids.

**Guaiacum Resin.** From the stem of Guaiacum officinale, partly exudating spontaneously, partly obtained by extracting with alcohol. Is greenish outside, inside reddish or greenish-brown, brittle, grey-white when pulverised, turns greenish at the air, has a balsamic odour, and a sweetish-bitter taste which is at the same time acrid and irritating to the throat; dissolves readily in alcohol, 90% in ether (the beta-resin is insoluble); readily in alkalies and re-precipitable by acids; fuses easily, and burns afterwards with a strong aromatic smell. The tincture turns blue with many oxydising substances (nitrous acid, chlorine, &c.). It consists of about $70\%$ guaiaconic acid, $10\%$ acid guaiacum-resin, $10\%$ beta-resin, and of guaiacic acid, and guaiacum-yellow.

**Guaiaretic Acid** $= C_{40} H_{26} O_8 + HO$. In guaiacum-resin, constituting $10\%$ of it. Boil the pulverised resin with half its weight of caustic lime and sufficient water for half an hour, strain, dry the residue and exhaust with hot alcohol, distil off the alcohol from the tincture, dissolve the remnant in soda-ley, allow the solution to stand cold, press the soda salt, recrystallise in water with a little caustic soda, decompose with hydrochloric acid, and crystallise in alcohol.—Soft, small warty and scaly crystals, of a faint odour of vanilla, when crystallised in acetic acid inodorous brittle needles, when crystallised in diluted alcohol shining laminæ; fuses at 75° to 80° under loss of water, by rapidly heating mostly volatilised without decomposition; insoluble in water, dissolves in 1·8 parts alcohol of $90\%$, likewise in ether, also in chloroform, sulphide of carbon, acetic acid, benzol; in concentrated sulphuric acid with purple colour, and reprecipitable by water with white colour; little soluble in liquor of ammonia. Its compounds with the fixed alkalies dissolve readily in water.

**Gum** $= C_{12} H_{11} O_{11}$. Just as scarcely any part of a plant is without fibre, so in all likelihood no plant is without gum; at all events, gum is obtained in every phyto-chemical analysis, though the quantity obtained is sometimes exceedingly small. Being insoluble in ether, and as alcohol dissolves scarcely traces of it, the gum always occurs in aqueous extracts, being extracted completely on account of its ready solubility in the solvent. The qualitative and quantitative determination is effected by boiling the extract (prepared with cold water) for a short time, filtering off a flocky turbidity (albumen), concentrating to a small bulk at a gentle heat, and mixing the residue with alcohol of 95% in small quantities as long as any cloudiness is produced. The viscid, dough-like precipitate is washed with alcohol, re-dissolved in as little water as possible, the solution precipitated as before with alcohol, the deposit washed with alcohol, dried at 100°, and weighed.

The gum, as obtained by this process, contains generally more or less foreign matters, as inulin, sugar, dyeing substances, mineral salts, &c., to remove which completely is most difficult or impossible. Should the gum, thus repeatedly precipitated with alcohol and then dried perfectly, be soluble in an equal weight of cold water, then it contains either no inulin or only so little (one-fifth per cent. or less) that this can be left out of consideration. But, as inulin is convertible into gum with comparative ease, it is also possible that a portion of the gum obtained may have been originally inulin.

Another contamination of the gum, obtained as above, consists of pigments of various kinds. They are almost never absent, and are indicated by the more or less darkened colour of the gum. They are so obstinately attached to it as to be removable only by bleaching, which would destroy also part of the gum. Fortunately, the quantity of the adhering pigment is always so minute as to be practically of no consequence in the quantitative analysis.

By the repeated precipitation with alcohol any sugar present in the gum may be expected to be thoroughly removed. Any remaining traces may be discovered as follows:—Dissolve from 10 to 20 grains of the dried and pulverised gum in a little water, add one drop of a solution of sulphate of copper (1 part to 9 parts water), and enough of solution of caustic potash or soda to give it a decidedly alkaline reaction. It must yield with pure gum a clear, blue liquid, and must remain so when heated to the boiling point; while in the presence of sugar the colour will be changed, and a red powder (suboxyd of copper) subsides. Gum free from sugar has no disoxydising effect upon the salts of oxyd of copper, but effects only a blue solution with excess of potash or soda. The best kinds even of commercial gum arabic contain traces of sugar, which may be thoroughly removed from the finely pulverised gum by treating with alcohol.

Another and very frequent impurity of gum obtained as above, or exudating spontaneously, consists in mineral substances (alkalies and alkaline earths), which are combined either with the gum itself, while this acts as a weak acid, and has, therefore, an acid reaction when dissolved in water; or, less frequently, these alkalies are fixed to stronger acids (phosphoric acid, &c.) forming compounds which dissolve slowly or not at all in alcohol. Their presence is best ascertained by incinerating a portion of the gum at the air. If a remnant is left, mineral substances were present, and the weight of the ashes thus obtained has, therefore, to be subtracted from that of the gum. To prepare a gum free from mineral substances, its aqueous solution has to be strongly acidified with hydrochloric acid and precipitated with alcohol; the deposit is washed with alcohol and redissolved in water, again acidified with hydrochloric acid and precipitated with alcohol, and these operations are repeated several times.

Different gums have different reactions with the same reagents. Gum Arabic (from Acacia Arabica, A. Seyal and other species) is precipitated from its aqueous solution by subacetate of lead, chloride of iron, and silicate of potash; it becomes thick with borax solution, and is not affected by acetate of lead; whereas many other gums, on the contrary, are precipitated by acetate of lead, yet are not affected by subacetate of lead, chloride of iron, silicate of potash and borax. A cold concentrated sulphuric acid colours pulverised gum, only after several hours, but renders it black instantly on heating; when boiled with diluted sulphuric acid it is converted into grape sugar. Nitric acid produces oxalic and mucic acids. Iodine shows itself inactive to gum.

**Gum-Resins.** Genuine mixtures of resin, volatile oil and gum; soft or hard at a mean temperature.

**Gurjunic Acid**$= C_{44} H_{34} O_8$. Forms an ingredient of the gurjun-balsam or so-called wood-oil, an exudation from various species of the genus Dipterocarpus, especially Dipterocarpus turbinatus, and remains together with other substances when the balsam is distilled with water, while a volatile oil$= C_{40} H_{32}$ passes over. Dissolve the residue in hot patash-ley, add to the red-brown solution chloride of ammonium in excess; filter and precipitate the liquid with hydrochloric acid. The acid subsiding in dense yellow flocks, is dissolved in ether, and yields after evaporation of the latter, a crusty mass, which has to be recrystallised in alcohol.—Forms colourless, crumbly crusts of slightly acid reaction, fuses at 220°, boils at 260° and decomposes afterwards; is insoluble in water; dissolves difficultly in weak, readily in strong alcohol and in ether, slowly in benzol and sulphide of carbon, readily in alkalies forming soaps. The salts of other bases are insoluble in water.

**Gutta**$= C_{40} H_{32}$, the main ingredient of gutta-percha. Dissolve the gutta-percha, purified by treating with water and hydrochloric acid in hot ether, press quickly the mass formed on cooling, dissolve again in hot ether and wash the re-separated portion with cold ether and alcohol, whereby it becomes a jelly-like mass. The pressed substance is immediately to be heated to 100°, in order to prevent oxydation, and is then dried.—White, fine powder, when free of air bubbles heavier than water, fuses at 150° to a viscid mass, and is destroyed in higher temperatures; is insoluble in alcohol and in cold ether, readily soluble in sulphide of carbon and chloroform, less readily in benzol and oil of turpentine.

**Gutta-Percha,** the hardened milky juice of Isonandra Gutta, Sideroxylon attenuatum, Ceratophorus Leerii, Payena macrophylla, Bassia sericea, Mimusops Elengi, Mimusops Manilkara, Imbricaria coriacea, and probably other saponaceous trees. Is pale-yellow, grey-white or reddish, almost as hard as wood, tough and flexible at + 25°, at 48° plastic and of a doughy consistence, consequently mollifying in hot water; yields to water only a little acid and an extractive substance, dissolves in absolute alcohol partially (22%), little in cold ether, readily in warm ether, sulphide of carbon, and chloroform, less readily in benzol and oil of turpentine, not in acids and alkalies. Consists mainly upwards to about $80°/_o$ of a hydrocarbon (gutta), and contains besides casein an organic acid, a resin soluble in ether and oil of turpentine, another resin soluble in alcohol, and an extractive substance.

**Gyrophoric Acid**$=$ LECANORIC ACID.

**Hæmatoxylin**$= C_{32} H_{14} O_{12} + 6 HO$. In the wood of Hæmatoxylon Campechianum. Mix the commercial extract with much sand, macerate the whole with several changes of ether, evaporate the brown-yellow solutions to a syrup-like consistence, mix with water and allow to crystallise. Wash the crystals with cold water, press and recrystallise in water containing a little sulphurous acid.—Colourless, snow-white, or pale straw-yellow quadratic prisms, lose by keeping 4 equivalents, at 100° the last 2 equivalents of water without melting when slowly heated. It has a strong liquorice-like very lasting taste, not bitter or acerb; is destroyed in high temperatures; dissolves little and slowly in cold, abundantly in boiling water, more readily in alcohol than in ether, in liquor of ammonia with at first pink, then beautiful purple-red colour under formation of hæmatein$= C_{32} H_{12} O_{12}$; fixed alkalies and their carbonates have a similar effect, likewise lime-water and solution of baryta; is precipitated pure white by acetate and by subacetate of lead, the precipitate assuming quickly a blue colour at the air.

**Hagenic Acid.** Supposed peculiar acid of the flowers of Hagenia Abyssinica, occurring in combination with ammonia. Explicit statements are wanting.

**Harmalin**$= C_{26} H_{14} N_2 O_2$. Alkaloïd of the seeds of Peganum Harmala. After the solution of the chlorides of harmin and Harmalin, as stated under Harmin, has been mixed with a little ammonia, harmin subsides, while Harmalin remains dissolved. Precipitate this solution with excess of ammonia, triturate and mix the deposit with water, add acetic acid until it is almost completely dissolved, throw down the filtrate with nitrate of soda, with chloride of sodium or with hydrochloric acid, wash the subsiding salt with a diluted solution of the precipitating agent, and purify by treating its aqueous solution with animal charcoal. Precipitate the solution with potash-ley; wash the deposit with water, afterwards with absolute alcohol, dissolve in boiling absolute alcohol, and leave to cool under exclusion of the air.—Forms colourless, klinorhombic crystals; is almost tasteless by itself; when dissolved of a pure bitter taste; loses at 190° nothing of its weight, fuses afterwards and decomposes. Newly precipitated or moist Harmalin assumes at the air, more especially if under the influence of vapours of ammonia, a brown colour, dissolves very little in water, little in cold, readily in boiling alcohol, little in ether, slighly in volatile oils, is converted into harmin by heating its nitrate with alcoholic hydrochloric acid. Neutralises the acids and forms readily soluble, crystallisable salts. These and their solutions are yellow, taste bitter, and are precipitated by caustic alkalies; they dissolve more readily in pure water than in water containing acids or salts, and are precipitated in aqueous solutions by acids and by salts.

**Harmin**$=C_{26} H_{12} N_2 O_2$. Alkaloïd of the seeds of Peganum Harmala. Exhaust with cold water, containing sulphuric or hydrochloric acid, neutralise the free acid of the extracts, and add a large quantity of a concentrated solution of chloride of sodium to throw down the hydrochlorides of Harmin and harmalin. Wash these with a solution of chloride of sodium, dissolve in cold water, which leaves behind dyeing matter, decolourise with animal charcoal, and drop into the filtrate, heated to 50° to 60°, and under stirring, ammonia, until a precipitate begins to form, which rapidly increases under continued stirring without the addition of more ammonia and usually contains the whole of the Harmin, but no harmalin. Collect the deposit obtained hereby, precipitate, if the filtrate contains any more Harmin, the latter by carefully adding ammonia, or remove any harmalin present from the precipitate by dissolving the whole of it in an acid, and partially precipitating as above. The presence or absence of harmalin for the above operation may be ascertained under the microscope, for Harmin forms needle-like crystals, while harmalin crystallises in lamellary form. The Harmin is afterwards purified by recrystallising and decolourising with charcoal.—Forms colourless, brittle prisms of great lustre and light-refracting power, tasteless, bitter in solution, permanent at the air; is almost insoluble in water, slowly soluble in cold, more readily in boiling alcohol, slightly in ether, less so in volatile and fixed oils. Forms with acids colourless, or slightly yellowish, crystalline salts, the concentrated solutions of which have a yellowish colour, while in the diluted state (especially in alcohol) they exhibit a bluish fluorescence. The salts dissolve mostly more copiously in pure than in acid water, and their aqueous solutions are precipitated by hydrochloric and nitric acids and by chloride of sodium and nitrate of soda. Caustic alkalies throw down the base.

**Hazelnut Oil**, obtained by pressing the seeds of Corylus Avellana and other species of that genus. Pale-yellow, thickish, mild, inodorous, of 0·924 density, solidifies at 19°. Belongs to the non-drying oils.

**Hederic Acid** $= C_{30} H_{26} O_8$. In the seeds of Hedera Helix. Free the seeds from fat by means of ether, boil the remnant with several changes of alcohol, distil off one quarter of the alcohol from the tinctures and allow the impure acid to separate. It is difficult to procure it in the pure state, and has been only once obtained pure by keeping the ether-alcoholic solution at rest for some time.—Fine, white, soft crystalline needles and lamellæ, inodorous, of very acrid taste and slightly acid reaction, not fusible by heat, insoluble in water and ether, soluble in alcohol, in concentrated sulphuric acid with beautiful purple-colour. Yields amorphous, jelly-like salts with alkalies and alkaline earths, soluble in alcohol, but scarcely or not in water. Nitrate of silver produces in the alcoholic solution of hederate of ammonia a white precipitate, which dissolves in hot alcohol and subsides on cooling in a crystalline state.

**Hedera-Tannic Acid.** In the seeds of Hedera Helix. Is obtained from the seeds, exhausted successively with ether and with alcohol (to be used for the preparation of hederic acid), by boiling with water. Mix the decoction with acetic acid and acetate of lead, remove the precipitate and throw down the liquid with ammonia. The beautiful yellow precipitate is slightly washed (being soluble), afterwards decomposed under water with sulphuret of hydrogen and the solution filtered. The liquid yields on evaporating the acid but in an impure state. — Inodorous, amorphous, acid substance, the solution of which colours the salts of iron-oxyd dark-green, but does not precipitate glue.

**Helenin** $= C_{16} H_{14} O_5$. In the root of Inula Helenium. Exhaust with alcohol, and mix the tincture hot with three times its volume of water. It becomes cloudy, and the Helenin subsides slowly in a crystalline form.—Forms white, friable, quadrangular prisms and needles of faint smell and taste, insoluble in water, readily soluble in alcohol and ether; fuses at 75°, boils at 275° to 280° under partial decomposition, dissolves in concentrated sulphuric acid with red colour, which becomes slowly darker.

[According to Kallen, Helenin is resolvable into two crystallisable substances, for one of which he retains the name Helenin $= C_{12} H_8 O_2$. It is devoid of odour and taste, fusible at 110°. The other is Alant-camphor $= C_{20} H_{16} O_2$, with a smell and taste suggestive of peppermint; fuses at 64°; yields cymol by distillation with phosphorus-pentasulphide.]

**Helianthic Acid** $= C_{14} H_9 O_8$. In the seeds of Helianthus annuus. Exhaust the seeds, freed from the husks, with hot alcohol, distil off the latter from the tincture, filter the residue, precipitate the liquid with acetate of lead, decompose the washed precipitate under water with sulphuret of hydrogen, filter and evaporate the liquid.—Brownish-yellow, amorphous mass, friable to a slightly coloured powder, fusible by heat; dissolves readily in water and alcohol, not in ether; imparts a splendid dark-green colour to chloride of iron, which changes to violet on addition of ammonia; does not precipitate glue.

**Helleborein** $= C_{52} H_{44} O_{30}$. One of the two glucosids of the root of Helleborus niger and H. viridis, and present in larger quantity than the other. It is obtained by boiling with water, precipitating with subacetate of lead, removing the excess of lead by sulphate or phosphate of soda, evaporating, precipitating with tannic acid, mixing the washed precipitate with alcohol, adding oxyd of lead, drying, extracting with hot alcohol, precipitating the Helleborein from the highly concentrated solution with ether, separating the H. before it conglutinates, and drying in a vacuum. By repeatedly dissolving in alcohol and precipitating with ether, it is obtained completely pure.—Crystallises from the concentrated alcoholic solution slowly in translucid warty masses of about the size of peas, composed of microscopic needles, turning readily to a chalky white at the air and yielding a yellowish-white, very hygroscopic powder. Tastes sweetish, dissolves most readily in water and aqueous liquids, less readily in alcohol; insoluble in ether. The aqueous solution has a slightly acidulous reaction, and dries to a yellowish resin. At 160° it becomes straw-yellow and conglutinates, at 220° to 230° it turns brown and becomes paste-like, at 280° viscid and is charred. Concentrated sulphuric acid dissolves it with brown-red colour, passing slowly into violet. Alkalies and alkaline earths do not affect it. Diluted acids quickly decompose it on boiling under formation of sugar and of another product appearing in violet-blue flocks.

**Helleborin** $= C_{72} H_{42} O_{12}$. The other of the two glucosids of the root of Helborus niger and H. viridis; only occurring in small quantity. To prepare it, boil with alcohol, concentrate the decoctions to a small volume; shake repeatedly with much hot water, and evaporate the aqueous solutions after removing the supernatant fixed oil. The H., which then separates, is collected, washed with water, and recrystallised repeatedly in boiling water until it is snow-white. — Dazzling white, concentrically arranged needles, almost tasteless when dry, but of an extremely acrid and burning taste when dissolved in alcohol. It is insoluble in water, slightly soluble in ether and fixed oils, but readily so in boiling alcohol and in chloroform. Above 250°, it fuses and becomes carbonised. Concentrated sulphuric acid imparts to it a magnificent crimson-red colour, and dissolves it slowly with the same colour. This reaction is much more intense and sensitive than the similar one of salicin. By immediately diluting the solution with water, most of the H. subsides unaltered, only a small part having undergone a decomposition into sugar and a resin. The same decomposition takes place on boiling with diluted acids, while aqueous alkalies have no effect.

**Helonin** $= C_{14} H_9 NO_3$. Resinous substance, not belonging to the alkaloïds, from the seeds of Schoenocaulon officinale. Is obtained by treating the impure veratrin, precipitated by caustic potash after Couerbe's method, with water in order to remove sabadillin and sabadillin-hydrate. Extract the veratrin from the residue by means of ether, dissolve the insoluble part in alcohol, and evaporate the solution.—It is brown, solid, insoluble in water, ether, and alkalies, soluble in alcohol, fuses at 185°, combines with acids, but does not neutralise them.

**Hempseed Oil** = Oil of Cannabis sativa.

**Hesperidin.** Bitter ingredient of the unripe bitter oranges, and of the white spongy parts of the fruit-peelings of bitter oranges and lemons. Exhaust with water, concentrate, treat the extract with strong alcohol, evaporate, add water and allow to crystallise slowly.—Delicate, silky, bitter, needle-like crystals, united to warty concrescences; dissolves very little in cold, in six parts boiling water; slightly in cold, most readily in hot alcohol; not in ether and oils. According to recent investigations it appears to be a glucosid, said to contain instead of sugar dulcit $= C_{12} H_{14} O_{12}$, a non-fermentable kind of sugar.

**Huanokin** = $C_{20} H_{12} NO$ (isomeric with Cinchonin). Alkaloïd occurring in the Quina huanoko plana (from Cinchona micrantha), distinguishable by the following characteristics. It crystallises in small colourless prisms; is tasteless, of slightly alkaline reaction, which becomes more decided in an alcoholic solution, but then possessing a slightly bitter taste; fuses readily, and sublimates by a stronger heat; is insoluble in water; dissolves in 400 parts cold, and in 110 parts boiling alcohol of 80%, in 600 parts cold, and in 470 parts boiling ether.

**Hurin.** Acrid, crystalline substance of the milky juice of Hura crepitans. It is obtained after the milky juice has been evaporated and boiled with alcohol, the alcoholic solution evaporated and exhausted with water, the residue extracted with ether, and the etherous solution evaporated; as an oily, afterwards crystalline substance of acrid and burning taste, of alkaline reaction, fusing above 100°, boiling afterwards and evaporating in extremely pungent vapours; readily soluble in alcohol, ether, and oils, not in water; not changed by alkalies.

**Hydrastin** = $C_{44} H_{23} NO_{12}$. Alkaloïd, associated with berberin, in the root of Hydrastis Canadensis. The mother-ley remaining from the preparation of berberin is freed from alcohol, is diluted with water and cautiously mixed with ammonia until the precipitate, consisting of resins, remains permanent. The filtered liquid, when mixed with a little excess of ammonia, forms a drab-coloured precipitate, which has to be washed and re-crystallised in alcohol with aid of charcoal.—White, very glossy, quadrangular prisms, turning opaque on drying, undissolved tasteless, when dissolved of a bitter and acrid taste; narcotic; of alkaline reaction; fuses at 135°, decomposes afterwards; scarcely dissolves in water, readily in alcohol, ether, chloroform and benzol, not in alkalies. Its salts are not crystallisable, but readily soluble and very bitter.

**Hydrocarotin** $= C_{36} H_{30} O_2$. Besides carotin in the cultivated root of Daucus Carota. The alcoholic solution of hydrocarotin and mannit, obtained in the preparation of Carotin, throws down on cooling a red brown mucilaginous deposit, and forms, after the latter has been removed and the liquid allowed to rest for eight days, crystals, consisting of Hydrocarotin and mannit. Remove the mannit by dissolving in water, and purify the H. by recrystallising several times in the least possible quantity of boiling alcohol, and lastly by boiling with water.—Forms colourless, large, flexible, rhombic lamellae, without taste or smell; floats on water like a fat, without being wetted; becomes at 100° hard and brittle, a little above 100° yellowish and soft, then dark yellow; fuses at 126° without loss of weight, and consolidates again resinlike; is destroyed by a higher temperature; is insoluble in water, soluble in boiling alcohol and crystallising almost entirely on cooling; readily soluble in ether, sulphide of carbon, chloroform, benzol, oils, in concentrated sulphuric acid, with vividly red colour, and thrown down by water in the amorphous state; not soluble in alkalies.

**Hydroelaterin.** *See* ECBOLIN.

**Hydrocyanic Acid** $= HC_2 N$ or HCy., does probably not exist as such in the living vegetable organism, but appears always, when amygdalin (*see* "Amygdalin") is decomposed under access of water. It is easily detected by its odour of bitter almonds, or by distilling the substance in question with water, adding to the distillate potash-ley in sufficient quantity to blue litmus-paper, then a stale solution of subsulphate of iron, and after agitation hydrochloric acid in excess, when a deep blue precipitate or a similar colouration will indicate the presence of Hydrocyanic acid. [Very minute traces of H. acid are, according to Almén, detected by adding to the colourless distillate one drop each of diluted soda-ley and of hydrosulphide of ammonium, evaporating the whole, on the water-bath, acidifying with 1-2 drops of hydrochloric acid, and adding a little chloride of iron. A more or less blood-red colouration ensues through the formation of sulphocyanide of iron.]

Quantitatively the amount of Hydrocyanic acid is determined by mixing, the distillate or a certain fraction of it, with nitrate of silver, then with ammonia in excess, and after agitation with excess of nitric acid. The precipitate is collected, washed and dried at 100°. After noting down its weight, it is heated in a porcelain crucible, until reduced to the metallic state. The remaining silver is dissolved in nitric acid, and any insoluble portion (chloride of silver), filtered off, washed, dried at 100°, and its weight deducted from that of the first precipitate. The rest gives the weight of the pure cyanide of silver, which, divided by five, represents the amount of Hydrocyanic acid.—Or, the resulting filtered solution of nitrate of silver is mixed with hydrochloric acid, the deposit collected, washed, dried, and weighed. This weight, divided by 5.35, also gives the quantity of H. acid.

Volumetrically, the same object of estimation is completed with greater expedition and unaffected by the presence or absence of hydrochloric or formic acids. For this purpose, dissolve 3·15 grammes of fused nitrate of silver in water, until the whole amounts to 100 cubic centimeters. On the other hand, add to the distillate in question, or to a certain fraction of it, potash-ley in excess, then a few drops of chloride of sodium solution, and at last cautiously, and under continual stirring, just enough of the nitrate of silver solution to obtain a slight precipitate, which does not redissolve. Every cubic centimeter of the silver solution, used for this purpose, indicates 0·01 gramme Hydrocyanic acid.

To compute from the quantity of H. acid the amount of amygdalin, originally present, the following equation will serve as a guide:—One equivalent Amygdalin $= C_{40} H_{27} NO_{22} + 4 HO$ is equal to 1 eq. Hydrocyanic acid $= HC_2 N + 2$ eq. Grape-sugar$= 2 C_{12} H_{12} O_{12} + 1$ eq. Oil of bitter almonds$= C_{14} H_6 O_2$. It follows herefrom, that by multiplying the weight of the Hydrocyanic acid with 17, we obtain the weight of the amygdalin as required.

**Hygrin.** Alkaloïd, besides cocain (*see this*), in the coca-leaves. When, in the preparation of cocain, more of the soda is added to the slightly alkaline liquid, from which the cocain has been removed by means of ether, Hygrin and a neutral oil of tobacco-odour are dissolved by once more shaking with ether, and remain behind after the ether has been distilled off. By heating this residue to the boiling point, the temperature rises quickly to 280°, a brown alkaline oil passes over and a black resin remains. The distillate when kept at 140° for several hours in a current of hydrogen gas, allows the greater part (*a*) of a yellow colour to pass over, the rest being only volatilized at 140° to 230°, and condensing to a thick brown oil (*b*). Both portions contain Hygrin, but contaminated in *b* with a neutral oil, and in *a* with other volatile substances. To remove any ammonia present, *a* is converted into the oxalate and dissolved in absolute alcohol, the liquid is evaporated, and the remnant is mixed with potash-ley, which separates the H. as an oil. After heating this alkaline solution to the boiling-point in a current of hydrogen gas, Hygrin, dissolved in water, passes over. It is separated from the distillate by ether and remains after the ether has been distilled off. The neutral oil, present besides Hygrin in *b*, is removed by dissolving *b* in water acidulated with hydrochloric acid, shaking with ether and decanting the ethereous liquid; the solution, containing the hydrochloric acid, is then oversaturated with soda-ley and the H. withdrawn from it by ether.—Thick, pale-yellow oil of strongly alkaline reaction, of burning taste and of an odour of trimethylamin; forms white vapours with volatile acids; does dissolve in water but not in every proportion, readily in alcohol and in ether; the aqueous solution throws down subchloride of tin white, sulphate of iron yellowish, sulphate of copper pale-blue, chloride of mercury and nitrate of silver white (the latter deposit soon changing to brown). Hygrin forms with hydrochloric acid a crystalline, deliquescent salt; this is precipitated by bi-iodide of potassium red-brown, by subchloride of tin white, by chloride of mercury white, partly in flocks and partly in oily drops; by chloride of platinum dirty white-yellow, by picric acid yellow, by tannic acid white.

**Hyoscyamin** $= C_{20} H_{17} NO_2$. Alkaloïd of the genus Hyoscyamus. Exhaust the seeds at 50° with alcohol, containing 2% sulphuric acid, render the tincture slightly alkaline with baryta; filter after a short digestion, precipitate the baryta with sulphuric acid, distil the alcohol from the acid liquid, neutralise the remnant as exactly as possible with carbonate of potash, filter, render alkaline with carbonate of potash, and shake with ether. Decant the ethereous solution, distil off the ether, dissolve the remnant in water, filter, mix the liquid with a mixture of 1 part kaolin, 1 part pulverised charcoal, and 2 parts ivory-black, so as to form a pulp, spread over porcelain plates, let dry at the sun, triturate, exhaust with ether, evaporate the solution, fuse the remnant carefully and recrystallise in ether.—It forms tuftily united, colourless, silky needles, is inodorous (smells narcotic in the impure state), has a nauseous, poignant, tobacco-like taste and a strongly alkaline reaction; dissolves slowly in water, but more readily than atropin, more readily in alcohol, ether, and acids. When heated with pure soda-ley under a pressure of $1\frac{1}{2}$ atmospheres, and for a rather long time, it evolves vapours of strongly alkaline reaction, and the crystalline remnant yields, by means of hydrochloric acid, a white crystalline body, stated by Kletzinsky to be santonin.

[HYOSCYAMIN $= C_{30} H_{23} NO_6$, is, according to Hoehn and Reichardt, prepared in the following manner from the seeds of Hyoscyamus niger. The seeds are freed from fat-oil by means of ether, then extracted with alcohol and sulphuric acid, the liquid evaporated, freed from a resinous mass; nearly neutralised with soda, and precipitated by tannic acid. The well-washed precipitate is spread on porous clay, and still moist mixed with excess of lime and exhausted with strong alcohol. The filtrate is acidified with sulphuric acid; evaporated; freed by ether from fat and colouring matter; mixed with excess of soda-ley, and the alkaloïd extracted by ether. The ethereous solution is purified by shaking with water and evaporated, leaving the H. as an oily liquid, solidifying over sulphuric acid to a crystalline warty mass of wax consistence, fusing at 90°C. The aqueous solution has a strongly alkaline reaction. Precipitates are produced by alkalies (in concentrated solutions), by tannic acid white; iodine-water, kermes-brown; chloride of mercury, white; chloride of gold, yellow-brown, easily soluble in excess, and becoming decomposed by keeping; chloride of platinum gives a resinous precipitate. Heated with alkalies the H. separates into Hyoscic acid and Hyoscin.]

**Hypogaeic Acid**$= C_{32} H_{29} O_3 + HO$. In the oil of the earth-nut (the fixed oil of the seeds of Arachis hypogaea). Saponify with soda-ley, decompose the soap with sulphuric acid, dissolve the fat-acids in alcohol, throw down with acetate of magnesia and ammonia the arachic and palmitic acids, filter and mix the liquids with an alcoholic solution of acetate of lead and ammonia. Press the precipitate, dissolve in ether, shake the solution with hydrochloric acid, remove the chloride of lead, shake the filtered liquid with water previously boiled, decant the ether, evaporate, press the remaining crystalline mass and recrystallise in alcohol. —Forms colourless, concentrically arranged needles, tasteless and inodorous, fusing at 34°; becomes yellow-red, rancid, and uncrystallisable at the air, and this even in very low temperatures ; is readily soluble in alcohol and ether ; is converted by nitrous acid into an acid of the same composition, but fusing at 38° (analogous to elaidic acid).

**Iatrophic Acid.** *See* CROTON OIL.

**Igasuric Acid.** *See* MALIC ACID.

**Igasurin.** Alkaloïd, supposed to occur in the seeds of Strychnos Nux vomica, but which requires further investigations. It is said to be nearly related to brucin.

**Ilicic Acid.** In the leaves of Ilex Aquifolium [and doubtless in other species of the genus.—F. v. M.]. Only known in combination with bases. Precipitate the aqueous decoction with subacetate of lead, free the filtrate from lead by sulphuret of hydrogen, warm with hydrated lead oxyd, remove again from the filtrate any dissolved lead by sulphuret of hydrogen, and evaporate to the consistence of syrup ; purify the lamellae, which will have formed after several days, by pressing, dissolving in water, precipitating with alcohol and recrystallising, whereby colourless ilicate of lime is obtained. This salt contains 18% lime, dissolves readily in water, not in alcohol, does not precipitate the salts of manganese, zinc, iron, copper and silver, but throws down sub-chloride of tin and acetate of lead.

**Ilicin.** Bitter ingredient of the leaves of Ilex Aquifolium, only known in the impure state.

**Ilixanthin** $= C_{34} H_{22} O_{22}$. Yellow pigment of the leaves of Ilex Aquifolium, scarcely present during winter, but copiously in the hotter part of the summer. The alcoholic extract is distilled, the remnant filtered and allowed to rest cold; the grains which will have formed after a few days are dried, washed with ether (to remove chlorophyll) dissolved in alcohol, freed from the latter by evaporation and addition of water, and recrystallised in hot water.—Forms straw-yellow, microscopic needles, fusible at 198°, destroyed in higher temperatures; dissolves scarcely in cold, readily with yellow colour in hot water, also in alcohol, not in ether; the aqueous solution turns orange with alkalies and becomes colourless on addition of sulphuric acid, no further change being observable even on boiling; assumes a sap-green colour with chloride of iron.

**Imperatorin** = Peucedanin.

**Indican** $= C_{52} H_{31} NO_{34}$. Substance forming the indigo-blue of Isatis tinctoria. Exhaust the leaves in a displacement apparatus with cold alcohol, precipitate the green tincture with a solution of acetate of lead in alcohol and a little ammonia, wash the pale-green deposit with cold alcohol, and decompose under water by means of carbonic acid gas. The deposit becomes decolourised and yields a yellow solution, which has to be freed from any dissolved lead by means of sulphuret of hydrogen and evaporated over sulphuric acid.—Yellow or light brown syrup-like liquid, which can not be obtained dry without decomposition, has a slightly bitter, unpleasant taste and an acid reaction, dissolves in water, alcohol and ether, is decomposed even by a gentle heat, even cold by diluted acids, under formation of blue flocks. The formation of the blue indigo is represented by the following equation:—

$$C_{52} H_{31} NO_{34} + 4HO = C_{16} H_5 NO_2 + 3C_{12} H_{10} O_{12} \text{ (indiglucin)}.$$

**Indigo-blue** $= C_{16} H_5 NO_2$. Contained in many plants, especially those of the genera Indigofera, Isatis and Polygonum, but is only formed on drying. Plants, turning blue on drying, are known also among the genera Asclepias, Croton, Galega, Marsdenia, Mercurialis, Nerium, Phytolacca, and Pimelea, and should be tested for indigo-pigment. To ascertain the presence of indigo, the respective green parts are extracted with warm water, and the clear solution allowed to rest at a temperature not below 15°; the original indigo compound is decomposed by a kind of fermentation, and the blue pigment subsides.—Pure Indigo-blue is deep blue, assumes on rubbing a copper-red colour, is inodorous and tasteless, fuses with heat, and sublimates at 228°, mostly undecomposed, in purple-red fumes, which condense to crystalline masses; is insoluble in water, alcohol, ether, diluted acids, and alkalies, dissolves unaltered in fuming sulphuric acid, is destroyed by nitric acid and chlorine, dissolves in alkalies in the presence of a reducing substance, as subsulphate of iron, grape-sugar, &c. The solutions contain indigo-white $= C_{16} H_5 NO + HO$, which forms again Indigo-blue on contact with the air.

**Inosit** $= C_{12} H_{12} O_{12} + 4 HO$. Found in the green fruit of Phaseolus vulgaris, Pisum sativum and Robinia Pseudacacia, in the leaves of Brassica oleracea, Digitalis purpurea, and Taraxacum officinale, in the shoots of Solanum tuberosum, in the herb and green fruit of Asparagus officinalis, in Agaricus piperitus, A. croceus, and others. Bruise the husks of French beans, press, evaporate the juice to a syrup, and mix with alcohol, sufficient to produce a permanent turbidity; the crystals, which will have formed, are purified by repeatedly recrystallising in water, with aid of animal charcoal.—Rhombic, tabular crystals, one inch long and about a quarter inch thick, or cauliflower-like conglomerations, of a pure, sweet taste, without rotation, turn opaque in dry air, over sulphuric acid, or at 100°, and lose all their water of crystallisation, fuse only above 210°, afterwards decomposed, dissolve in 6 parts cold water, slightly in strong alcohol, not in absolute alcohol and in ether, yield with nitric acid oxalic acid, dissolve in sulphuric acid, when cold or heated to 100°, without colouration, but become black in higher temperatures. Is not altered on boiling either with diluted sulphuric acid or with alkalies and alkaline earths, does not reduce the alkaline tartarate of copper, and does not ferment with yeast.

**Inulin** $= C_{12} H_{10} O_{10}$. Occurs throughout the whole order of Compositæ, replacing the starch and especially contained in the roots, but has also been met with in other plants, and seems to be widely distributed. In the living plant it exists dissolved in the cellular juice, from which it may be precipitated by means of water-absorbing agents as alcohol, glycerin, calcium-chloride, &c., under the form of white, tasteless granules similar to starch, but assuming a brown instead of a blue colour with solutions of iodine. [According to Sachs, Inulin is obtained in "sphæro-crystals" by immersing the roots of Compositæ in alcohol or glycerin.] In the dried plant it appears (as observed under the microscope) under the form of brittle, pellucid, amorphous masses, soluble by access of water but reprecipitated by alcohol.—It is prepared by boiling the substance in question, after exhaustion with ether and alcohol, with the least possible quantity of water for a quarter of an hour, straining or filtering hot, and allowing to cool, when a portion of the Inulin will separate, another portion being obtainable by evaporating and cooling as before, while the rest becomes converted into gum and subsequently into sugar.

Cold water dissolves only one-fifth per cent. Inulin, while boiling water takes up large quantities, but readily converts the Inulin, especially by prolonged boiling, into gum. It is not possible to separate the two bodies completely, though they are distinguishable by gum yielding mucic acid on treating with nitric acid, whereas Inulin forms different products.

[ALEURON, a substance closely related to Inulin, has been discovered by Hartig, in the seeds of a great many and widely different plants, either replacing or associated with starch. It resembles the latter in size, form and colour, but differs by its easy solubility in water, dilute acids and alkalies. With iodine it becomes brown, and with acid nitrate of mercury it acquires a brick-red colour. The surface of its granules presents under the microscope a dotted appearance. It is insoluble in volatile and fixed oils, alcohol and ether, and may, by means of these liquids, be washed out of the respective parts and collected like starch. Aleuron, in some cases, has a characteristic colour, which is green in the seeds of Pistacia, indigo-blue in Cheiranthus annuus, rose-red in Hibiscus, brown in Arachis, yellow in Ailanthus, Frangula, Myristica, Lupinus luteus.]

**Inverted Sugar.** *See* FRUIT SUGAR, *also* GRAPE SUGAR.

**Ipecacuanhic Acid** $= C_{14} H_9 O_7$. In the root of Cephaelis Ipecacuanha. Precipitate the aqueous extract with subacetate of lead, wash the deposit with alcohol, dissolve in diluted acetic acid, mix the solution with subacetate of lead, afterwards with ammonia; wash the deposit obtained with the strongest alcohol, divide under ether, and decompose with sulphuret of hydrogen. The liquid, separated from the sulphide of lead, has to be evaporated in a current of carbonic acid gas; the remnant is mixed with water, filtered off from the fat, and the filtered liquid digested with animal charcoal and evaporated.—Amorphous, reddish brown mass, very hygroscopic, very bitter, fusible; dissolves readily in water, also in alcohol, less readily in ether; colours salts of oxyd of iron green, which becomes violet on addition of a little ammonia, and with more ammonia ink-black.

**Iris-Steraopten** $= C_{16} H_{16} O_4$, obtained by distilling the root of Iris Florentina with water. White scales, of a pleasant odour of violets, lighter than water, readily soluble in alcohol. [Is identical with Myristic acid. Flueckiger and Hanbury].

**Isocetic Acid** $= C_{30} H_{29} O_3 + HO$ (therefore of the same composition as cetic acid from spermaceti). Forms the glycerid of the solid part of castor-oil. It fuses at 55°, crystallises from alcohol in lamellæ.

[**Ivain.** Indifferent bitter substance, contained, according to v. Planta-Reichenau, in the herb of Achillea moschata, together with

I

Iva oil, achillein, and moschatin. The alcoholic extract of the plant is mixed with acetate of lead, filtered, the filtrate treated with sulphuret of hydrogen, filtered and evaporated. The brown residue is freed from achillein and moschatin by means of acetic acid, which leaves behind the Ivain.—The I. is yellow, of the consistence of turpentine, easily soluble in alcohol, and of an intensely bitter taste.]

**Ivy Resin.** Exudation of the stem of Hedera Helix. Red-brown or greenish, transparent, brittle mass, of a faint aromatic smell, and likewise acrid taste. Contains about 23% resin, 7°/$_0$ gum, and many impurities.

**Jalapin**$=C_{68} H_{56} O_{32}$. In the tubers of Ipomœa Orizabensis and Convolvulus Scammonia. The crude resin is obtained by extracting with alcohol, mixing with water, distilling off the alcohol and drying the remaining mass. Dissolve this in much alcohol, mix the solution with water until turbid, boil the mixture repeatedly with animal charcoal, precipitate the filtrate, coloured still, with acetate of lead and a little ammonia, which forms a small quantity of a green-brown deposit; filter, free the liquid by means of sulphuret of hydrogen, heating and subsequent filtering from lead; distil off the alcohol, knead the remaining resin repeatedly in hot water, dissolve in ether and evaporate.—Colourless, transparent in thin layers, amorphous, brittle, even at 100°, softens at 123°, fuses at 150°; inodorous, tasteless, has a scarcely perceptible acid reaction when dissolved in water; dissolves very little in water, most readily in wood spirit, alcohol, ether, chloroform, acetic acid, benzol, oil of turpentine, in cold concentrated sulphuric acid with a beautiful red colour, changing into brown-black under formation of sugar; splits, when heated with diluted acids, into sugar and jalapinol, $=C_{32} H_{31} O_7$, a white mass, shaped like the particles of cauliflower. [Is, according to Flueckiger and Hanbury, identical with Convolvulin].

**Jamaicin**$=$BERBERIN.

**Japanese Wax.** Obtained from the leaves, branches, and fruits of Rhus succedanea by boiling with water.—Yellowish-white, smells and tastes a little acrid, and affects the throat; softer and more unctuous than beeswax, also more friable; has an acid reaction, fuses at 45°, contains palmitin but no olein; is, according to Berthelot, perhaps di-palmitin.

**Jervin**$=C_{60} H_{45} N_2 O_5$. Alkaloïd besides veratrin of the root of Veratrum album. Treat the alcoholic extract with water acidulated with hydrochloric acid, precipitate the solution with carbonate of soda, dissolve the deposit in alcohol, digest with charcoal, distil off the alcohol, press the crystalline residue in order to remove most of the difficultly crystallisable veratrin, and rinse with a little alcohol.—Forms white crystals, fusible, insoluble in water, readily soluble in alcohol; yields with acetic acid a readily soluble salt, with hydrochloric, nitric and sulphuric acids salts which dissolve sparingly in water and in acids.

**Juniperin.** Resinous substance of the fruits of Juniperus communis and other species of Juniperus. Effect a distillation with water, after the fruits have been macerated with cold water; strain the remaining mass hot, let cool, collect the sediment, treat it first with cold, then with boiling alcohol, and distil off the alcohol from the united tinctures. The remaining liquid throws down successively wax, resin, at last Juniperin as a yellow powder, which conglomerates like a resin. This has to be washed and ground up with water, whereby it dissolves in 60 parts of water and passes, when this solution is shaken with ether, into the latter, remaining after the evaporation of the solvent as a light-yellow, brittle, tasteless mass, which burns with the odour of juniper, dissolves in ammonia with gold-yellow, in concentrated sulphuric acid with light-yellow colour.

**Kaempherid.** In the root of Alpinia Galanga. Exhaust with ether, evaporate the liquid and recrystallise in alcohol.— Light-yellowish lamellæ of mother-of-pearl lustre, without smell or taste, fusible by heat and decomposing afterwards; scarcely dissolving in water, in 25 parts cold ether, in 50 parts cold alcohol, better in both when hot, also a little in acids, readily in caustic alkalies without change. It is decomposed by concentrated sulphuric acid under various changes of colour. Consists of 65·32 C, 4·40 H, and 30·28 O.

**Kawahin**=Methysticin.

**Kinotannic Acid.** In kino, an induration of the sap exudated from incisions of the stem of Pterocarpus Marsupium. It is regarded by some as different from, by others as identical with, gallotannic acid.

[Kinotannic Acid has been found by J. Wiesner to exist in combination with a kind of gum in the Eucalyptus kino, a spontaneous exudation of the stem of Eucalypts. Pyrocatechin was invariably found to be present in small quantities, and in a few instances also Catechin.]

**Kokum-Butter**=Brindonia Tallow.

**Koussin** = $C_{26}H_{22}O_5$. The active ingredient (vermifuge) of the kousso-flowers (from Hagenia Abyssinica). Digest with alcohol, containing lime; press, let subside, filter, draw off the alcohol, allow the residue to rest warm and uncovered for some time, in order to remove the last traces of alcohol, filter, oversaturate slightly with acetic acid, collect the deposit, wash with cold water, and dry with a very gentle heat.—White, often a little yellowish powder of crystalline appearance under the microscope, easily friable, inodorous, of an acrid bitter taste. At 140° it turns grey, at 150° brown, at 194° it fuses, and becomes carbonised with more heat. It dissolves slightly in water, at 17° in 1300 parts alcohol of 45%, and in 12 parts alcohol of 90%, in hot alcohol and in ether almost in every proportion. The solutions have an acid reaction. It is also rather soluble in hydrated alkalies. Chloride of iron forms in the aqueous as well as in the alcoholic solution a strong brown, acetate of lead a grey-yellow precipitate. Concentrated sulphuric acid dissolves Koussin with yellow-brown colour; the solution, at first clear, becomes spontaneously turbid after a few minutes, and throws down white flocks.

**Labdanum** = LADANUM.

**Laburnin.** Alkaloïd of the unmatured seeds of Cytisus Laburnum. It is obtained from the aqueous extract, which has been purified by subacetate of lead, by precipitating with phosphomolybdic acid and boiling the deposit, after mixing with chalk and drying, with alcohol. The base is afterwards purified by isolating it from the platinum salt. It does not combine with acids, and forms large concretions of crystals, consisting of klinorhombic prisms containing water of crystallisation; dissolves most readily in water, difficultly in absolute alcohol, scarcely in ether, and evolves with caustic potash ammonia even when cold.

**Lactic Acid** = $C_6 H_5 O_5 + HO$. As this acid is easily formed during the spontaneous decomposition or fermentation of vegetable extracts, its presence in the living plants must be traced with great caution; especially it must be ascertained, if no kind of fermentation takes place while the analysis is carried on. However, Lactic acid is no uncommon ingredient of plants, and I, for my part, have encountered it not rarely in the examination of the constituents of plants, though it may often have been overlooked on account of its not being endowed with any striking properties, and because it yields no precipitate with metallic salts. The Lactates of the alkalies being very deliquescent, it is probable that the hygroscopicity of vegetable extracts is not occasioned by malates, as generally assumed, but by alkaline Lactates. The presence of Lactic acid is ascertained by the following process:— Free the aqueous liquid from all precipitable matters by means of acetate and subacetate of lead, remove excess of lead by carbonate of ammonia, evaporate to a liquid of syrup consistency, add a concentrated solution of acetate or chloride of zinc, and leave the mixture to stand cold for several days. In the presence of Lactic acid crystalline crusts will be formed of Lactate of zinc, slowly soluble only in 50 parts cold water. This salt has the formula $Zn O + \bar{L} + 3 HO$ (in 100: 27·13 $Zn O$, 54·69 $\bar{L}$ and 18·18 $HO$); it loses at 100° all its water and contains then 66·83% acid. From it the acid may be obtained by admitting sulphuret of hydrogen and evaporating, towards the end of the operation, in a vacuum.— It forms a colourless syrup-like liquid of 1·21 density, is inodorous, has a strong and pure acid taste, mixes with water and alcohol in every proportion, dissolves less readily in ether, becomes anhydrous $= C_6 H_5 O_5$ at 130° and solidifies on cooling to a pale-yellow mass of very bitter taste; it is decomposed in a higher temperature. All the Lactates dissolve in water, though many of them only sparingly in cold.

**Lactucerin** $= C_{30} H_{24} O_2$. In Lactucarium (the hardened milky juice of Lactuca virosa and other species). Boil with alcohol, filter hot and recrystallise the warty masses obtained, on cooling, in alcohol with aid of animal charcoal; it is well to withdraw first from the lactucarium the bitter substance by means of water.—Forms fine, colourless, concentrically united columnar prisms, without taste or smell; neutral; fuses between 150° and 200°, sublimates mostly undecomposed in a current of carbonic acid gas; is insoluble in water, soluble in alcohol, ether, and oils; is not altered by potash dissolved in alcohol; is not precipitable from the alcoholic solution by metallic salts.

**Lactucin** $= C_{22} H_{12} O_6 + HO$. The bitter ingredient of Lactucarium. Treat and press a few times with cold water and boil repeatedly with water, evaporate the united liquids until equal in weight to half the lactucarium employed, separate the substance, settled to a granular mass, from the mother-ley, and dissolve in hot water, precipitate with subacetate of lead, wash the deposit with hot water, adduce to the filtered liquid sulphuret of hydrogen, filter again, evaporate and allow to stand cold. The Lactucin forms in crystals, and by concentrating the mother-ley still further an additional quantity will be obtained. Recrystallise in hot alcohol with aid of animal charcoal.—Forms white, pearly scales, similar to boric acid; fusible; becomes charred in higher temperatures; has a strong and pure bitter taste; dissolves scarcely in cold water, in hot water less readily than in alcohol, not in ether, readily in acetic acid; is decomposed by alkalies and loses its bitterness. Is not glucosid.

**Lactucon** $=$ LACTUCERIN.

**Ladanum.** Exudation of Cistus Creticus, and to some extent also from C. ladaniferus, C. Ledon, C. laurifolius, and C. monspeliensis.—Black-brown, soft, of pleasant smell and of bitter taste. Contains 86% resin, 7% wax, and some volatile oil.

**Laetia-Resin.** Exudation of the stem of Laetia apetala.— Small, yellow-white, translucid, brittle grains of concheous fracture, and of faintly aromatic smell; slowly soluble in alcohol, yielding a volatile oil when distilled with water.

**Laevo-glucose** = Fruit Sugar.

**Lapathin** = Chrysophanic Acid.

**Laricin** = $C_{14} H_{12} O_4$. The purgative ingredient of the Larch-agaric, Polyporus officinalis. It is separated with difficulty from the accompanying resin.—Is in the pure state a white, amorphous powder of bitter taste, dissolves readily in alcohol and oil of turpentine; forms with boiling water a paste.

**Laserpitin.** = $C_{48} H_{36} O_{14}$. Bitter substance of the root of Laserpitium latifolium. Treat with alcohol of 80%, distil off the alcohol from the tinctures, separate the upper resinous layer of the residue from the aqueous lower one; allow the former to stand at the air until it is converted into a crumbly crystalline pulp; collect it in a filter, remove most of the resin by washing with weak alcohol; next dissolve in alcohol, precipitate the rest of the resin with subacetate of lead dissolved in alcohol; remove from the filtered liquid the excess of lead by sulphuret of hydrogen, and leave the liquid to evaporate spontaneously. At first a flocky matter is formed, the pure substance crystallising later.

The Laserpitin crystallises in colourless prisms, has neither smell nor taste, only the resinous L. has a bitter taste; is insoluble in water, readily soluble in alcohol, ether, chloroform, sulphide of carbon, oil of turpentine, benzol, and fixed oils; the alcoholic solution has a neutral reaction and strongly bitter taste. It fuses at 114° without loss of weight, and sublimates in higher temperatures undecomposed. It does not dissolve in alkalies and diluted acids; dissolves in concentrated sulphuric acid with cherry-red colour, also in fuming nitric acid. The alcoholic solution is not precipitated by acetate of lead, nitrate of silver, chloride of mercury, iodide of potassium, and alkalies. It separates when heated with a concentrated aqueous or better alcoholic solution of caustic potash, into angelic acid, and into a brown resin (Laserol).

**Lauric Acid** = $C_{24} H_{23} O_3$ + HO. In the fat of the fruit of Laurus nobilis, Ocotea Pichurim, Tetranthera calophylla, Irvingia Barteri, Cocos nucifera, Croton Tiglium, combined with glyceryl oxyd. Saponify the solid fat of the bay-berries (the laurostearin) with soda-ley, separate the soap with common salt, and decompose with tartaric acid. The Lauric acid, which rises to the top, is purified by repeatedly melting with water and by recrystallising in alcohol. Appears in white, tuft-like, brittle, crystalline needles of 0·883 density at + 20°; fuses at 43·5°; volatile with the vapours of boiling water; is insoluble in water, readily soluble in alcohol and in ether, the solutions having an acid reaction. The Laurate of baryta contains 71·38%, the Laurate of lead 63·12% acid.

**Laurostearin** = $C_{54} H_{50} O_8$. As to occurrence see Lauric Acid. Expose the fixed bay-oil to the sunlight, and press off the solid fat after the green colour has disappeared; dissolve in warm alcohol, and precipitate by evaporating and mixing with water.— Snow-white, voluminous, easily friable, inodorous mass, consisting of concentrically united needles; fuses at 45°, dissolves sparingly in cold, better in hot alcohol, readily in ether.

**Lecanoric Acid** = $C_{32} H_{13} O_{13} + HO$. In various lichens of the genera Evernia, Lecanora, Roccella, Variolaria. Extract with ether, evaporate, wash the remnant with cold ether until the latter passes off uncoloured, boil with water to remove orsellic ether, and recrystallise in alcohol.—White, radially united crystalline needles without taste or smell, of acid reaction; not volatile; dissolve in 2500 parts of boiling water and crytallise on cooling; yield when boiled with water, orsellic acid = $C_{16} H_7 O_7 + HO$, the latter yielding orcin by continued boiling under evolution of carbonic acid; dissolve in 150 parts cold alcohol of 80%, in 15 parts of boiling alcohol under formation of orsellic ether= $C_4 H_5 O + C_{16} H_7 O_7$; in 80 parts of ether. Lecanoric acid forms with alkalies easily crystallising salts.

**Ledum-Tannic Acid** = $C_{14} H_6 O_6$. In the leaves of Ledum palustre. Add to the aqueous decoction solution of acetate of lead by drops, until a sample of the precipitate dissolves completely in acetic acid, filter and precipitate with subacetate of lead. The deposit, after washing, is decomposed with sulphuret of hydrogen, the sulphide of lead is removed and the liquid evaporated.—Reddish powder, dissolves in water and in alcohol, the solutions becoming dark-green by chloride of iron; throws down on boiling with diluted acids a yellow or red powder (Ledoxanthin).

**Legumin.** Peculiar protein substance, occurring principally and in large quantity (up to 20% or 30%) in the seeds of the leguminous plants. To prepare it, treat peas, &c., with warm water, precipitate with acetic acid, wash the deposit with a little cold water, treat with alcohol and ether, redissolve in potash-ley, precipitate with acetic acid and wash again with water, alcohol and ether.—White powder with a yellowish tinge, dissolves in cold and hot water. The solution becomes covered with a pellicle on evaporating, and the Legumin is thereby converted into the insoluble modification. Acetic acid throws down Legumin, and does not redissolve it when added in excess; sulphuric acid behaves in the same manner, while oxalic, tartaric, citric and malic acids do not precipitate the Legumin from solutions. For other properties see Protein substances.

**Lepidin.** Peculiar substance, occurring in every species of the genus Lepidium, prevailingly in the younger parts and in the seeds. It is obtained by boiling with water acidulated with sulphuric acid, saturating the decoction with carbonate of lime, filtering, evaporating the liquid to a honey consistence, extracting with alcohol and evaporating the solution.—Brown, transparent mass, yielding a yellow powder; permanent at the air; of a faint smell, and of very bitter taste; softens when warm; dissolves readily in water and alcohol, slightly in oils, not in ether, combines neither with bases nor with acids.

**Lichenic Acid**=Fumaric Acid.

**Lichenin**=$C_{12} H_{10} O_{10}$. In lichens and in algae, observed for instance in the following genera:—Cetraria, Cladonia, Evernia, Parmelia, Ramalina, Sticta, Usnea; Delesseria, Fucus, Helminto-chorton. Is not distributed in isolated grains like starch, but as a turgid mass uniformly embedded between the cells. Free, for instance Cetraria Islandica, from the bitter substance by macerating with a weak solution of carbonate of soda and by washing; boil with water for two hours, strain hot, press, collect the jelly, separated after cooling, on a cloth; press, redissolve in little boiling water; precipitate with alcohol and dry the deposit.—Colourless or yellowish, hard, brittle, transparent mass of vitreous fracture and difficult to pulverise, inodorous and tasteless, not volatile, swells up considerably in cold water, dissolves in boiling water to a thick slime, of a jelly-like appearance when concentrated; is insoluble in alcohol and ether, becomes blue with iodine, dissolves in warm nitric acid to a thin liquid, and yields on heating oxalic but no mucic acid; is converted into sugar when boiled with water containing sulphuric acid. The aqueous solution, when boiled by itself, loses its property of gelatinising. Alkalies and alkaline earths dissolve the Lichenin likewise.

**Lichenstarch**=Lichenin.

**Licheno-stearic Acid** = $C_{28} H_{23} O_5 + HO$. In Cetraria Islandica, also in Agaricus muscarius, and, doubtless, in numerous other lichens and algae. Boil the Cetraria for half an hour with alcohol and a little carbonate of potash, strain, mix the liquid with an excess of hydrochloric acid and four to five times its volume of water, wash the deposit with water and boil it several times with alcohol of 42 to 45 %. The alcoholic solutions deposit, on cooling, a mixture of Licheno-stearin, Cetraric acid, and another substance from which the Licheno-stearin acid is dissolved by boiling petroleum and subsides after cooling, or better when the petroleum is distilled off partly. It is purified by re-crystallising in alcohol under aid of animal charcoal.—Loose, white mass, consisting of fine, pearly, crystalline lamellae, inodorous, of a rancid taste and affecting the throat; not bitter; fuses at 120° without loss of weight; is not volatile; insoluble in water, readily soluble in alcohol, ether, and oils. The Licheno-stearates are permanent at the air, only soluble in water with aid of alkalies, the solutions yielding a froth on boiling.

**Ligustrin** } = Syringin.
**Lilacin**

**Limonin** = $C_{44} H_{26} O_{14}$. In the seeds of oranges and lemons (Citrus Aurantium and C. medica). Bruise, draw out with cold alcohol and let the solutions evaporate.—White powder, consisting of microscopic crystals; has a strong and pure bitter taste; is not altered at 200°; fuses at 244°; dissolves very little in water, readily in alcohol; very little in ether, in concentrated sulphuric acid with blood-red colour, and percipitable from it by water; not soluble in liquor of ammonia, readily in potash-ley, and precipitable from it by acids.

**Linoleic Acid** = $C_{32} H_{27} O_3$ + HO. In linseed oil, poppy oil, perhaps also in other drying oils. Saponify with soda-ley, salt out, dissolve the soap in much water, precipitate with chloride of calcium, wash, press and treat the lime-soap with ether, which dissolves the Linoleate of lime, but not the other oleates. The ethereous solution is decomposed with cold hydrochloric acid, the L. acid remaining dissolved in the ether; this solution is decanted and the ether distilled at a temperature as low as possible, in a current of hydrogen gas; the remnant is dissolved in alcohol and precipitated with ammonia and chloride of baryum; the precipitate is dissolved in ether and repeatedly recrystallised in ether; the Linoleate of baryta is decomposed by shaking with ether and hydrochloric acid, and the solution evaporated in a vacuum.— Slightly yellowish, thin oil of 0·926 density at 14°, of a highly light-refracting power; tastes at first mild, afterwards acrid; has an acidulous reaction; is still liquid at—18°; dissolves not in water, readily in ether, less so in alcohol, becomes viscid by keeping at the air; forms with nitrous acid no elaidic acid, but a glutinous resin, suberic acid and very little oxalic acid. Its neutral salts show a great propensity for forming acid salts, they become coloured at the air and odorous.

**Linseed Oil.** Obtained by pressing the seeds of Linum usitatissimum. Is yellow, smells and tastes peculiarly, has 0·934 density, does not congeal at—15°, separates at—18° a little solid fat, dissolves in 32 parts alcohol of 0·820, in 1.6 parts ether. Forms with alkalies very soft soaps, consists of about nine-tenths linoleate of glyceryl, and one-tenth palmitin. Dries at the air.

**Liriodendrin.** Bitter aromatic substance of the bark of the root of Liriodendron tulipifera. Is obtained by extracting with alcohol, evaporating the tincture, washing the impure L. which has separated, with diluted potash-ley, in order to remove resin and dyeing matter; dissolving in alcohol; diluting the solution with water, until it turbifies, and crystallising.—Forms colourless scales, similar to boric acid, or concentrically arranged needles,

smells faintly aromatic, tastes warming, bitter; fuses at 83°, sublimates partly undecomposed, partly yielding ammoniacal products; is
almost insoluble in water, readily soluble in alcohol and ether, not
in aqueous alkalies and diluted acids, is not decomposed by concentrated nitric, but by hydrochloric and concentrated sulphuric
acids.

[**Lobelacrin.** Isolated by Enders from the herb of Lobelia inflata
by exhausting the drug with alcohol and distilling the liquid in
presence of charcoal, which then retained the acrid principle.
The charcoal was washed with water and then treated with boiling
alcohol. This, on evaporation, yielded a green extract, which was
further purified by means of chloroform. Warty tufts were thus
finally obtained, yet always of a brownish colour. The tufts are
readily soluble in ether and chloroform, but only slightly in
water; they possess the acrid taste of the herb. Lobelacrin is
decomposed by merely boiling with water; by the influence of
acids or alkalies it is resolved into sugar and Lobelic acid. The
latter is soluble both in water and alcohol, and is nonvolatile; it
yields a soluble salt with baryta, whereas the lead-salt is insoluble
in water.]

**Lobelin.** Alkaloïd-like substance of the stalks and leaves of
Lobelia inflata, said to act similarly to nicotin, but as yet only
obtained in the impure state, namely, by extracting with water,
acidulated with hydrochloric acid, evaporating, treating the extract
with alcohol, and evaporating the tincture.—Shining yellow, gummous, hygroscopic mass of an acrid tobacco-like taste; dissolves
readily in water and in alcohol, not in ether; is precipitable by
tannic acid, bi-iodide of potassium, iodide of potassio-mercury,
nitrate of silver, chloride of gold and chloride of platinum.

**Lupinin.** Bitter substance of the seeds of Lupinus albus, and
probably of many other species. Draw out with hot alcohol, evaporate the tinctures to dryness, treat the mass with water, digest the
aqueous solution with charcoal and evaporate to the consistence of
a syrup, which throws down the L. in small, white, amorphous
grains.—It is transparent, brittle like gum arabic, deliquesces at
the air, dissolves readily in water and in weak alcohol, not in
absolute alcohol and in ether, is not perceptibly altered by acids
and by alkalies.

**Lupulic Acid** $= C_{32} H_{25} O_7$. The bitter ingredient of hops
(Humulus Lupulus), in the purest state. Exhaust with ether,
distil off the ether from the extracts, treat the remaining thick
mass with cold alcohol of 90°/$_o$, concentrate the alcoholic tincture,
dissolve the residue again in ether, shake this ethereous solution
repeatedly with strong potash-ley in order to remove resinous
bodies, shake then with water, which takes up mostly the bitter

substance, and precipitate this aqueous solution with sulphate of copper. The precipitate of fine, blue, microscopic needles, being a combination of the bitter substance with oxide of copper, has to be washed with a little ether. Dissolve in more ether, decompose the solution with sulphuret of hydrogen, filter off from the sulphide of copper and evaporate in a current of carbonic acid gas. The remaining brown, crystalline mass is freed from the adhering mother liquor by means of nitro-benzol.—Colourless, rhombic prisms of great lustre, brittle and breaking under a gentle pressure; tasteless, but of a pure and pleasant bitter, when dissolved in alcohol; insoluble in water, most readily soluble in alcohol, ether, chloroform, sulphide of carbon, benzol, oil of turpentine, &c.; of a decidedly acid reaction.

[**Lupulin.** A liquid, volatile alkaloïd contained, besides Trimethylamin, in hops. It has. the odour of Coniin, and assumes a violet hue when treated with chromate of potash and sulphuric acid.—Griessmayer.]

**Luteolin**=$C_{24}$ $H_8$ $O_{10}$. Yellow pigment of weld (Reseda luteola). Draw out the herb with alcohol of 80°/$_o$. evaporate the tinctures, collect the L. formed on keeping and dry, wash with a little cold ether, dissolve in alcohol, pour the solution into water, heat to the boiling point, filter, and let crystallise.— Pure yellow, silky needles, inodorous, of a slightly bitter and acrid taste, and of acidulous reaction; fuse above 320° under partial decomposition; dissolve in 14,000 parts cold and in 5000 parts boiling water, in 377 parts alcohol, and in 625 parts ether, readily in warm concentrated acetic acid.

**Lycoctonin.** Peculiar alkaloïd of the root of Aconitum Lycoctonum. Its preparation is partly given under "Acolyctin." The ether, employed for the purification of acolyctin, and containing Lycoctonin, is allowed to evaporate, the opaque white, usually warty, crystals obtained are washed with ether, and afterwards with cold water and dried.—It tastes strongly bitter, has an alkaline reaction, dissolves readily in alcohol, less readily in ether, slightly in water; the alcoholic solution does not become turbid with ether, and scarcely so with water. Concentrated sulphuric acid colours it yellow. In several of its properties it shows a similarity with narcotin, but is distinguished by its ready solubility in alcohol, by its decided alkalinity and by the form of its crystals. The neutral sulphates and the hydrochlorids of L. are precipitated white by tannic acid.

**Lycopodium Bitter.** Found in Lycopodium Chamæcyparissias. Prepare, according to Boedeker and Kamp, from the herb at first an alcoholic, afterwards from the remnant an aqueous extract, precipitate the aqueous solution of the latter with acetate and subacetate of lead, and evaporate the filtered liquid, freed from the lead by sulphuret of hydrogen. The remnant, when extracted with alcohol, yields grape-sugar to it, while the bitter substance remains undissolved, and has to be dissolved in water and precipitated with subacetate of lead. The deposit is decomposed under water with sulphuret of hydrogen, the solution, separated from the sulphide of lead, is required to ferment with yeast, then dried and extracted with absolute alcohol. The alcohol dissolves the Bitter and leaves after evaporating a syrup-like liquid undermixed with colourless needles.—Neutral, non-nitrogenised, of very bitter, nauseous taste, dissolves in water, alcohol and ether, reduces the alkaline solution of copper, after being boiled with diluted sulphuric acid.

**Lycopodium-Resin** = $C_{36} H_{32} O_4$. In Lycopodium Chamæcyparissias. Evaporate the mother-ley, left from the preparation of the lycopodium-stearon (see this), treat the remnant with water and boil the insoluble portion with a little soda-ley. After cooling, this resin separates and has to be recrystallized in boiling alcohol.— Is insoluble in water, readily soluble in alcohol and in ether, fuses at 170° under decomposition, is slightly soluble in cold alkalies, but becomes decomposed by heating.

**Lycopodium-Stearon** = $C_{30} H_{30} O_4$. Found in Lycopodium Chamæcyparissias. Separates from the alcoholic tincture of the herb on evaporating, and is obtained by washing with cold alcohol and water and by repeatedly dissolving in boiling alcohol, after cooling as a jelly, which dries to a starch-like mass.—Amorphous, inodorous and tasteless mass, fusing at 100°, burns on heating with an odour of fat, has a neutral reaction, is insoluble in cold, slightly soluble in boiling water, in cold alcohol and in ether, abundantly so in these liquids at boiling heat.

**Mace Balsam.** Obtained by pressing the covering of the seed (mace) of Myristica fragrans. Is of a rather thin fluidity, when newly drawn, smells and tastes like mace, deposes on keeping a white, granular substance.

**Madarin** = MUDARIN.

**Madder Orange** = RUBIACIN.

[**Magnolin.** Bitter substance found by Wallace Procter in the fruits of Magnolia umbrella. The M. crystallises from weak alcohol in needles, and from strong alcohol or petroleum ether in prisms; is less bitter than Liriodendrin, but produces a scratching sensation in the throat; is almost insoluble in cold, slightly soluble in boiling water, abundantly in alcohol, chloroform, sulphide of carbon, and petroleum ether, especially when warm; is also easily soluble in fat-oils, and less copiously in hot glycerin; has a neutral reaction; evolves no ammonia with potash; is precipitated from the potash or soda solution by acids; becomes red with concentrated sulphuric acid, with concentrated nitric acid brown and resinous. Iodine has no action on it. The crystals of M. melt at 80° to 82°, becoming amorphous; at 125° they emit white vapours which condense to oily drops and consist partly of unaltered M. and partly of a resinous mass.]

**Maleic Acid** = Equisetic Acid.

**Malic Acid** = $C_4 H_2 O_4 + HO$. A widely-distributed substance, and perhaps the commonest of all vegetable acids. It has been observed in all parts of plants, principally and most abundantly in fruits, especially in unripe and acid ones. Several of the acids that have been found in plants and are described under different names, are probably nothing but Malic acid, as for instance Igasuric acid of the seeds of Strychnos nux vomica, and Braconnot's Fungic acid. When a vegetable extract yields with lime-water no turbidity either cold or hot, the presence of Malic acid may be inferred; yet it may even possibly be present when a turbidity does ensue, because one plant often contains two and more organic acids. Malic acid is distinguished and prepared on a small scale in this manner. Neutralise the respective liquid when acid with ammonia, precipitate with acetate of lead, leave to subside for one day in a cold place and collect the deposit on a filter, wash with cold water, mix with more water under stirring, boil and filter boiling hot. Boil again with water all that has been left undissolved until the whole is either dissolved or the remnant has been exhausted. The Malate of lead crystallises from the liquids, when kept cold, and more of it is obtained by concentrating the mother-ley. The whole of the Malate of lead is ground up with water to a fine pulp, the latter is decomposed with sulphuret of hydrogen, and the liquid is evaporated first with a gentle heat and afterwards in a vacuum, when it will yield the pure acid in crystalline needles, united to warty masses.—It is inodorous, of a pure and strongly acid taste, deliquesces at the air, dissolves most readily in water, alcohol and ether; the aqueous solution is not clouded by lime-water either cold or hot. It fuses at 83° and becomes decomposed in a higher temperature. The Malates dissolve nearly all in water, most of them readily so.

The quantitative estimation of Malic acid cannot be effected through precipitation with metallic salts, because those Malates are either not insoluble in water or are decomposed during washing. If the quantity of Malic acid has to be determined in a liquid which contains no other acid, the solution is warmed with carbonate of baryta, until the acid reaction has disappeared. It is then filtered off from the excess of carbonate of baryta, is evaporated to dryness and heated for some time at 100°; 100 parts of this anhydrous Malate of baryta contains 43·09 parts acid.

**Mangostan-Resin** $= C_{36} H_{22} O_{10}$. Exudation of the stem of Garcinia Mangostana. Of a beautiful lemon-yellow colour; brittle; inodorous and tasteless; fuses at 110°; dissolves readily in alcohol and in ether. Is decomposed by liquor of ammonia into two resins, the one of which fuses at 80°, and the other at 115°.

**Mangostin** $= C_{40} H_{22} O_{10}$. In the fruit-peels of Garcinia Mangostana. Boil the peels, after exhausting with hot water, with alcohol, evaporate the tincture until amorphous, yellow masses of resin and Mangostin are formed, dissolve the latter in alcohol and add to the solution, heated to the boiling-point, water in small quantities, until it becomes turbid. On cooling, the resin subsides and the liquid, after being poured off and kept for some time, throws down the Mangostin, which has to be purified by dissolving in alcohol and precipitating with sub-acetate of lead. The precipitate after washing is mixed with alcohol and decomposed by sulphuret of hydrogen, the filtered alcoholic liquid is mixed with water and allowed to rest until crystals are formed, which have to be recrystallised in diluted alcohol.—Thin laminæ of a beautiful gold colour, inodorous and tasteless, fusible at 190°, partly sublimating at a higher heat, neutral, insoluble in water, readily soluble in alcohol, ether, and alkalies.

**Mannit** $= C_{12} H_{14} O_{12}$. In considerable quantity in the manna, a sweet exudation obtained by incisions into the stem of Fraxinus Ornus and Fr. rotundifolia, and also occurring in many other plants, being contained in their roots, stalks, leaves, barks, seeds, and also nearly in all fungi. To prepare it from the manna, allow the aqueous solution to ferment with yeast, in order to destroy the sugar, decolourise with animal charcoal, evaporate so as to form crystals, and recrystallise. Or, boil with alcohol, filter hot, let crystallise and recrystallise.—Forms long rhombic, concentrically or tuftily united needles of a slight and pleasant sweet taste, without rotating power; loses at 120° nothing of its weight; fuses at 166°, volatilises when kept in a fused state in small quantity and sublimates unaltered; begins to boil at 200°, while a portion volatises and another portion loses 2 eq. water being converted into mannitan, but the largest portion remains undecomposed; above 250° it swells up and is destroyed. It dissolves in 6 parts cold and in any quantity of hot water, almost insoluble in cold absolute alcohol, in hot aqueous alcohol in large quantity (in 1430 to 1660 parts absolute alcohol at 14°, in 84 to 90 parts alcohol of 0.898 at 15°, not soluble in ether. Yields with nitric acid, oxalic, but no mucic acid, dissolves in concentrated sulphuric acid without colouration, is not altered on boiling with diluted sulphuric acid, with aqueous alkalies and with alkaline tartarate of copper; is not able to ferment with yeast, is not precipitable by acetate and subacetate of lead, but by the ammoniacal acetate of that metal.

**Marrubin.** Bitter substance of Marrubium vulgare. Draw out with hot water, digest the liquid with animal charcoal, which takes up the bitter substance; treat the coal with alcohol, distil off the alcohol from the tincture, warm the remaining liquid, until every trace of alcohol is removed, withdraw the Marrubin from the thick mass by means of ether, and allow the solution to evaporate slowly.—It crystallises in colourless, rhombic, tabular crystals, from the alcoholic solution in needles, tastes strongly bitter, fuses at 60°, decomposes in higher temperatures, is almost insoluble in cold water, slightly soluble in hot water, readily in alcohol and in ether; has a neutral reaction; dissolves in concentrated sulphuric acid with brown-yellow colour, in hot concentrated nitric acid with yellowish colour; hydrochloric acid and alkalies have no effect.

**Marum Camphor,** passes over in the distillation of Teucrium Marum with cold water.—White leaflets, heavier than water, of a disagreeable aromatic smell and taste.

**Masopin**$=C_{44} H_{36} O_2$. A crystalline resin, main ingredient of the dschilte (the hardened sap of a Mexican tree, Achras Sapota). To prepare it, boil the dschilte with water, and treat the remaining viscid, elastic body with absolute alcohol, which leaves the adherent caoutchouc undissolved. Precipitate the Masopin from the alcoholic solution with water, and recrystallise in ether.—White needles, of silky lustre, devoid of taste and smell; fuse at 155° without loss of weight, exhaling a pleasant odour, and solidify afterwards to an amorphous, vitreous mass, the fusing point of which is only 69° to 70°.

**Mastich.** Exudation of Pistacia Lentiscus. Yellowish white, transparent grains, on the fracture of glass-like lustre, of faint smell, of aromatic and somewhat bitter taste, softens on masticating, fuses at 80°, readily soluble in absolute alcohol, ether and oils, in alcohol of 80°/₀ to the extent of four-fifths, and leaving a soft resin (masticin). The readily soluble portion of the resin has the formula $C_{40} H_{31} O_4$, the slowly soluble $C_{40} H_{31} O_2$.

**Maynas-Resin**$=C_{28} H_{18} O_8$. From incisions in the stem of Calophyllum Caloba and C. longifolium, in the South American province of Maynas.—Crystallises from boiling alcohol in beautifully yellow rhombic prisms, fuses at 105°, is afterwards decomposed, is insoluble in water, readily soluble in alcohol, ether, acetic acid, fixed and volatile oils, also in alkalies; in concentrated sulphuric acid with beautiful red colour.

**Mecca Balsam.** Exudation of the stem of Balsamodendron Opobalsamum. Of thin fluidity, pale yellow, of 0·95 density, smells pleasantly, tastes bitter and warming, becomes thicker by age, afterwards solid and darker, dissolves readily in alcohol and ether. Contains about $10\%$ of volatile oil and two resins, the one of which dissolves readily in cold alcohol, the other sparingly. Some kinds also contain gum.

**Meconic Acid** $= C_{14} H_4 O_{14} + 6 HO$. As yet only found in opium. Mix the alcoholic solution of opium with a solution of chloride of baryum, and decompose the precipitate of Meconate of baryta with sulphuric acid. It crystallizes in colourless, mica-like laminæ, is inodorous, of acid taste, cooling afterwards a little bitter; loses at 120° all its water of crystallisation; fuses at 150 to 200° and becomes decomposed, yielding a new acid (pyromeconic acid $= C_{10} H_3 O_5$); dissolves moderately in water, readily in alcohol; the aqueous solution is decomposed on boiling under formation of carbonic acid and another new acid (comenic acid $= C_{12} H_4 O_{10}$). Salts of oxyd of iron impart to Meconic acid a vividly red colour.

**Meconin** $= C_{20} H_{10} O_8$. Indifferent substance of opium (the hardened milky juice of the green capsules of Papaver somniferum). Precipitate the extract, obtained by maceration with water, with ammonia; filter, evaporate the liquid to a syrup consistence and leave to stand cold for a few weeks; press the brown crystals thus obtained, boil with alcohol of 36° B., concentrate the solution, crystallise what has formed first in boiling water with aid of animal charcoal, next in hot alcohol.—Colourless needles, inodorous; of no perceptible taste at first, afterwards acrid or bitter. Meconin fuses at 90°, distils at 155° unaltered; dissolves in 700 parts cold and in 20 parts boiling water, in alcohol, ether, acetic acid and volatile oils.

**Melampyrit** $=$ DULEIT.

**Melezitose** $= C_{12} H_{11} O_{11}$. Peculiar sweet substance of the manna of Briançon, effusing from the stem of Pinus Larix. Extract the above manna with boiling water, evaporate the extract to a syrup consistence and keep in a cold place. The Melezitose then forms slowly and has to be recrystallised in alcohol.—Minute, short, microscopic crystals, constituting a white mealy powder about as sweet as grape-sugar; effloresce readily at the air, lose in heating about 4 per cent. of water, fuse below 140° without further change; are decomposed at 200°, dissolve readily in water, scarcely in cold, slightly in boiling alcohol, not in ether; become carbonised by cold concentrated sulphuric acid; yield with nitric acid oxalic acid, with diluted sulphuric acid on heating grape-sugar (more readily than trehalose, but less so than cane-sugar). Melezitose generally undergoes with beer-yeast a slow alcoholic fermentation; is not altered by heating with alkalies or alkaline tartarate of cooper; is precipitable by ammoniacal acetate of lead.

**Melin** $=$ RUTIN.

**Melitose** $= C_{12} H_{11} O_{11} + 3 HO$. Distinct sweet matter of the peculiar manna of various kinds of the genus Eucalyptus. Crystallises, on evaporating the aqueous solution of the manna, and is purified with animal charcoal. From water Melitose crystallises in felted needles, from alcohol in well-formed but small crystals; it has a slightly sweet taste, loses at 100° 2 equivalents water, at 130° another eq., while evolving a peculiar odour and becoming anhydrous and of the appearance of a pale-yellow translucent mass; it fuses on rapidly heating to 94° to 100° under loss of $11·23°/_5$ (5 eq.) water, smells like caramel, when heated to a higher temperature, and is afterwards carbonised; dissolves in water almost like mannit; in hot alcohol more extensively than mannit, yields with nitric acid a little mucic and more oxalic acid, separates on heating with diluted sulphuric acid into equal parts fermentable and unfermentable sugars (the latter called by Berthelot Eucalin $= C_{12} H_{12} O_{12}$); yields with yeast half as much alcohol and carbonic acid as grape-sugar; is not altered on boiling with alkalies, alkaline earths or alkaline tartarate of copper; is precipitated by ammoniacal acetate of lead.

**Melonemetin.** Emetic ingredient of the root of Cucumis Melo, obtained in the impure state by treating the aqueous extract with alcohol and evaporating the tincture to dryness.— Brown, hard, glossy, deliquescent mass; tastes acrid and somewhat bitter; dissolves most readily in water, alkalies and alcohol, but not so well in strong alcohol; not in ether, acetic acid and oils.

**Menispermin** $= C_{36} H_{24} NO_4$. Alkaloïd of the seed-husks of Cocculus indicus of Anamirta paniculata. Exhaust the contused husks with boiling alcohol of 36° B., filter, distil off the alcohol, withdraw from the residue picrotoxin by means of boiling water and afterwards the Menispermin and paramenispermin by acidulated water; precipitate the latter two bases with ammonia, dissolve the deposit in diluted acetic acid, precipitate again with ammonia, dry the deposit, draw out with alcohol and allow the solution to evaporate spontaneously, whereby a yellow alkaline resin with crystals of Menispermin and a yellowish slimy mass are obtained. Pick out the crystals as well as possible, remove from the yellow gelatinous mass the resin by means of cold alcohol, afterwards, by rinsing with cold ether, the rest of the Menispermin, which remains after the evaporation of the ether, and purify all the crystals obtained by rinsing with cold alcohol. The yellowish slime, dissolved in absolute alcohol, yields, after evaporating, paramenispermin. — Forms white, half-translucid, quadratic prisms, similar to cyanide of mercury; tasteless; fuses at 120°, and is destroyed in higher temperatures; is insoluble in water, soluble in alcohol, ether and diluted acids, forming salts with the latter, which are precipitable by alkalies.

K

**Menyanthin**$=C_{60} H_{46} O_{23}$. In the leaves of Menyanthes trifoliata. Prepare with little water a very concentrated aqueous extract, and digest with roughly pounded boneblack, to absorb the bitter substance. Wash the boneblack with cold water and extract with boiling alcohol; filter hot, distil the solution, evaporate the residue to treacle consistence, and treat repeatedly with ether, to remove a substance of rancid taste; dissolve the extract in water; precipitate the solution with tannic acid; knead the precipitate after it has become of a tough plasticity repeatedly with pure water, dissolve in six times its weight of alcohol, filter, mix the filtrate with carbonate of lead, add an equal volume of water to the alcoholic solution, and heat over the water-bath under continual stirring, until carbonic acid ceases to be evolved. Treat the dry mass with hot alcohol, shake the filtrate with boneblack, filter, distil off the alcohol, throw down the M. from the residue with tannic acid, and proceed as before.— White amorphous mass, yielding after trituration a pure-white permanent powder; neutral, of a strong and pure bitter taste; becomes soft at 60° to 65°, but is completely fused only at 115°; carbonises in a higher temperature; dissolves somewhat slowly in cold, readily in hot water and in alcohol; not in ether, but in concentrated sulphuric acid, with yellow-brown colour, which by access of the air gradually changes to a violet colour, in concentrated nitric acid with yellowish tinge, in concentrated hydro-chloric acid colourless, in alkalies unaltered: is not precipitable by metallic salts; changes on heating with diluted sulphuric or hydrochloric acid into sugar and a volatile oil. The latter (Menyanthol$=C_{16} H_8 O_2$) has an agreeable smell like bitter almonds, and shows also in other respects similarity with oil of bitter almonds; has a very burning taste; solidifies after a few days' rest to a white crystalline mass, probably a new acid, called provisionally menyanthic acid.

**Mercurialin.** Volatile alkaloïd of Mercurialis annua. Is obtained by distilling the herb and fruits with lime and water, saturating the distillate with sulphuric acid, evaporating to dryness, shaking the salty mass with absolute alcohol, &c., and proceeding as with coniin, to which also it greatly resembles.—Limpid, colourless, oily liquid of a very penetrating narcotic odour, similar to coniin and nicotin, of a strongly alkaline reaction; becomes converted at the air into a soft, resinous body; begins to boil at 140°; yields with platinum-chloride a double salt in beautiful mother-of-pearl-like laminæ, combines also with carbonic acid.

**Metacetonic Acid**$=$PROPIONIC ACID.

[**Metacopaivic Acid**$=C_{44} H_{34} O_8$. Discovered by E. G. Strauss in the balsam of Maracaibo, a kind of copaiva-balsam, exported from Columbia. Obtained by treating the balsam with soda-ley, filtering, mixing with chloride of ammonium, filtering again and neutralising with excess of hydrochloric acid. The precipitate is dissolved in hot alcohol and crystallised.—The M. acid crystallises in leaflets, insoluble in water, soluble in alcohol, easily soluble in ether, soluble also in warm petroleum, in potash-ley and ammonia. Fuses at 205°-206°. Is, according to Flueckiger and Hanbury, probably identical with Gurjunic acid, which itself may be regarded as Hydrated Abietic acid.]

**Metamorphin.** Obtained only once in the preparation of morphin (after the method, indicated there, with lime, &c.), as hydrochloride of Metamorphin. This salt was decomposed with an equivalent weight of silver-sulphate, the chloride of silver removed by filtering, the liquid digested with carbonate of baryta and the Metamorphin withdrawn by alcohol from the mixture consisting of sulphate of baryta, carbonate of baryta and Metamorphin.—Forms flat, concentrically united prisms; has at first no perceptible, but afterwards a very slightly pungent, not bitter taste; becomes at 100° opaque, grey-brown at 130° without fusing, but fuses on rapidly heating to a colourless liquid; has a neutral reaction when dissolved in water, and a slightly alkaline one when dissolved in alcohol; dissolves in 6000 parts cold and in 70 parts boiling water, in 330 parts cold and 9 parts boiling alcohol of 90%, not in ether, in the hydrates and carbonates of alkalies; gives the same reactions with sulphuric, nitric, and iodic acids as morphin; becomes grey-blue with chloride of iron.

**Methysticin.** Peculiar crystalline body of the so-called kawa-root (from Piper methysticum). Evaporate the alcoholic tincture to honey-consistence, treat with alcohol, let the solution crystallise and purify the crystals by recrystallisation.—Colourless, inodorous and tasteless, neutral crystals, fusible at 130°, decomposing in a higher temperature, insoluble in water (according to other statement: slightly soluble in cold, more readily in hot water), scarcely in cold alcohol, ether and volatile oils, abundantly in hot alcohol and in hot volatile oils; in hydrochloric acid with yellow, in nitric acid with red, in pure sulphuric acid with beautiful violet, in commercial sulphuric acid with blood-red colour.

**Morin** $= C_{18} H_8 O_{10}$ (isomeric with morus-tannic acid). The yellow pigment of the wood of Maclura tinctoria. Boil the wood with water, concentrate the decoction to a small bulk, keep in a cold place, collect the yellow sediment which will have formed, press, dissolve in hot alcohol, dilute with ten times its quantity of water, collect the deposit of morin-lime, boil with a solution of oxalic acid in alcohol, filter hot and let crystallise.—White crystalline powder, passing exposed to ammoniacal vapours into a faint yellow; of slightly bitter, not acerb taste; of acidulous reaction; loses water at 180°, remains so unaltered up to 250°, decomposes in still higher temperatures yielding pyrocatechuic acid; dissolves in 4000 parts cold and in 1000 parts boiling water, readily in alcohol with dark-yellow colour, also readily in ether; is precipitable by glue.

**Morindin $= C_{28} H_{15} O_{15}$.** Yellow dye of the root of Morinda citrifolia. Boil with alcohol, filter hot, recrystallise what has been separated on cooling in alcohol, and afterwards in alcohol containing some hydrochloric acid, in order to remove anorganic salts.— Forms sulphur-yellow, fine needles, of silky lustre, fusible by heat and decomposing in higher temperatures; dissolves little in cold with yellow colour, and readily in boiling water; forming a jelly-like substance after cooling; slightly soluble in cold absolute, copiously in diluted boiling alcohol; not in ether; in alkalies with orange-red, in concentrated sulphuric acid with purple-red colour; precipitates the salts of the alkaline earths and of the earth-metals, likewise sub-acetate of lead.

**Moringic Acid.** In the oil of ben (from Moringa oleifera), is probably identical with oleic acid.

**Morus Tannic Acid $= C_{18} H_8 O_{10}$.** In the wood of Maclura tinctoria, as secretion in the interior of the logs. Recrystallise these secretions repeatedly in boiling water, dissolve in more boiling water containing hydrochloric acid, filter the solution, which has become turbid after cooling by the formation of a reddish resin, and allow the solution to stand quiet, when the acid will slowly separate.—It crystallises in light-yellow, microscopic needles, has a sweetish astringent taste and an acid reaction; fuses at 200° and decomposes afterwards; dissolves in 6·4 parts cold, and in 2·14 parts boiling water with yellow colour; readily in alcohol, ether and wood-spirit, not in volatile and fixed oils; is precipitable by glue, likewise with dark-green colour by salts of oxyd of iron.

**Morphin $= C_{34} H_{19} NO_6 + 2 HO$.** In opium, in the ripe and in the green capsules and probably in all the other parts of Papaver somniferum, also in the capsules of Papaver Rhoeas, and other species of that genus [and in the herb of Argemone Mexicana.] Boil the opium with water, press, and repeat these operations twice; evaporate to one-half, add milk of lime; boil for a quarter of an hour, strain, press, boil the remaining mass twice with water, concentrate the whole of the calcareous liquids to a small bulk, add chloride of ammonium, heat for an hour or as long as ammonia is evolved, leave to stand cold for eight days, collect the sediment, wash with cold water, dissolve in hydrochloric acid, treat the solution again with lime and chloride of ammonium, dissolve the deposit again in hydrochloric acid, digest the solution with coal, filter, precipitate with ammonia and dry the deposit.—

Forms fine, white needles of silky lustre, or, when obtained from the alcoholic solution and by slow evaporation, considerably long, colourless, half-translucid, klinorhombic prisms, inodorous, slightly bitter, in solution very bitter; loses at 120° its water and turns opaque, fuses afterwards and becomes decomposed; dissolves in 1000 parts of cold and in 500 parts boiling water, in 30 parts cold and in 20 parts boiling alcohol of 80°/₀, with decidedly alkaline reaction; not soluble in ether, but in 60 parts chloroform, also in amylalcohol; most readily in diluted acids, in concentrated sulphuric acid yellowish, in concentrated nitric acid red afterwards yellow, in concentrated sulphuric acid containing a little nitric acid, violet-red; separates from iodic acid instantly iodine with brown colour, blues the salts of oxyd of iron; dissolves in fixed caustic alkalies and in alkaline earths, little in ammonia; neutralises the acids completely. Its salts are for the most part crystallisable, taste very bitter, dissolve in water and in alcohol, not in amylalcohol, are only precipitable by tannic acid when quite neutral.

[**Moschatin.** *See* IVAIN and ACHILLEIN. Contained, besides achillein, in the alcoholic solution of the aqueous extract. Dissolve the flocculent mass, obtained by treating the alcoholic residue with water in absolute alcohol, evaporate to dryness on the water-bath, heat with water, and wash with cold water, until the mass becomes brittle under water.—A slightly hygroscopic powder of aromatic, bitter taste, soluble in absolute alcohol, scarcely in water.]

**Moss-starch** = LICHENIN.

**Mucilage** = VEGETABLE MUCILAGE.

**Mucin.** Peculiar protein substance contained in crude gluten; but the complete isolation of which has not been achieved yet. *See* Gluten and Glutin.

**Mudarin.** Peculiar bitter substance of the root of Calotropis gigantea and C. procera. Is obtained by treating the alcoholic extract with water and evaporating the solution.—Light-brown, pellucid, brittle, inodorous, of nauseous bitter taste, insoluble in ether and in oils. The cold concentrated aqueous solution becomes turbid when gently heated, gelatinises and splits up at last completely into water and a pitch-like coagulum which does not redissolve on cooling, but does so slowly on addition of fresh water. The alcoholic solution does not coagulate.

**Mycose** = $C_{12} H_{11} O_{11} + 2 HO$. Peculiar kind of sugar of ergot (the mycelium of Cordiceps purpurea). Precipitate the aqueous extract with subacetate of lead; remove any lead from the filtrate by means of sulphuret of hydrogen; evaporate to a syrup consistence, and purify the crystals which will have formed after some time by rinsing with alcohol and recrystallising in water.—Crystallises in rhombic prisms, tastes sweet, but less so than cane-sugar, fuses at 100° without a perceptible loss of weight, intumesces at 130°, and loses 2 eq. water; becomes solid again and loses no more water; fuses at 210° again, becomes next brown, and smells unmistakably like caramel. Dissolves in less than its equal weight of water, in boiling alcohol not quite to the extent of 1%; is not altered by alkalies or by alkaline tartarate of copper; yields with nitric acid, oxalic, but no mucic acid; is converted, by heating with diluted sulphuric acid, into grape-sugar. Rotates three times more to the right than cane-sugar, and even more than dextrin.

**Myrica Wax.** Obtained by boiling the berries of Myrica cerifera, M. cordifolia, M. quercifolia, M. serrata, with water.— Pale green, transparent, brittle and friable when cold, smells and tastes aromatic; has 1·00 density; fuses at 44 to 49°. Contains much palmitic, little myristic acid, mostly in the free state; no oleic, nor any volatile acids.

**Myristic Acid** $= C_{28} H_{27} O_3 + HO$. As for distribution, see Myristin. Saponify myristin with soda-ley; decompose the soap with a mineral acid and crystallise in alcohol what has separated.— White, wart-like groups or fine leaflets, fusing at 53.8°; insoluble in water; soluble in 545 parts alcohol of 50% at 17°; in 7 parts absolute alcohol at 16°; and in $\frac{1}{4}$-part at the boiling heat; in 2 parts ether at 16°. The Myristates of the alkalies dissolve undecomposed in water.

**Myristin** $= C_{90} H_{86} O_{12}$ $(C_6 H_5 O_3 + 3 C_{28} H_{27} O_3.)$. Principally in the fat of the seeds of Myristica fragrans, M. Otoba, M. sebifera, and of other kinds of this genus, in the fat of Cocos nucifera, in the fat of Bassia species, of Myrica cerifera and other congeners, of the Dica-bread (the fruit of Irvingia Barteri). Dissolve the nutmeg-balsam in 4 parts boiling alcohol, let the mixture cool, and wash what has crystallised with alcohol.— White, opaque, or silky shining mealy mass, of granular appearance under the microscope, inodorous and tasteless, fuses at 52°; is insoluble in water, soluble in 4458 parts absolute alcohol at 17°, in 3 parts boiling alcohol, in $7\frac{1}{4}$ part ether at 17°, and in $3\frac{1}{4}$ parts boiling ether.

**Myronic Acid.** As Myronate of potash $= KO + C_{20} H_{18} NS_4 O_{20}$ in the seeds of Brassica nigra. Press the seeds, draw out with alcohol of 85% at first cold, then at 50° to 60°, treat afterwards with cold or warm water, neutralise the free acid in the aqueous extract with carbonate of baryta, evaporate to a syrup consistence, remove mucous substances by digesting with weak alcohol, and evaporate the filtrate to the formation of crystals. The Myronate of potash crystallises in limpid, short, rhombic prisms of glass lustre, is permanent at the air, neutral, has a refreshing bitter taste, dissolves readily in water, decomposes in aqueous solutions and in the presence of myrosin (emulsin of the white mustard) into $KO + 2 SO_3 + HO$, $C_8 H_5 NS_2$ and into $C_{12} H_{12} O_{12}$, the latter body being identical with grape-sugar. From the Myronate of potash the Myronic acid is obtained by removing the potash with tartaric acid, or by converting the potash-salt into the baryta compound, and decomposing the latter with sulphuric acid.—The Myronic acid forms a colourless and inodorous syrup-like liquid, has a bitter taste and an acid reaction; decomposes readily with heat; dissolves readily in water and in alcohol, scarcely in ether, forms with potash, soda, ammonia and baryta crystallisable, with lime, oxyd of lead and oxyd of silver amorphous salts of bitter taste.

**Myrosin.** The albuminous or emulsin-like constituent of the black and white mustard seeds, Brassica alba and B. nigra, and which gives rise to the formation of volatile mustard oil from the myronate of potash, contained in black mustard seeds, and by the concurrence of water. From the black mustard it cannot be obtained, as it is instantly decomposed through the myronate of potash when brought into contact with water. From the white mustard seeds it is obtained by treating with cold water; the filtered liquid is evaporated to the consistence of syrup at a temperature not exceeding 40° and mixed with alcohol; the Myrosin, precipitated thereby, is redissolved in water, and the solution brought to dryness with a gentle heat.—It is in physical properties very similar to emulsin of almonds; yields with water a slimy solution which coagulates at 60°, and likewise readily through alcohol and acids. It has not been possible as yet to separate it from albumin.

[According to Will and Koerner, Myronate of potash (Sinigrin), when dissolved in water and brought into contact with myrosin, splits into Sulphocyanide of Allyl (mustard oil), Bisulphate of potash, and grape-sugar.]

**Myroxocarpin** $= C_{48} H_{35} O_6$. Peculiar, crystalline matter of the white balsam of Peru, the latter being obtained by pressing the inner parts of the fruit and the seeds of Myroxylon Pereiræ. Digest the balsam with alcohol of moderate strength, and leave the clear liquid to evaporate; the crystals which have formed are purified by animal charcoal and by recrystallising.—The crystals are colourless, inodorous, and tasteless, hard, glossy, flat, thin, more than one inch long, insoluble in water, soluble in alcohol and in ether, neutral, fusible at 115°, and mostly decomposing in higher temperatures. Acids and alkalies have scarcely any effect; nitric acid converts it slowly into oxalic acid and an amorphous resin.

**Myrrh.** Exudation of the stem of Balsamodendron Myrrha. Yellow-brown, of a wax or resin-like gloss, brittle, of balsamic odour, and of bitter aromatic taste. Contains about 45% resin, 41% gum, 2.5°/₀ volatile oil. The resin (myrrhin) is red-brown, brittle, has (warm) a myrrh-like odour, fuses at 90-95°, dissolves in alcohol, ether, acetic acid, only partially in hot potash-ley, and consists of $C_{48} H_{32} O_{10}$.

**Napellin.** Besides Aconitin the other peculiar alkaloïd of the root of Aconitum Napellus, and some other kinds of the same genus of Ranunculaceæ. To prepare it, withdraw the aconitin from the raw aconitin with the least possible quantity of pure ether, dissolve the residue in absolute alcohol, filter, add acetate of lead as long as it causes any turbidity, agitate and let digest, filter, throw down the excess of lead by sulphuret of hydrogen, leave to digest warm, filter, evaporate the alcohol, add carbonate of potash, bring to dryness, exhaust with absolute alcohol, filter through animal charcoal and bring to dryness.—Triturated, a white, electric powder of bitter and afterwards burning taste, dissolves with some difficulty in ether, considerably more readily in water and in alcohol than aconitin, has a decidedly alkaline reaction, saturates the acids completely; is not (on account of its solubility in water) precipitated by ammonia from the aqueous solutions of its salts.

[According to later researches of Huebschmann, Napellin is probably identical with Acolyctin.]

**Narcein** $= C_{46} H_{29} NO_{18}.$ In opium. Mix the aqueous extract of opium, after it has been freed from meconin, morphin, narcotin, and meconic acid, by oversaturating it with ammonia, cooling, filtering, concentrating the filtrate, and precipitating with solution of baryta:—with carbonate of ammonia, in order to remove the excess of baryta, and evaporate to a liquid of syrup thickness.—The impure Narcein crystallises after a few days, and has to be purified by pressing off the mother-ley, and recrystallising in alcohol. Traces of codein and of meconin are removed by ether. It crystallises in white, silky, shining, delicate needles, tastes slightly bitter with an almost metallic aftertaste, fuses at 92°, and is decomposed in higher temperatures; dissolves in 375 parts cold and in 230 parts hot water (according to Anderson it dissolves readily in hot water), readily in alcohol, not in ether, has no alkaline reaction, is dissolved more readily by ammonia and by the diluted fixed alkalies than by water, but is separated as an oily liquid by addition of much concentrated potash-ley even from the hot solution. With diluted nitric acid a yellow liquid is obtained on heating, which evolves the odour of a volatile base with caustic potash; with concentrated nitric acid oxalic acid is obtained. It dissolves in concentrated sulphuric acid with an intensely red colour, and turns green on heating. With strong hydrochloric acid it does not strike the blue colour observed by Pelletier. The sulphate, nitrate and hydrochloride of Narcein are crystallisable, yet of acid reaction.

**Narcotin** $= C_{46} H_{25} NO_{14}$. In the unripe and the matured capsules of Papaver somniferum; also alleged to exist in the root of Aconitum Napellus, and called at first Aconellin. Exhaust opium with cold water, treat it afterwards with water containing hydrochloric acid, precipitate the latter solution with bicarbonate of soda, exhaust the deposit with alcohol of 80%, distil one-half to two-thirds of the alcohol, and pour the remnant boiling hot into a flat vessel. The Narcotin will soon crystallise and has to be washed with cold alcohol and recrystallised in hot alcohol. Or, begin by drawing out the opium with water and hydrochloric acid, and precipitate the Narcotin from the liquid by means of chloride of sodium; dissolve the deposit in hydrochloric acid, precipitate with potash and recrystallise in alcohol.— Forms colourless needles of pearly lustre, inodorous and tasteless, neutral; fuses at 170° and decomposes afterwards; is insoluble in cold water, not or very slightly soluble in boiling water, in 100 parts cold and in 20 parts boiling alcohol of 85%, in 126 parts cold and in 40 parts boiling ether, most readily in chloroform, little in oils, not in alkalies, readily in acids; becomes yellow and then orange with concentrated sulphuric acid; with nitric acid yellow, and on warming red; with sulphuric acid, containing a little nitric acid, beautifully blood-red. Its salts have an acid reaction, are mostly uncrystallisable, have a more bitter taste than the salts of morphin; dissolve in water, alcohol and ether; those formed with weak acids are decomposed by much water, those formed with volatile acids partly lose the acid in evaporating. The hydrates and the carbonates of alkalies precipitate the Narcotin completely.

**Narthecic Acid.** Peculiar crystalline acid of the herb of Narthecium ossifragum. Draw out with water, containing a little soda, precipitate the extract, acidified by acetic acid, with acetate of lead and afterwards with subacetate of lead, decompose the latter deposit with sulphuret of hydrogen, evaporate the filtrate to honey consistence, draw out with ether and evaporate the latter.— White, needle-shaped crystals of acid taste, not volatile, soluble in water, alcohol and ether, yields with the pure and with the earthy alkalies readily soluble salts, which are precipitable by most of the salts of heavy metals.

**Narthecin.** Peculiar rancid constituent of the herb of Narthecium ossifragum. Exhaust with water, draw out the remnant with alcohol, precipitate the tincture with acetate of lead, decolourise with animal charcoal, remove the excess of lead by sulphuret of hydrogen, distil off most of the alcohol, allow the remnant to evaporate spontaneously, and purify by recrystallising in ether.—White, wart-shaped mass, of a very rancid taste, fuses at 35°, and remains then in the amorphous state; has an acid reaction; dissolves little in water, readily in alcohol and in ether, also in alkalies, and reprecipitable by acids; is not volatile.

[**Nataloin**$=C_{34}$ $H_{19}$ $O_{15}$. Crystallised bitter substance, discovered by Flueckiger from Natal-aloes. By triturating the drogue with an equal weight or less of alcohol at 49° or below, the Nataloin is left behind in pale-yellow crystals.—The Nataloin dissolves at 15·5° in 70 parts common alcohol, in 60 parts of a mixture of 1 ether and 3 alcohol, in 35 parts methyl-alcohol, in 50 parts acetic ether, in 1236 parts ether, and in 230 parts absolute alcohol. It is of a pure bitter taste, gives out no water at 100°, fuses at 180° to 189°. It dissolves in concentrated sulphuric acid, the solution becoming beautifully green on the addition of nitre or chlorate of potash. Nitric acid yields with Nataloin oxalic acid; alkaline liquids dissolve Nataloin under darkening. The three varieties of aloin, obtained respectively from Barbadoes, Natal, and Zanzibar, are named by Flueckiger Barbaloin (Aloin), Nataloin, and Socalcin. They are distinguished, according to Histed, by the following reactions:—A drop of nitric acid gives, with a few particles of Barbaloin or Nataloin (not with Socalein), a vivid crimson. Nataloin (not Barbaloin or Socalein), assumes a fine blue colour by adding a minute quantity to a drop or two of concentrated sulphuric acid, then allowing the vapour of a rod touched with nitric acid to pass over the surface.]

[**Ngaï-camphor**$=C_{20}$ $H_{18}$ $O_2$. Obtained from Blumea balsamifera. White crystals, precisely like Borneo-camphor, which they also resemble in odour and hardness, as well as in being a little heavier than water, and not so volatile as camphor. The alcoholic solution rotates to the left as powerfully as Borneo-camphor to the right. Boiling nitric acid converts Ngaï-camphor into a laevogyric camphor$=C_{20}$ $H_{16}$ $O_2$, identical with the camphor of Chrysanthemum Parthenium, and distinguished from common camphor only by the optical properties.—PLOWMAN.]

**Nicotianin**$=C_{40}$ $H_{32}$ $N_2$ $O_6$. Separates on the surface of the aqueous distillate of tobacco.—White laminæ, smelling like tobacco smoke, have not an acrid, but a warm and bitter aromatic taste; soluble in water, alcohol, and ether, also in potash-ley, the latter solution yielding nicotin by distillation.

**Nicotin** $= C_{10}$ $H_7$ N. Volatile alkaloïd of Nicotiana species. Draw out the comminuted tobacco-leaves with hot water, containing $1°/_o$ sulphuric acid, strain, press, concentrate the liquid to syrup thickness, add one-twentieth of the weight of the leaves pulverised charcoal, evaporate to dryness, pulverise, treat warm with alcohol of $90°/_o$, allow to cool, filter, drive off the alcohol, add water, filter off the resin that has formed, distil the liquid with caustic potash, saturate the distillate with sulphuric acid, bring almost to dryness, and shake the mass with absolute alcohol, which dissolves only the sulphate of Nicotin, leaving the sulphate of ammonia. Drive off the alcohol from the solution, pour the remaining concentrated aqueous solution of sulphate of Nicotin into a glass-stoppered bottle, add a few pieces of caustic potash, add ether after solution, shake, pour the ether, containing the Nicotin, into another stoppered bottle, and allow to evaporate spontaneously.—Colourless oil of a nauseous odour of tobacco, of an extremely acrid and burning taste, fluid at —10°; density = 1·033; boils at 240-250°, distils under partial decomposition, has a strongly alkaline reaction, is coloured brown by the light or the air, dissolves in water in every quantity, likewise in alcohol and in ether, readily in fixed oils, less in volatile oils; saturates the acids completely.

**Nigellin.** Bitter substance of the seeds of Nigella sativa, as yet only obtained as extract.

**Nucin.** In the green fruit-shells of Juglans regia. Draw out with benzol or with sulphide of carbon for only a brief time, evaporate, mix the remnant with quartz-sand and sublimate at 60° to 80°.—Red-yellow, very glossy, very brittle needles, sublimate unaltered; non-nitrogenised; insoluble in water, sparingly soluble in alcohol, readily in ether. Nucin acquires a beautiful purple colour with the hydrates, the carbonates, the borates and the phosphates of alkalies and with subacetate of lead; dissolves copiously in the above alkalies and alkaline salts, and is reprecipitated in brown-red flocks from these solutions by acids.

[**Nucitannin.** Contained, according to T. L. Phipson, in the epidermis of the walnut (Juglans regia), and procurable therefrom by means of water or alcohol. N. is related to the tannic acids, and separates with mineral acids into sugar, acetic acid and a new acid, Red acid. The latter dissolves easily in alcohol and ammonia, little in cold water; with alkalies it forms dark-red salts, with lead a brown amorphous salt.]

**Nutmeg-balsam.** Obtained by pressing the seeds of Myristica fragrans.—Brownish-yellow, with a great many white, granular, crystalline veins, presenting a marble appearance; smells and tastes strongly of nutmegs, is moderately hard, of 0·956 density, fuses at 47°, dissolves slowly in cold alcohol, readily in ether, chloroform, sulphide of carbon and benzol. Contains volatile oil (6°/₀), myristin (70°/₀), olein (20°/c), an acid resin, butyrin, and traces of other volatile acids.

**Oenolin** or **Oenolic Acid** = $C_{20} H_9 O_9 + HO$. The red dyeing matter of the skins of grapes (from Vitis vinifera). Wash the skinny parts with water, and draw out with water containing acetic acid, precipitate the solution with acetate of lead, and decompose the deposit, after washing and while still moist, with sulphuret of hydrogen. The sulphide of lead, after washing, yields up the dyeing matter to alcoholic acetic acid, impure Oenolin remaining after evaporating, and which is freed from fat by boiling with ether.—An almost black mass, yielding a beautiful violet powder, or a red-brown one when dried at a temperature of 100° to 120°; permanent at the air; intumesces in the heat and becomes decomposed, is insoluble in water, yet dissolves in water containing acetic or tartaric acid, likewise in wood spirit, not in pure alcohol, but in alcohol containing even very little acetic acid, with blue, with more acetic acid with red colour; not in ether, benzol, chloroform, sulphide of carbon, volatile and fixed oils. Acetate of lead occasions in the alcoholic solutions a purely blue precipitate, nitrate of lead a violet one, subacetate of lead blue and a little brownish, subsulphate of iron blue-violet, sulphate of iron dark nut-brown, chloride of iron yellowish, acetate of copper nut-brown, subchloride of tin violet-red, subnitrate of mercury a precipitate of the colour of wine-yeast, nitrate of mercury light-brown, nitrate of silver brown-red; alum causes no precipitate.

**Oil of Achillea Millefolium** (milfoil). Obtained by distillation with water. The oil of the flowers is dark-blue, of subacid reaction and 0·92 sp. gr. The oil of the herb is also blue and of a deeper colour than oil of chamomile, thick and of almost butter consistence when cold, of strong smell, tastes similar to the herb, afterwards a little burning, has a density of 0·852–0·917. The oil of the fruits is greenish. The oil of the root is colourless or slightly yellowish, smells peculiar and disagreeable, somewhat like valerian, has an unpleasant, but not penetrating or burning taste, is lighter than water.

[**Oil of Achillea moschata** (oil of iva). Obtained by distilling the herb, before flowering, with water.—Clear yellowish liquid, of a very pleasant, strongly ethereous odour, and bitter, warming taste. Boils at 180° to 210°.]

**Oil of Achillea nobilis** (showy milfoil). Obtained by aqueous distillation of the herb, the flowers or the fruits.—Pale-yellow, thick; of a very strong smell, similar, but more refined, than oil of milfoil, and at the same time somewhat camphor-like; of an aromatic, camphoraceous and somewhat bitter taste; of 0·97-98 density; dissolves readily in alcohol.

**Oil of Acorus Calamus** (sweet flag). Obtained by distillation with water from the rhizome.—Pale to dark yellow; of a strong penetrating odour similar to the root, and of an aromatic bitter, burning, slightly camphoraceous taste; of 0·89-98 density; dissolves readily in alcohol; boils at 195°, after the more volatile part (probably a hydrocarbon isomeric with oil of turpentine) has passed over.

**Oil of Aleurites triloba** (kekune, the candle-nut tree). Fixed oil, obtained from the seeds by boiling with water.—Is thin, inodorous, and tasteless. Acts as a mild purgative.

**Oil of Allium sativum** (garlic)$= C_6 H_5 S$. Obtained by distillation with water from the bulbs; also from the leaves of Sisymbrium Alliaria, and mixed with oil of mustard from the herbs and seeds of Thlaspi arvense and other Cruciferæ; seems, like oil of mustard, not to exist ready formed (certainly not in cruciferous plants), but to be produced by the action of water. In the raw state it is brownish yellow; of the most intense garlic odour; heavier than water and slowly soluble in it; it is partly decomposed on rectifying and an oil $= C_6 H_5 S$ passes over as a pale-yellow or colourless liquid of great light-refracting power and of a less nauseous odour; lighter than water; without action on metallic potassium; soluble in cold concentrated sulphuric acid, with purple-red colour, changing to a deep indigo blue with hydrochloric gas, and precipitating much sulphide of silver from a solution of the nitrate.

**Oil of Alpinia Galanga**, A. officinarum (galingal). Obtained by distillation with water from the tubers.—Smells similar to cajeput oil; is lighter than water; dissolves readily in alcohol; of a similar constitution as oil of cajeput.

**Oil of Andropogon** (lemon grass). Obtained by aqueous distillation of A. Iwarancusa, A. Calamus, A. citratus, A. Martini, A. Schoenanthus, A. muricatus.—Colourless or yellowish; thin; smells penetrating, aromatic, similar to roses, but fainter; has an acrid taste, similar to oil of citron; of neutral reaction; lighter than water; boils at 147° and above. Is a mixture of different oils. [It yields a solid compound with bisulphide of soda solution.—Flueckiger and Hanbury.]

**Oil of Anemone.** *See* ANEMONIN.

**Oil of Anime.** Obtained by the distillation of the resin. —Limpid; of a strong but not unpleasant odour, and of hot taste.

**Oil of Anthemis nobilis** (chamomile). Obtained by aqueous distillation from the flowers.—Blue or greenish; is a mixture of a hydrocarbon $C_{20} H_{16}$ with angelic and valerianic acids, which remain in rectifying the oil with potash-ley. The hydrocarbon has a pleasant, lemon-like smell and boils at 175°.

[According to Demarçay, oil of chamomile is a mixture of several compound ethers, principally the angelates and valerates of butyl and amyl.]

**Oil of Apium graveolens** (celery). Obtained by aqueous distillation of the herb and fruits.—Colourless or pale yellow; of penetrating odour; of a sweetish, warming taste, of 0·881 density; dissolves readily in alcohol.

**Oil of Archangelica officinalis** (angelica). Obtained by aqueous distillation of the root.—Colourless, lighter than water, of a penetrating odour, and tastes like the root.

**[Oil of Argemone Mexicana.** Fat-oil, obtained by pressing the seeds.—Light-yellow, still liquid at 5° C., of a slightly nauseous odour and raw taste, dries, dissolves in five to six times its volume of alcohol of 90°/$_0$, becomes dirty-brown with concentrated sulphuric, red with nitric acid, is easily saponified.]

**Oil of Aristolochia Serpentaria.** Obtained by distillation with water from the root.—Light-brown, lighter than water; has the smell and taste of valerian and of camphor.

**Oil of Arnica montana.** Obtained by distilling with water the flowers and the root. The former is blue or brownish-green, the latter brownish-yellow.

**Oil of Artemisia Absynthium** (wormwood)$=C_{20} H_{16} O_2$. Obtained by aqueous distillation from the leaves and flowers.—Dark-green, smells and tastes of the plant, of 0·973 density, boils at 205°, and dissolves readily in alcohol. When distilled with another volatile oil, it yields to it the colouring matter.

**Oil of Artemisia Cina and A. Sieberi** (worm seed)$=C_{24} H_{20} O_2$. Obtained by aqueous distillation from the flowers.—Colourless to yellowish, smells of the drug; has an acrid, burning, aromatic taste; is neutral; of 0·925-945 density: boils at 175° after rectification; is a mixture of cinaeben$=C_{20} H_{16}$, cinaeben-stearopten$=C_{20} H_{18} O_2$, and a little propionate of propyl.

**Oil of Artemisia Dracunculus** (tarragon). Obtained by distilling the leaves with water.—Of 0·935 density; boils at 200° to 206°; consists entirely or wholly of anethol.

**Oil of Artemisia vulgaris** (mugwort). Obtained by distilling the root with water.—Pale greenish-yellow; butter-like; crystalline; of penetrating peculiar odour; tastes nauseous, somewhat bitter, at first burning, afterwards cooling; neutral; lighter than water; readily soluble in alcohol.

**Oil of Asa-fœtida.** Obtained by distillation with water.—Yellowish; heavier than water; of a strong garlic-odour; dissolves to a considerable extent in water, readily in alcohol; boils at 130° to 140°; is a mixture of two oils $C_{12} H_{11} S$ and $C_{12} H_{11} S_2$.

**Oil of Asarum Europæum.** Obtained by distilling the root with water.—Yellowish, thickish, lighter than water (according to other observations, of 1·018 sp. gr.); smells valerian-like; of burning, acrid taste, neutral; contains oxygen.

**Oil of Aspidium Filix mas** (male fern). By treating the ether-extract of the tubers with water containing ammonia, filicic acid passes into the water, while the oil remains in the ether, to be obtained by evaporation.—It is dark grass-green, thicker than olive oil; of a mild, afterwards rancid taste ; smells like the root ; remains liquid at temperatures much below 0°.

**[Oil of Atherosperma moschatum** (Australian sassafras). Obtained by aqueous distillation from the bark.—Thin, unctuous, pale yellow when fresh, becomes yellowish-brown by age. Resembles in odour sassafras oil with an admixture of caraways. Taste aromatic, bitter, prickling on the tongue. Sp. gr. 1·04. Boils at 230° to 245°. Report of Exhibition of 1862].

**Oil of Atropa Belladonna** (deadly night-shade). Obtained by pressing the seeds.—A little thicker than linseed oil, inodorous, of mild taste, of 0·925 density, dries slowly, becomes very thick and turbid at — 16°, congeals completely at — 27°.

**Oil of Balsam of Copaiva** $= C_{20} H_{16}$. Obtained by distillation with water. Is colourless, thin ; smells similar to the balsam ; has an acrid, lasting, bitter taste, of 0·88 to 0·91 density ; boils at 245°.

**Oil of Balsam of Peru** $= C_{32} H_{14} O_4 (C_{14} H_7 O + C_{18} H_7 O_3)$, Cinnamcïn or Cinnamate of Benzyl. Boil the balsam with a solution of soda and wash with water. The remnant separates into a resin and a yellow-brown liquid. Heat the latter to 170°, distil with steam of the same temperature, and desiccate the distillate with chloride of calcium.—A colourless oil of great light-refracting power, liquid at — 12°; has a faint, pleasant smell and an acrid, aromatic taste ; of 1·098 density ; of neutral reaction ; boils at 340° to 350° and distils under partial decomposition ; dissolves scarcely in water, readily in alcohol and in ether ; is decomposed by potash-ley into benzyl alcohol and cinnamate of potash.

**Oil of Bassia.** *See* BASSIA FAT and GALAM-BUTTER.

**Oil of Betula alba** (birch). Obtained by distilling the leaves with water.—Colourless, thin ; of a pleasant balsamic odour, similar to young birch-leaves or roses; tastes at first mild, sweetish, afterwards peculiarly balsamic, acrid, and hot; becomes a little turbid and thickish at 0°, but not hard or crystalline even at — 10°; is lighter than water; dissolves in 8 parts alcohol of 0·850. [Not to be confounded with the oil of birch-bark, obtained by dry distillation, and utilised in the preparation of the fragrant Russian leather.—F. v. M.]

**Oil of Brassica alba and B. nigra** (mustard). Fixed oil, obtained by pressing the seeds. Is yellow; of 0·917-920 density ; of mild smell and taste; thickens at—12°; is not drying. Contains the glycerides of erucic and sinapoleic acids.

**Oil of Brassica nigra** $= C_8 H_5 NS_2$. Obtained by distillation with water, but does not pre-exist. Oils of similar, but as yet undefined constitution, are also obtained from the root of Cochlearia Armoracia and of Sisymbrium Alliaria; from the herb of Cochlearia officinalis; from the herb and seeds of Iberis amara; mixtures of the oils of mustard and of garlic are obtained from the herb and seeds of Thlaspi arvense; from the seeds of Nasturtium officinale; Thlaspi Bursa pastoris; peculiar oils, containing sulphur, are furnished by Lepidium sativum, Raphanus sativus, &c. —It is colourless or yellowish; of an intensely penetrating smell and taste of mustard; of $1 \cdot 01$ density; dissolves slightly in water; readily in alcohol and in ether; attacks potassium metal violently, forming another oil and sulphocyanide of potassium; decomposes, on heating with monosulphide of potassium, into the sulphocyanide and oil of garlic; absorbs much ammonia and is converted into a white crystalline mass of thiosinamin $= C_3 H_8 N_2 S_2$.

**Oil of Brassica oleracea** = Rape-oil.

**Oil of Buphthalmum.** *See* Buphthalmum-Stearopten.

**Oil of Bursera gummifera** = $C_{20} H_{16}$. Obtained by distilling the Gomart-resin (an exudation of the stem) with water. Is similar to oil of turpentine.

**Oil of Camelina sativa.** Obtained by pressing the seeds. Thicker than hempseed-oil, inodorous and tasteless; of $0 \cdot 925$ density; freezes at $-19°$; yields soft soaps. Dries at the air.

**Oil of Canella alba** (white cinnamon). Obtained by distilling the bark with water. It is a mixture of caryophyllic (eugenic) acid, an oil similar to that of cajeput and an oxygenised oil.

**Oil of Cannabis sativa** (hemp). The fixed oil is obtained by pressing the seeds.—At first greenish or brownish-yellow; turns yellow at the air; has a mild taste; specific gravity $= 0 \cdot 927$; dissolves in thirty parts cold and in any proportion of boiling absolute alcohol; saponifies with difficulty, the soaps not being so soft as those of linseed oil; belongs to the non-drying oils.

The volatile oil is obtained by distilling the herb with water. Pale-yellow; smells like the green herb; tastes aromatic, not burning, but penetrating; is lighter than water. [According to Personne, it consists of two hydrocarbons—one liquid (Cannaben $= C_{18} H_{20}$), the other solid, forming platy crystals (Hydride of Cannaben $= C_{18} H_{22}$).]

**Oil of Carapa Guianensis.** Obtained by distilling the seeds with water.—Colourless, unctuous; of very bitter taste; becomes solid at $4°$.

**Oil of Carum Carvi** (caraway). Obtained by distilling the fruit with water.—Colourless or pale-yellow; thin; of a strong taste and smell of the fruit; of $0 \cdot 91$–$97$ density; begins to boil at $175°$, Carven passing over, while Carvol distils up to $232°$.

**Oil of Carum Petroselinum** (parsley)$=C_{20} H_{16}$. Obtained, besides a less volatile stearopten (see Parsley-stearopten), by distilling the fruits with water.—Is freshly distilled greenish-yellow, rectified colourless; thin; smells like the fruit; of $1·01$–$1·04$ density; solidifies between $2°$ and $8°$; boils between $160°$ and $170°$; dissolves readily in alcohol.

**Oil of Chenopodium ambrosioides** (Mexican tea). Obtained by distilling the herb with water.—Pale to greenish-yellow, rectified colourless; very thin; of great light-refracting power; smells strongly of the herb; tastes strongly aromatic and cooling similar to peppermint; of $0·902$ density; boils at $179$–$181°$; dissolves readily in alcohol.

**Oil of Chrysanthemum Parthenium** (pyrethrum). Obtained by distillation with water from the flowering herb.—Greenish; deposits on keeping stearopten; distils between $165$-$220°$.

**Oil of Cinnamomum Burmanni** (massoy). Obtained by the aqueous distillation of the bark.—Consists of a light, of a heavy oil, and of stearopten. The first is colourless, mobile; smells similar to sassafras. The heavy oil is thicker, like the first of acrid and pungent taste, but of fainter smell; is less volatile; becomes thick at $— 10°$, but not crystalline. The stearopten is white, pulverulent, inodorous, almost tasteless, heavier than water.

**Oil of Cinnamomum Camphora.** *See* CAMPHOR.

**Oil of Cinnamomum Culilaban.** Obtained like the preceding oil.—Colourless; smells like the oils of cajeput and cloves mixed; is heavier than water.

**Oil of Cinnamomum Zeilanicum** and **C. Cassia** (cinnamon) $=C_{18} H_8 O_2$. Obtained by distilling with water the bark and flowers of the first and the bark of the latter.—Yellow, rectified colourless, of the pleasant odour and burning taste of cinnamon; of $1·008$ density; boils at $220°$.—A steropten $= C_{56} H_{29} O_{10}$ subsides slowly from the oil and appears, after re-crystallisation in alcohol, under the form of colourless and inodorous, highly shining, brittle prisms. The oil of the leaves of C. Zeilanicum is very similar to that of cloves; of $1·053$ density; contains caryophyllic and benzoic acids and a hydrocarbon $C_5 H_4$.

**Oil of Citrus Aurantium** (orange). The oil of the flowers is obtained by aqueous distillation, and consists of an oil, easily soluble in water—constituting the so-called Aqua Naphæ—and another sparingly soluble, which is colourless or yellowish, of extremely pleasant odour, of neutral reaction, of $0·85$-$90$ density; forms stearopten on keeping.—The oil of the fruit$=C_{20} H_{16}$, is obtained by pressing the rind, and is purified by rectification. It has a pleasant smell, $0·830$–$0·880$ density, and boils at $180°$. The oil of C. Aurantium var. sinensis is obtained from the rind of the fruit in the same manner, and has the same composition as the preceding. It resembles oil of lemon.

**Oil of Citrus Bergamia** (bergamot). Prepared like oil of C. Aurantium, from the fruit.—Colourless, thin, of very pleasant odour; of 0·860–0·870 density; boils at 183°. Is a mixture of a hydrocarbon $= C_{20} H_{16}$, and a hydrate of the same.

**Oil of Citrus Limetta** $= C_{20} H_{16}$. Obtained like oil of lemon, and greatly resembling it.

**Oil of Citrus medica** (lemon) $= C_{20} H_{16}$. Obtained like the preceding.—Colourless or yellowish, thin; of pleasant lemon-odour, and 0·840–0·860 density; boils at 160°–175°.

**Oil of Cochlearia officinalis** (scurvy grass) $= C_6 H_5 SO$. Obtained by distillation with water from the herb of Cochlearia officinalis, C. Danica and C. Anglica.—Possesses the pungently acrid smell and taste of the green herb in the highest degree; has a density $= 0·942$.—The so-called Cochlearia stearopten, which forms in the aqueous and spirituous distillates of the herb, is a substance crystallising in small iridescent laminae and needles of a faint odour, but acrid aromatic taste; of 1·248 density; fuses at 45°, sublimates unaltered, and consists of $C_6 H_7 O_2$ ($C_6 H_5 + 2 HO =$ hydrate of allyl).

**Oil of Convolvulus scoparius and C. floridus** (rosewood). Obtained by distilling the root and stem with water.—Pale-yellow, thin, lighter than water; smells of roses and cubebs, a little rancid; has a bitter, aromatic taste.

**Oil of Coriandrum sativum** (coriander) $= C_{20} H_{18} O_2$. Obtained by aqueous distillation from the fruits.—Colourless or yellowish, of the smell and taste of the fruits; neutral; of 0·859–0·871 density; boils at 150°, but not constantly.

**Oil of Corylus Avellana.** *See* HAZELNUT-OIL.

**Oil of Crocus sativus.** Obtained by distillation with water from the stigmata.—Yellow, thin, lighter than water, of the specific odour of saffron; becomes slowly converted into a solid mass, which sinks in water. [The oil is, according to Rochleder, a decomposition product of crocin.]

**Oil of Croton Eluteria and C. Sloani** (cascarilla). Obtained by distillation with water from the bark.—Dark yellow; smells of camphor, lemons, and thyme; has an aromatic, somewhat bitter taste; of 0·938 density; begins to boil at 180°, the boiling point rising afterwards. Consists of, at least, two different oils, the more volatile of which is probably a hydrocarbon.

**Oil of Croton Tiglium** $=$ CROTON OIL.

**Oil of Cucurbita Pepo** (pumpkin). Obtained from the seeds by pressing.—Pale-yellow, thick, inodorous, and tasteless; of 0·923 density; solidifies at — 15°; dries slowly.

**Oil of Curcuma longa.** Obtained by distillation with water from the root.—Citron-yellow, thin, of penetrating smell and hot taste.

**Oil of Curcuma Zedoaria** (zedoary). Obtained, like the foregoing, from the tubers.—Pale-yellow, turbid, thick, heavier than water; has a peculiar fragrant, camphor-like odour, and a somewhat bitter, hot, camphoraceous taste.

**Oil of Cyperus esculentus** (earth-nut). Obtained by pressing the tubers.—Yellow, inodorous, mild; of 0·919 density; solidifies at 0°; is readily saponified; hardens with hyponitric acid.

**Oil of Dahlia purpurea.** Obtained by aqueous distillation from the tubers.—Yellowish; smells very strongly like the tubers; has a sweetish, afterwards sub-acrid taste; sinks slowly in water; becomes thick like butter, somewhat crystalline, and separates benzoic acid.

[**Oil of Daphne Mezereum.** Fat-oil obtained by pressing the fruits.—Yellowish, drying oil of at first sweet, afterwards burning-sharp and acrid taste, and of 0·8903 density at 15° C. It dissolves in ether, sulphide of carbon and benzol, less in strong alcohol; does not solidify at — 16°. It forms with concentrated sulphuric acid a deep-red, heavy liquid; with concentrated nitric acid a red mass of a bitter-almond-like odour. It consists, according to A. Casselmann, of about 10% stearin, palmitin and myristin, and of 90% linolein and olein, with traces of volatile fat acids and coccognin.]

**Oil of Daucus Carota** (carrot). Obtained by distilling the root with water.—Has a peculiar strong, penetrating smell, and a similar taste, warming and somewhat disagreeable; of 0·886 density.

**Oil of Dicypellium caryophyllatum** (clove-bark). Obtained by distilling the bark with water.—Is heavier than water; resembles in odour oil of cloves, and is somewhat similarly constituted.

**Oil of Dryobalanops.** *See* BORNEEN and BORNEOL.

**Oil of Elais Guineensis.** *See* PALM-OIL.

**Oil of Elemi**$=C_{20}$ $H_{16}$. Obtained by aqueous distillation.—Colourless, mobile, smells like elemi-resin; has an acrid taste; of 0·850 density; boils at 166°–174°.

**Oil of Elettaria Cardamomum** (cardamom). Obtained by distilling the seeds with water.—Pale-yellow, of the odour and taste of the seeds, of 0·92–0·94 density; neutral; deposits a stearopten $= C_{20} H_{22} O_6$, which is isometric, or identical with hydrate of oil of turpentine.

**[Oil of Eriostemon squameus.** Obtained by distilling the leaves with water.—Pale-yellow, lighter than water; of a taste and odour similar to, but milder than, oil of rue. *Rep. of Exh. of* 1862.]

**Oil of Excæcaria.** *See* TALLOW, CHINESE.

**[Oil of Eucalyptus.** Obtained by distillation with water from the leaves and branchlets of various species of Eucalyptus, especially E. amygdalina, the first samples of which were exhibited by Baron F. v. Mueller in 1854, and subsequently in 1862 prepared by Mr. Johnson, and especially by Mr. J. Bosisto.—The oils, obtained from different species, often vary very considerably in physical properties, as will be seen by the following short description of a few of the oils :—*E. amygdalina :* Pale-yellow, thin; of pungent odour, resembling, but coarser than, lemons; tastes rather mild, and cooling, afterwards bitter; of 0·881 sp. gr. at 15°; boils at 165°–188°; deposits stearopten at — 18°, which melts at — 3°; becomes resinous at the air. *E. oleosa :* Thin, mobile, pale-yellow, of mild taste; flavour camphoraceous, suggestive of turpentine; odour mint-like; sp. gr. = 0·911; boils at 161°–177°. *E. sideroxylon :* Thin, limpid, very pale-yellow, taste and smell like E. oleosa; sp. gr. $=$ 0·923; boils at 155°–178°. *E. goniocalyx :* Pale-yellow; of pungent, penetrating, rather disagreeable odour; taste exceedingly unpleasant; sp. gr. $=$ 0·918; boils at 152°–175°. *E. globulus :* Very pale-yellow, thin; of cajeput-like odour, but of less disagreeable, more cooling, and mint-like taste; of 0·917 sp. gr.; boils at 149°–177°. *E. corymboza :* Smells slightly of lemons and roses; tastes slightly bitter, somewhat camphor-like; colourless; of 0·881 sp. gr. at 15°. *E. obliqua :* Reddish-yellow, of mild odour and bitter taste; of 0·899 sp. gr.; boils at 171°–195°; becomes turbid at — 18°. *E. fissilis :* Pale reddish-yellow; smells similar to the preceding oil; of 0.903 sp. gr.; boils at 177°–196°. *E. odorata :* Pale-yellowish with a greenish tinge; smells aromatic; sp. gr. $=$ 0·899–922; boils between 157° and 199°. *E. longifolia :* Of oily consistence; taste aromatic, cooling; odour fragrant, camphor-like; of 0·940 sp. gr.; boils at 194°–215°. *E. rostrata :* Pale yellow to reddish-amber; smells and tastes like E. odorata; of 0·918 sp. gr.; boils at 137°–181°. *E. viminalis :* Pale yellowish-green; of disagreeable, but not penetrating smell; of 0.921 sp. gr.; boils at 159°–182°. All these oils are now manufactured on a large scale by Mr. J. Bosisto, who, by untiring energy, by expenditure of capital, and by perseverance, has succeeded in establishing the Eucalyptus oils amongst the prominent articles of commerce. (*Report of the Exhibition of* 1862.)

The Eucalyptus oils dissolve resins, &c., with great ease. These substances, arranged according to their diminishing solubility, form the following series :—Camphor, rosin, mastich, callitris-sandarac, elemi, sandarac, kaurie-gum, dammara, asphalt, xanthorrhœa-resin, dragon's-blood, benzoin, copal, amber, anime, shellac, caoutchouc, beeswax. Gutta-percha is not dissolved. (*Rep. of Exh. of* 1862.) According to Cloez, Eucalyptus oil contains hydrocarbons and a distinct compound, Eucalyptol$=C_{24} H_{20} O_2$, obtained by repeated fractional distillations. Eucalyptol is a very mobile, colourless liquid, boiling at 175°, and of 0·905 sp. gr. at 8°. It rotates polarised light to the right ( $[a]=+10·42°$), remains liquid at — 18°; is little soluble in water, completely in alcohol, the solution having a rose-like odour when highly diluted. Eucalyptol is slowly acted upon by nitric acid ; converted into a tarry mass by concentrated sulphuric acid ; forms, when distilled with phosphoric anhydride, Eucalyptin$=C_{24} H_{18}$, a liquid boiling at 165°, and of 0·836 density.]

**Oil of Eugenia caryophyllus** (cloves). Obtained by distilling the flower-buds and flower-stalks with water.—Colourless, but becomes coloured by age; of the smell and taste of cloves; of 1·030–1·060 density. Is a mixture of eugenic acid and a hydrocarbon$=C_{20} H_{16}$. The latter is obtained by distilling the crude oil with potash-ley, washing the oily product, desiccating with calcium-chloride, and rectifying. It is colourless, of great light-refracting power; smells like oil of turpentine, but dissolves more sparingly in alcohol; is of 0·910 density; boils at 142° to 151°.

**Oil of Euonymus Europæus.** Obtained by pressing the seeds.—Pale-yellow; smells like rape-oil, tastes bitter, afterwards rancid; has 0·938 density; solidifies at — 12°; yields the bitter principle to warm water; dissolves with difficulty in alcohol with an acid reaction. Contains olein, palmitin, acetin, and free benzoic acid.

**Oil of Fagus sylvatica** (beech). Obtained by pressing the seeds.—Yellow, of a mild and pleasant taste ; of 0·920 density; becomes thick and turbid at — 10°, solidifies at — 17°; turns easily rancid.

**Oil of Fœniculum officinale** (fennel). Obtained by distilling the fruits with water.—Pale-yellow ; smells and tastes like the fruit, sweetish and aromatic; solidifies below 10°; is of 0·968 density at 20°; consists of a hydrocarbon isomeric with oil of turpentine, boiling at 185° to 190°, and of liquid and solid anethol.

**Oil of Galipea Cusparia** (angustura). Obtained by distilling the bark with water.—Pale-yellow; smells peculiarly aromatic, similar to Ligusticum ; tastes at first mild, afterwards acrid ; has a density of 0·934; boils at 266°. Is a mixture of a hydrocarbon and an oxygenised oil.

**Oil of Galbanum.** Obtained by distilling the gum-resin with water.—Colourless, of galbanum-odour; of 0·904 density; boils at 160°; is isomeric with oil of turpentine.

**Oil of Garcinia.** *See* BRINDONIA TALLOW.

**Oil of Gaultiera.** *See* SALICYLATE OF METHYL.

**Oil of Geum urbanum** (avens). Obtained by distilling the root with water.—Greenish-yellow, of butter-consistence, smells of cloves.

**Oil of Hedwigia balsamifera.** Obtained by distillation with water from the balsam.—Yellow, has a pleasant turpentine-like odour and a hot taste; is lighter than water.

**Oil of Helianthus annuus** (sunflower). Obtained by pressing the seeds.—Pale-yellow, thicker than hempseed oil, of 0·926 density, dries slowly; becomes turbid even at an ordinary temperature; solidifies completely at — 16°.

**Oil of Hesperis matronalis.** Obtained by pressing the seeds. —Greenish, turns brown on keeping; of 0·928 density; almost inodorous, drying readily; still completely liquid at — 15°.

**Oil of Humulus Lupulus** (hops)$=C_{20} H_{16}$, mixed with $C_{20} H_{16}$ + 2 HO. Obtained by aqueous distillation from the female flowers.—Colourless or yellowish, thin; smells penetrating-narcotic of hops; tastes hot, slightly bitter like wild marjoram and thyme; is neutral, of 0·910 density. The hydrocarbon distils between 125°–175°, the hydrate at 210°.

**Oil of Hyoscyamus niger** (henbane). Obtained by pressing the seeds.—Pale green-yellow, rather thin, inodorous, mild, of 0·913 density, dissolves scarcely in 60 parts absolute alcohol.

**Oil of Hyssopus officinalis** (hyssop). Obtained by distilling the herb with water.—Colourless, of peculiar odour; of acrid, camphoraceous taste and of neutral reaction; has a density of 0·88–0·98; distils between 142° and 162°, the last portion coloured; is a mixture of at least two different oils.

**Oil of Iatropha Curcas.** From the pressed seeds, called physic-nuts.—Colourless, inodorous, mild, of 0·91 density, congeals at — 8° butter-like; dissolves scarcely in alcohol; contains the glycerids of ricinoleic and isocetic acids.

**Oil of Illicium anisatum** (star-anise). Obtained by distilling the capsules with water.—Yellowish, of a sweetly aromatic smell and taste, of 0·976 density at 20°; separates cold a stearopten consisting of anethol.

**Oil of Illipe.** *See* GALAM BUTTER.

**Oil of Iris.** *See* IRIS STEAROPTEN.

**Oil of Irvingia.** *See* DICA FAT.

**Oil of Juglans regia.** *See* WALNUT OIL.

**Oil of Juniperus communis** $=C_{20} H_{16}$. Obtained by distilling the berry-like fruits with water.—Colourless or yellowish, smells strongly of the fruits; of 0·847–0·870 density; dissolves little in alcohol. The oil of the ripe fruits boils at 205°; deposits on keeping or in a cold place a stearopten, which is heavier than water. The oil of the unripe fruits contains a more volatile oil, boiling at 155°, and another boiling at 205°, like the oil of the ripe ones.

**Oil of Juniperus Sabina** (savin) $=C_{20} H_{16}$. Obtained by distilling the branchlets with water.—Colourless, of a strong taste and smell of the shrub, of 0·89–0·94 density; boils at 155°–161°.

**Oil of Juniperus virginiana** (pencil or red cedar). Obtained by distilling the wood with water.—A soft, white, crystalline mass of a peculiar aromatic smell; solidifies at 27° after desiccation, distils on the main at 282°.—Mixture of a liquid hydrocarbon and an oxygenised stearopten. The hydrocarbon, Cedren $=C_{30} H_{24}$, is obtained by pressing the crude oil and rectifying the liquid portion, when cedren passes over at 264°–268°, and is re-distilled with metallic potassium. It is now colourless, of 0·948 density, and boils at 237°. The cedar-stearopten forms white needles of satin-lustre, and fuses at 79°.

**Oil of Laurus nobilis** $=$ BAY OIL.

**Oil of Lavandula** (lavender). Obtained by aqueous distillation from the flowers of L. Stœchas, L. angustifolia, and L. latifolia.—Pale-yellow, thin, of a pleasant smell of the flowers; has a burning, bitter, aromatic, acrid taste; is neutral, of 0·876–0·880 density, boils at 185°–188°, dissolves readily in alcohol. Its composition is $C_{30} H_{28} O_4$, but it consists of two portions, a liquid and a solid one, the latter being identical with common camphor.— The oil, distilled from the stalks and leaves or from the whole herb, is called oil of Spike. It smells not so pleasant as oil of Lavender, has a greater density and contains more stearopten.

**Oil of Ledum palustre** (wild rosemary). Obtained by distillation with water.—Consists in the main of a hydrocarbon $=C_{20} H_{16}$, and of an oxygenised oil. It has, fresh, the composition $C_{80} H_{63} O_5$. From the oil soon a stearopten crystallises, which forms colourless, fine prisms of a faint odour of the herb, and somewhat similar to roses and turpentine; of a warming, aromatic taste; fuses with a gentle heat, sublimates at a higher temperature, and contains a hydrocarbon $=C_{20} H_{16}$, and its hydrate $=5C_{10} H_8 + 3HO$.

**Oil of Lepidium sativum** (cress). Obtained by pressing the seeds. — Brown-yellow, of specific taste and odour, of 0·924 density; becomes thick and turbid at — 6°, solid at — 15°; dries slowly.

**Oil of Linum.** *See* Linseed Oil.

**Oil of Lucuma.** *See* Shea Butter.

**Oil of Lycopus Europæus.** Obtained from the herb by distillation with water.—Green, butter-like; smells like the herb; has an acrid taste.

**Oil of Madia sativa.** Obtained by pressing the seeds.—Deep-yellow, thick, mild, of 0·935 density; solidifies at — 10° to — 17°; dries slowly.

**Oil of Matricaria Chamomilla** (German chamomile). Obtained from the flowers by distillation with water.—Dark-blue, thick, smells strongly of the flowers, tastes aromatic hot, is lighter than water. Has nearly the same composition as common camphor$= C_{20} H_{16} O_2$.

[According to Kachler it contains capric acid, an oil isomeric or polymeric with common camphor, and boiling at 150°–165°; another oil of the same composition, deep-blue, of 281°–289° boiling point, and identical with blue oil of galbanum (of 281° boiling point); also a small quantity of hydrocarbon$= C_{20} H_{16}$. The distilled water contains, besides, propionic acid.]

**Oil of Melaleuca** (cajeput). Obtained by aqueous distillation from the foliage and branchlets of different species, especially M. Leucodendron.—Pale-green, rectified colourless, of a penetrating camphoraceous smell and burning taste, of 0·91–0·94 density; boils at 175°; dissolves readily in alcohol.

[M. ericifolia. Pale-yellow, taste and smell like cajeput-oil; thin, of 0·899–0·902 density, boils at 149°–184°.—*M. Wilsonii.* Resembles cajeput-oil; of 0·925 density.—*M. parviflora.* Of oily consistence and amber colour, 0·938 density; boils at 185°–209°; resembles the foregoing.—*M. uncinata.* Green, smells like M. ericifolia with an admixture of peppermint.—*M. genistifolia.* Pale greenish-yellow, mild in odour and taste.—*M. squarrosa.* Green, of disagreeable taste.—*M. linarifolia.* Light straw-coloured, mobile; of rather pleasant, cajeput-like odour; taste very agreeable, suggestive of mace, afterwards mint-like; of 0·903 density; boils at 175°–187°.—*Rep. of Exh. of* 1862.]

**Oil of Melissa officinalis** (balm). Obtained from the whole herb by aqueous distillation.—Pale-yellow, thin, of a pleasant lemon-like smell and of 0·85–0·92 density.

**Oil of Mentha.** Obtained by distilling the herb with water.—*M. piperita* (peppermint): Colourless, yellowish or greenish-yellow, of peculiar odour and burning, camphoraceous, then cooling taste; of 0·84–0·92 density; boils at 188°–193°; dissolves readily in alcohol.—*M. Pulegium* (penny-royal)$= C_{20} H_{16} O_2$. Of 0·927 density; boils at 183°–188°.—*M. viridis* (spearmint): Of 0·91-

0·93 density; after the stearopten has been removed by rectification, of 0·876 density; boils rather constantly at 160°. [Gladstone found in the crude oil a hydrocarbon almost identical with oil of turpentine, mixed with an oxydised oil $= C_{20}$ $H_{14}$ $O_2$, which is isomeric with carvol, of 0·951 density, 225° boiling point, and bearer of the peculiar smell of the plant.—*M. australis:* Like peppermint-oil of second quality.—*M. gracilis :* Odour between peppermint and penny-royal; of 0·914 density.—*M. laxiflora:* Of coarse smell; of fiery, bitter, very unpleasant, nauseous taste; of 0·924 density.—*Rep. of Exh. of* 1862.]

**Oil of Mercurialis annua.** Obtained from the (dried) herb by distillation with water.—Of a thickish consistence. The green herb is said to yield no oil.

**Oil of Moringa oleifera** (ben). Fat-oil of the seeds.—Pale-yellowish, of 0·912 density; fluid at 25°, thick at 15°, solid at lower temperatures; inodorous, of a pleasant mild taste (according to others, acrid and bitter); saponifies slowly; contains a peculiar, solid fat-acid, Benic acid $= C_{44}$ $H_{44}$ $O_4$.—The oil of *M. aptera* contains a peculiar but liquid fat-acid, Moringic acid, which scarcely differs from oleic acid.

**Oil of Myrica.** *See* MYRICA WAX.

**Oil of Myrica Gale** (gale). Obtained by distilling the leaves with water.—Brownish-yellow, at 12° thickish ; of a peculiar, pleasant, balsamic odour ; tastes at first mild, afterwards hot, and lastingly styptic ; is of neutral reaction, of 0·876 density ; dissolves in 40 parts alcohol of 0.875.

**Oil of Myristica fragrans.** Obtained by distillation with water.—(*a*) From mace, the covering of the seeds; yellowish, thin, smells strongly like mace, of a burning aromatic taste ; separates no solid at — 12°, begins to boil at 160°, the temperature slowly rising to 180°. The composition of the oil is $3C_{20}$ $H_{16}$ + HO, the more volatile portion being a hydrocarbon of thyme-like odour, 0·853 density.- -(*b*) From the seeds (nutmegs). Nearly colourless, thin, of a strong smell and taste of the seeds, neutral ; yields no sediment, even at — 7°; is of 0·850 density; commences to boil at 160°, the temperature rising slowly beyond 200°. Its composition is exactly like oil of mace.

**Oil of Myrrh** $= C_{20}$ $H_{14}$ $O_2$. Obtained by distillation with water. Pale-yellow, thickish, of the taste and smell of myrrh ; lighter than water ; of acid reaction.

**Oil of Myrtus communis** (myrtle). Obtained from the leaves, flowers, and fresh fruits by aqueous distillation.—Yellowish or greenish yellow, lighter than water.

**Oil of Narcissus Jonquilla** (jonquil). Extracted from the flowers by ether.—Yellow, of butter consistence, of pleasant smell of the flowers.

**[Oil of Nasturtium officinale.** Obtained by distillation.—Boils between 120–280°. Consists in the main of $C_{18} H_9 N$, which boils at 253·3°, and is of 1·0014 sp. gr. at 18°.—HOFFMAN.]

**Oil of Nicotiana Tabacum** (tobacco). Fixed oil of the seeds.— Gold-yellow, inodorous, mild, of 0·917 density, liquid even at — 15°; dissolves in 168 parts alcohol of 93 %; saponifies easily; is not drying. *See* also NICOTIANIN.

**Oil of Nigella sativa.** Obtained by distilling the seeds with water.—Colourless, with a bluish fluorescence ; lighter than water, smells like a mixture of fennel and oil of bitter almonds.

**Oil of Œnanthe Phellandrium.** Essential oil of the fruits.— Yellow or brownish, thin; smells and tastes penetrating, similar to the fruit ; is of 0·852 density, and neutral reaction.

**Oil of Olea Europæa** = OLIVE OIL.

**Oil of Olibanum.** Obtained by aqueous distillation.—Yellowish, thin; smells turpentine-like, but pleasanter ; of 0·866 density ; boils at 162°; consists of two oils, one of which contains oxygen, the other being a hydrocarbon = $C_{20} H_{16}$.

**Oil of Origanum Majorana** (sweet marjoram). Obtained from the herb by aqueous distillation.—Yellow-green, of a lighter hue after rectification; of a penetrating odour of the herb; of a warming, acrid, slightly bitter taste ; of sub-acid reaction and 0·89 density; boils almost constantly at 163°; forms by age or intense cold a stearopten = $C_{14} H_{15} O_5$.

**Oil of Origanum vulgare** (wild marjoram) = $C_{50} H_{40} O$. Obtained like the preceding.—Pale or brown-yellow, smells strongly of the herb, has an acrid aromatic taste, is neutral, of 0·86–90 density; forms a stearopten on keeping.

**Oil of Osmitopsis asteriscoides** = $C_{20} H_{18} O_2$. Obtained from the flowers by aqueous distillation.—Yellowish, rectified colourless, thin; smells penetrating, unpleasant of camphor and cajeput; has a burning, rancid taste; dissolves readily in alcohol; has a density of 0·921; boils at 178°.

**Oil of Papaver somniferum** (poppy). Fixed oil of the seeds. —Gold-yellow, thin, has a slightly acrid taste and a density of 0·924; solidifies at — 18°; dries at the air more readily than linseed oil; dissolves in 25 parts cold and in 6 parts hot alcohol; saponifies easily.

**[Oil of Pastinaca sativa.** Obtained from the fruits of aqueous distillation.—Colourless, clear, of not unpleasant odour and aromatic taste, of neutral reaction and 0·8672 density at 17·5°; consists for the greater part of butyrate of octyl (capryl). VON RENESSE.]

**Oil of Pelargonium.** Obtained by aqueous distillation from the leaves and flowers of P. odoratissimum, P. Radula and P. capitatum.—Colourless, of thick fluidity and roselike odour, becomes turbid at 0°. Consists of pelargonic acid and a neutral oil.

**Oil of Peucedanum.** Obtained by aqueous distillation.—(a) From the herb of P. Oreoselinum: Of a strong, aromatic, juniper-like odour and 0·840 density; boils at 163°; hydrocarbon $= C_{20} H_{16}$.—(b) From the root of P. Ostruthium: Colourless or pale-yellow, thin, of penetrating odour; of a warm, camphoraceous taste similar to oil of carrots. Is a mixture of various hydrates of a hydrocarbon $= C_{20} H_{16}$.

**Oil of Philadelphus coronarius.** Obtained from the flowers by extraction with ether.—Gold-yellow, in large quantities narcotic; of a delicious odour when diffused.

**Oil of Pimenta officinalis** (allspice). Obtained from the fruits by aqueous distillation.—Greatly resembles oil of cloves; has also a similar constitution; of 1·030 density.

**Oil of Pimpinella.** Obtained by distillation with water.—(a) From the fruits of P. Anisum (anis): Colourless or yellowish, possessing in a high degree the odour and the sweetish aromatic taste of the fruits; liquefies at 17°; has at 20° a density of 0·977; dissolves readily in alcohol; contains variable quantities of an elæopten and a stearopten, of the latter 25% to 80%. As for constitution, see Anethol.—(b) From the root of Pimpinella nigra: Light-blue, lighter than water, of a less penetrating odour than the following oil; of a burning taste of the root, afterwards irritating the throat.—(c): From the root of P. saxifraga (pimpernel); gold-yellow, thin, lighter than water, resembles in smell parsley-fruits; has a nauseous, bitter, afterwards rancid taste.

**Oil of Pinus.** The volatile oil (oil of turpentine $= C_{20} H_{16}$) pervades all parts of the numerous species of the above genus, and is mostly obtained from the resinous exudations by distillation with water. It is colourless, thin, of 0·850–0·880 density, boils at 150°–160°; has a strong, specific smell and taste; dissolves only by traces in water, sparingly in alcohol, readily in ether; is converted by hydrochloric gas into two compounds, a solid and a liquid one.

The fat-oils of Pinus are obtained by pressing the seeds. (a) From P. Abies: Brown-yellow, of turpentine-like smell and taste, and 0·928 density; remains liquid at — 15°.—(b) From P. Picea: Brown-yellow; of pleasant, balsamic odour and mild, aromatic taste; of 0·926 density. Is a mixture of resin, a volatile and a fat-oil, the latter more slowly drying and more readily soluble in absolute alcohol, than other drying oils.—*(c)* From *P. sylvestris:* Brown-yellow, of turpentine-like taste and smell, and of 0·931 density; becomes thicker at — 16°, at — 27° whitish and turbid, at — 30° solid; dries readily.

### Oil of Pinus Sabiniana = ABIETEN.

### Oil of Piper Cubeba.
Obtained from the fruits (cubebs) by aqueous distillation.—Colourless, thick; the portion which distils last in rectifying almost of butter-consistence; has a density of 0·936; distils mostly at 250°–260°; has a faint aromatic odour, and a warming taste like camphor and peppermint; is of neutral reaction, and consists of a liquid hydrocarbon and an oxygenised stearopten. The hydrocarbon, Cubeben = $C_{30} H_{24}$, distils first, is less thick than the crude oil, and has a density of 0·919.—The stearopten = $C_{30} H_{26} O_2$, is obtained by cold pressing and re-crystallising from alcohol. It forms colourless, klinorhombic prisms of glass-lustre; smells faintly like cubebs; tastes hot, after-wards cooling; is of neutral reaction; liquefies at 69° to an oil of 0·926 density; boils at 150°; sublimates unchanged in small quantities; dissolves readily in alcohol, ether, oils, and acetic acid, not in alkalies.

### Oil of Piper angustifolium (matico).
Obtained by distillation from the leaves.—Pale-green, somewhat thick, of a strong, cam-phoraceous smell and taste; is, when long kept, heavier than water, becomes thicker, and at last crystalline.

### Oil of Piper nigrum (pepper) = $C_{20} H_{16}$.
Obtained from the fruits by distilling with water.—Colourless, thin, of a hot peppery taste and smell, and 0·864 density; boils at 167·5° to 170°.

### [Oil of Pittosporum undulatum.
Obtained from the flowers by distillation.—Limpid, colourless, lighter than water; of an exceedingly agreeable, jasmine-like odour; taste disagreeably hot and bitter, slightly reminding of turpentine and rue.—*Rep. of Exh. of* 1862.]

### Oil of Populus (poplar).
Obtained by aqueous distillation from the leaf-buds of P. nigra and other species.—Colourless, lighter than water; of pleasant, balsamic odour.

### Oil of Primula Auricula.
*See* PRIMROSE-STEAROPTEN.

### [Oil of Prostanthera.
Obtained from the leaves by aqueous distillation.—*P. Lasianthos:* Greenish-yellow, of mint-like odour and taste, and 0·912 density.—*P. rotundifolia:* Of darker colour and 0.941 density, otherwise resembling the foregoing oil.—*Rep. of Exh. of* 1862.]

**Oil of Prunus Amygdalus.** The fat-oil (almond-oil) is obtained by pressing the seeds.—Yellowish, mild, thinner than olive-oil; of 0·920 density; becomes thickish at — 10°, at — 16° white and at — 21° of butter-consistence; contains more olein than olive-oil; dissolves in 25 parts cold and in 6 parts hot alcohol, in any quantity of ether; does not dry.

The volatile oil, Oil of Bitter Almonds$=C_{14} H_6 O_2$, does not exist ready formed, but is produced from amygdalin, when in contact with water and emulsin, together with hydrocyanic acid. Of plants, yielding those two products, many are known in the order of Rosaceæ (see Amygdalin). The oil, obtained by distillation, is freed from hydrocyanic acid by shaking with potash-ley or hydrated iron-suboxyd and water or mercury-oxyd and water; then poured off and rectified.—A colourless, thin oil of peculiar obour and burning aromatic taste; of 1·043 density; boils at 180°; dissolves in 30 parts water, readily in alcohol and ether; is, by keeping at the air, converted into benzoic acid; solidifies under the influence of ammonia slowly to a crystalline mass (hydrobenzamid).

**Oil of Quercus.** Obtained by distilling the fruit (acorns) of Q. robur with water.—Of butter-consistence and peculiar, strong smell; lighter than water.

**Oil of Reseda luteola.** Obtained by pressing the seeds.— Dark-green, thin, of nauseous taste and smell; of 0·935 density; remains liquid at — 15°; dries readily.

**Oil of Reseda odorata.** Obtained from the flowers by extraction with ether.—Is yellowish, thickish through the admixture of wax; lighter than water; of a most pleasant odour.

**Oil of Ricinus**$=$Castor Oil.

**Oil of Rosa.** Obtained by aqueous distillation from the flowers of various roses, especially from R. centifolia, R. Damascena, R. Indica, R. moschata, and R. sempervirens.—Colourless, at 11°–16° of lamellar-crystalline appearance, fuses at 20° to 30°, has a fragrant rose-odour and a mild, somewhat sweetish flavour; is of 0·870 density at 18°, and boils at 227°. It is a mixture of an inodorous stearopten$=C_{16} H_{16}$, fusing at 32° to 35°, distilling undecomposed at 280°–300°, and an oxygenised elæopten, which is the odoriferous principle. [According to R. Baur, the elæopten is convertible into the stearopten by treating with zinc, hydrochloric acid, and alcohol.]

**Oil of Rosmarinus officinalis** (rosemary). Obtained from the leaves and flowers by aqueous distillation.—Colourless or yellowish, tastes and smells of the herb, somewhat camphorlike; of 0·886–0·933 density; dissolves readily in alcohol. It is a mixture of hydrocarbon, isomeric with oil of turpentine, and an oxygenised oil.

**Oil of Rubus Idaeus.** *See* RASPBERRY-STEAROPTEN.

**Oil of Ruta graveolens** (rue). Obtained from the whole plant by distillation as usual. — Colourless, has a strong smell and taste of the herb, is of 0·831 density, congeals at — 1° to 2° completely to a mass of laminæ, boils at 228°–230°. Is in the main capryl aldehyd $= C_{20} H_{20} O_2$.

**Oil of Sagapenum.** By distilling the resin with water.— Thin, yellow, lighter than water; fresh of a nauseous, garlic-like odour, assuming by and by the smell of turpentine; yields, on drying, a translucid varnish; dissolves readily in alcohol.

**Oil of Salvia officinalis** (sage). Obtained from the herb by aqueous distillation.—Greenish-yellow, has the smell and taste of the herb; is of 0·864 density; boils between 130° and 160°. Is a mixture of several oxygenised oils, also of a stearopten, which forms spontaneously on keeping.

**Oil of Sambucus nigra** (elder). Obtained from the flowers by distillation.— Light-yellow, thin or of butter-consistence, smells strongly of the flowers; has a bitter, burning, afterwards cooling taste; is lighter than water.

**Oil of Sassafras officinalis** $= C_{18} H_{10} O_2$. Obtained from the root by distillation with water.—Brownish-yellow; of a fennel-like smell and taste, and 1·09 density; boils between 115° to 228°. [Consists, according to Grimaux and Ruotte, of Safren $= C_{20} H_{16}$, Safrol $= C_{20} H_{10} O_4$, and a very small amount of a phenol. Safren boils at 155°–157°, is dextro-rotating ($+ 17\frac{1}{2}°$ for 10 centimeters), and has at 0° a density of 0·8345. Safrol is optically inactive, of 1·1141 density at 0°, and boils at 231°–233°. It must be distilled in a current of hydrogen gas.]

**Oil of Sesamum orientale** (sesamé). Obtained by pressing the seeds.—Light-yellow, inodorous, mild, of 0·919 density at 23°, congeals at — 5°. Sesamé oil, either pure or mixed with oils of almond or olive, assumes a beautiful transient green colour, when agitated with a mixture of one-half volume of nitric and one-half volume of sulphuric acid.

**Oil of Spiraea Ulmaria** (meadow-sweet) $= C_{20} H_{18} O_4$. Obtained from the flowers by aqueous distillation. The crude oil contains also salicylous acid and a stearopten, from both of which it is purified by agitating with potash-ley and subsequent rectification.—Colourless, lighter than water, similar in smell to salicylous acid, of a slightly burning taste, partly solidifying by cold; readily dissolving in alcohol.

**Oil of Storax.** *See* STYROL.

**Oil of Syringa vulgaris** (lilac). Extracted from the flowers by ether.—Amber-yellow, smells similar to the flowers; deposits on keeping a hard, wax-like substance.

**Oil of Tanacetum vulgare** (tansy). Distilled with water from the herb and flowers.—Yellow, thin, of the specific odour of the plant, lighter than water.

**Oil of Teucrium Marum.** *See* MARUM-CAMPHOR.

**Oil of Thea Chinensis** (tea). Obtained by distilling Chinese tea with water, shaking the turbid distillate with ether, pouring the ethereous solution off, and evaporating.—Lemon-yellow, lighter than water, solidifies on keeping, is of a strongly narcotic tea-like odour and a similar taste, but without any astringency.

**Oil of Theobroma.** *See* CACAO-FAT.

**Oil of Thuya occidentalis.** Distilled from the green parts by water.—Colourless to greenish-yellow, of 0·925 density and camphoraceous smell and taste, boils at 190°, and for the greater part at 193°–197°; dissolves readily in alcohol. Is a mixture of at least two oxygenised oils.

**Oil of Thymus Serpillum.** Obtained by distillation with water.—Gold-yellow; of a pleasant odour of lemon and thyme, and of an aromatic, somewhat bitter taste ; of 0·89–0·91 density.

**Oil of Tilia Europæa.** Distilled by water from the flowers.— Colourless or yellowish, lighter than water, smells strongly and pleasantly of the fresh flowers, has a sweetish taste ; dissolves readily in alcohol.

**Oil of Tropacotum majus.** Obtained by distilling the fruits with water.—Yellow, heavier than water ; of a peculiar, strongly aromatic odour and acrid, burning taste; inflames the skin even more than mustard-oil; boils at 120°–130°; contains sulphur. [Contains, according to A. W. Hofmann, as chief constituent on oil$=C_{16}$ $H_7$ $N$, of aromatic odour, colourless, of 1·0146 density at 18°; boils at 226°. The same compound also occurs in the ethereal oil of Lepidium sativum, obtained from the herb, the aqueous distillate of which requires agitating with benzol, in order to deliver up the oil.]

**Oil of Turpentine.** *See* OIL OF PINUS.

**Oil of Valeriana officinalis.** Distilled from the root by water. —Thin, yellowish, neutral ; smells of the dried root; is of 0·90–0·96 density; dissolves readily in alcohol. It is a mixture of valerol (70%)$=C_{12}$ $H_{10}$ $O_2$, valeren or borneen$=C_{20}$ $H_{16}$, borneol $=$ $C_{20}$ $H_{18}$ $O_2$, and valerianic acid.

**Oil of Vitis vinifera.** Obtained by pressing the seeds (raisin-stones).—Colourless or yellowish, almost devoid of smell ; has a sweetish, aromatic taste; of 0·920 density; solidifies to butter-consistence at — 11°; turns readily rancid, yellow, and thick at the air; yields very soft soaps. [Contains, according to Fritz, mainly erucic, also stearic and palmitic acids as glycerids.]

**[Oil of Zieria Smithii.** Distilled from the leaves by water.— Pale-yellow, of the taste and odour of rue; of 0·950 density. —*Rep. of Exh. of* 1862.]

**Oil of Zingiber officinale** (ginger) $= C_{20} H_{16} + HO$. Obtained from the tubers by aqueous distillation.—Yellowish, very thin, smells strongly of ginger, tastes burning-aromatic; has a density of 0·893; boils at 246°.

**Oleandrin.** Alkaloïd of the leaves and branches of Nerium Oleander, accompanied by another alkaloïd named Pseudo-curarin. To prepare it, precipitate the concentrated aqueous decoction exactly with tannic acid, treat the deposit, after washing with a little cold water, for a short time only with an aqueous solution of tannic acid. Tannate of pseudo-curarin is hereby dissolved, and the tannate of Oleandrin remains. The latter, after dissolving in ether, is treated with hydrate of lime, in order to remove tannic acid and chlorophyll, and the filtrate is allowed to evaporate spontaneously.—Slightly yellowish, resin-like; very bitter, slightly soluble in water, readily so in alcohol and in ether, forms with acids uncrystallisable salts; is precipitable by chloride of gold and by chloride of platinum. It acts as a poison.

**Oleic Acid** $= C_{36} H_{33} O_3 + HO$. Contained as tri-olein in most of the (non-drying) liquid and solid fats. Saponify for instance oil of almonds or olive-oil with soda-ley, decompose the soap with hydrochloric acid, digest the separated fat-acids for several hours at 100° with lead-oxyd and treat the product with ether, the latter dissolving the Oleate of lead, while leaving behind the compounds of lead with the solid fat-acids. The ethereous solution, when agitated with an excess of aqueous hydrochloric acid, throws down chloride of lead, which sinks to the bottom of the lower aqueous fluid, while the Oleic acid remains with the ether and is obtained after the evaporation of the solvent, though contaminated by the products of oxydation and by dyeing matters. To remove these, allow the acid to solidify at — 6° to 7° and press between blotting paper, which absorbs the impurities.—Forms beautiful, snow-white needles, inodorous and tasteless; has a neutral reaction even when unchanged and dissolved in alcohol; fuses at 14°, congeals at 4° to a hard crystalline mass, is of 0·898 density at 15°, evaporates in a vacuum without decomposition, becomes brown above 100° and decomposes more and more; is in the solid state unalterable at the air, but absorbs, when liquid, much oxygen and turns rancid; is insoluble in water and mixes with alcohol and ether in every proportion. The Oleates are soft, frequently oily, or readily fusible to an oil; they dissolve more readily in alcohol, and especially in ether, than in water. The Oleic acid forms a solid mass with hyponitric acid.

**Olein** (Tri-olein) $= C_{108} H_{104} O_{12} = C_6 H_5 O_3 + 3 C_{36} H_{33} O_3$. Forms the main ingredient of the non-drying oils, and is in less quantity contained also in solid fats. Expose any fixed oil of this kind to a temperature of $- 5°$, press, treat the liquid portion again to a temperature of $- 10°$, and press again, removing the solid part.—Neutral oil, inodorous, of mild taste, and of $0·914$ density; remains liquid below $- 10°$; volatilizes in a vacuum, undecomposed, and partly at ordinary temperature; dissolves very little in alcohol, copiously in ether, is more readily saponified than the drying oils and the solid fats; becomes rancid at the air, and of a thicker consistence in thin layers, but not dry. Is solidified by hyponitric acid.

**Olibanum.** Gum-resinous exudation of Boswellia Carterii.— Yellowish or brownish grains, on the surface of a mealy appearance, of a faint aromatic smell, which becomes stronger on warming, of an aromatic, somewhat acrid, bitter taste; contains $56\%$ resin, $31\%$ gum, $6\%$ bassorin, and $5\%$ volatile oil. The resin is reddish yellow, brittle, tasteless, softens at $100°$, but fuses only at a much higher temperature, and consists of $C_{40} H_{32} O_6$.

**Olivamarin.** Bitter ingredient of all parts of Olea Europæa, occurring especially in the leaves, and still more in the unmatured fruits, but as yet only known as extract.

**Olive Oil.** Obtained by pressing the succulent part of the fruit of Olea Europæa.—Greenish yellow, of mild and pleasant taste, and of $0·916$ density; congeals partly at $10°$ to granules, at $0°$ completely; dissolves very little in wood spirit and alcohol. Consists in the main of $30\%$ palmitin, and of $70\%$ olein.

**Olivil** $= C_{28} H_{18} O_{10} + 2 HO$. In the gum of the olive-tree, associated with resin and with a little benzoic acid. Treat the gum with ether, which dissolves the resin; boil the residue with alcohol of $36°$ B.; filter hot and let cool. A crystalline pulp is obtained, which has to be washed with cold alcohol, and recrystallised in boiling alcohol.—Forms colourless crystals, inodorous, of a bitter-sweet and slightly aromatic taste, neutral, fuses at $120°$, losing its water; decomposes in a higher temperature; dissolves little in cold, in 32 parts boiling water, readily in alcohol, wood spirit, and concentrated acetic acid, little in ether and oils; turns blood-red with concentrated sulphuric acid, and becomes later carbonized; is not altered by diluted sulphuric or hydrochloric acids.

**Onocerin** $= C_{24} H_{20} O_2$. In the root of Ononis spinosa. Evaporate the alcoholic tincture of the root to the consistence of

M

syrup, press the crystals which will have formed after a few days; wash with cold water, and re-crystallise with aid of animal charcoal.—Small, delicate, very voluminous crystals of a beautiful satin-lustre ; inodorous and tasteless, highly electrified by friction, neutral, fusible, insoluble in water, soluble in boiling alcohol, very little so in ether, readily in warm oil of turpentine ; unaffected by alkalies and hydrochloric acid.

**Ononid** $= C_{36} H_{22} O_{16}$. Bitter-sweet substance of the dried root of Ononis spinosa, perhaps formed on drying from glycyrrhizin. Precipitate the aqueous decoction with diluted sulphuric acid; wash the brown flocks with cold water, dry, dissolve in absolute alcohol, evaporate the solution to dryness, and repeat dissolving and evaporating until the remnant is perfectly soluble in absolute alcohol.—Dark-yellow, amorphous, brittle mass of at first bitter, afterwards lasting sweet taste; has an acid reaction; fuses by heat and becomes decomposed; dissolves in water and alcohol ; is precipitable by metallic salts.

**Ononin** $= C_{60} H_{34} O_{26}$. Crystalline substance of the root of Ononis spinosa. Precipitate the decoction with acetate of lead; impregnate the filtrate with sulphuret of hydrogen; wash the deposit, consisting of sulphide of lead and of Ononin, dry, boil with alcohol, and recrystallise the Ononin which has been dissolved out, with aid of animal charcoal. — Forms colourless, quadrangular needles or leaflets, inodorous and tasteless; fuses at 235°, is decomposed by more heat; is insoluble in cold, little soluble in boiling water, sparingly in strong alcohol, almost insoluble in ether ; dissolves in concentrated sulphuric acid, with red-yellow colour, passing to cherry-red; separates, on boiling with diluted acids, into sugar and Formonetin (formate of Ononetin $= C_{46} H_{19} O_9 + C_2 HO_3$), the latter separating in a crystalline form ; by boiling with alkalies or with alkaline earths formic acid is formed and crystalline Onospin $= C_{58} H_{34} O_{24}$. Formonetin is tasteless, Onospin nearly so.

**[Ophelic Acid** $= C_{26} H_{20} O_{20}$. Discovered by Hoehn in the root and herb of Ophelia Chirata. Prepared by extracting the pulverised herb with alcohol of 60%, &c.—A deliquescent, syrupy, yellowish-brown substance, of at first slightly acid, afterwards intensely and lastingly bitter taste, of gentian-like odour, especially on warming ; dissolves in water, alcohol, and ether; reduces alkaline copper and silver solutions; becomes darker by alkalies, reddish-yellow by chloride of iron, dirty green by sulphate of copper, yellow by neutral and basic acetates of lead; yields with acids amorphous combinations.]

**Opian** $=$ NARCOTIN.

**Opianin** $= C_{66} H_{36} N_2 O_{21}$. In opium, but as yet only found in the Egyptian kind. Draw out with water, precipitate the liquid with ammonia, wash the deposit (mixture of morphin and Opianin) with water and alcohol, dry, dissolve in hot alcohol, decolourise by means of animal charcoal, and leave to crystallise, when the Opianin will separate first.—Forms colourless, klinorhombic crystals, inodorous, of a strong and lasting bitter taste when dissolved in alcohol; of strongly alkaline reaction; remains unaltered up to 100°, is decomposed by more heat; is insoluble in water, slightly soluble in boiling alcohol, in sulphuric acid, containing nitric acid, with blood-red colour. Is believed by some to be identical with narcotin.

**Opianyl** = Meconin.

**[Opium Alkaloids.** C. Hesse describes some new bases obtained in the following manner:—The aqueous extract of opium is precipitated by excess of soda or lime, and shaken with ether. From the latter the bases are removed by acetic acid; the acid solution is slowly added to a moderately diluted potash or soda ley; after 24 hours the precipitate is filtered off, and the filtrate, after being acidified by hydrochloric acid, is again supersaturated with ammonia. The liquid, together with the precipitate, is treated with chloroform, the latter solution with acetic acid, and the acid solution with ammonia, which produces a precipitate containing *Lanthopin* $= C_{46} H_{25} NO_8$. After 24 hours the liquid is added to potash-ley, and becomes turbid by *Codein*, which is removed by ether. The solution contains yet *Meconidin, Codamin, Laudanin*, and the base *X*, all of which pass into the ether on addition of chloride of ammonium. On evaporating the ethereous solution as slowly as possible, *Laudanin* $= C_{40} H_{25} NO_6$, crystallises first. After washing the mother-ley with bicarbonate of soda under addition of ether, *Codamin* $= C_{35} H_{23} NO_6$, crystallises. The yet remaining *Meconidin* $= C_{42} H_{23} NO_8$ is separated from the base *X* by acidifying the mother-ley and saturating with chloride of sodium; the precipitate is dissolved in weak acetic acid, and thrown down again by chloride of sodium. The precipitate is dissolved in water, mixed with bicarbonate of soda, shaken with ether, and decolourised by animal charcoal. On evaporating the ether, the Meconidin remains as a yellowish amorphous mass.]

**Opobalsamum siccum** = Tolu Balsam.

**Opobalsamum verum** = Mecca Balsam.

**Opoponax.** Gum-resinous exudation of Opopanax Chironium. From red-yellow to nearly white, opaque, of a disagreeable balsamic odour, and of a bitter and acrid taste. Contains 42% resin, and besides, gum, caoutchouc, wax, and volatile oil. The resin is red-yellow, fuses at 50°, dissolves in alcohol, ether, and alkalies, and has the formula $C_{40} H_{25} O_{14}$.

**Orcin** $= C_{14} H_8 O_4 + 2 HO$. Peculiar, sweet substance; occurs in Lecanora, Roccella, and Variolaria, lichens used for the preparatio of archil and litmus. To obtain it, draw out with alcohol, evaporate the solution, keep cold, pour off from the resin which will have formed, evaporate the liquid to honey-consistence, draw out with water, evaporate the aqueous solution to the consistence of syrup, keep cold, and purify the crystals which will have formed, by recrystallising.—Forms colourless crystals, which become anhydrous when dried in a vacuum and recrystallised in anhydrous ether; of a very sweet but disagreeable taste, of neutral reaction; fuses with a gentle heat, losing its water; boils at about 280°, and distils undecomposed; dissolves most readily in water, alcohol and in ether. The aqueous solution yields, with chloride of iron, a dark-red precipitate, from which the Orcin is dissolved by ammonia; sub-acetate of lead gives a white precipitate, turning quickly red at the air. The solution, mixed with a little ammonia, becomes slowly brown-red when exposed to the air, under formation of Orcein $= C_{14} H_7 NO_6$. Nitrate of silver, chloride of mercury and sulphate of copper produce no precipitates with Orcin. When mixed with nitrate of silver, and afterwards with ammonia, a flocky precipitate is formed; on boiling the silver is deposited as a mirror, and the liquid turns red.

**Oreoselon.** *See* PEUCEDANIN.

**Otoba-fat,** from Myristica Otoba, is similar to nutmeg-balsam.

**Oxalic Acid** $= C_2 O_3 + HO + 2$ eq. water. It occurs as an acid Oxalate of alkalies in the stalks and leaves of Oxalideæ, Polygoneæ, and Chenopodeæ, and is widely distributed as Oxalate of lime, especially in roots and woods. The Oxalate of lime, being insoluble in water and in vegetable acids (acetic acid, &c.), does not pass into the aqueous vegetable extract. Oxalic acid was formerly produced from species of Oxalis and Rumex, these herbs containing a large quantity of the acid; but at present it is produced more economically by the action of nitric acid on sugar, or by treating certain organic substances (sawdust, &c.) with caustic alkalies at rather high temperatures.—The pure Oxalic acid crystallises in colourless and inodorous klinorhombic prisms and needles of a strongly acid taste; it loses in a gentle heat its water of crystallisation (2 eq.), and volatilizes in a stronger heat (at 150°) mostly undecomposed in white, pungent vapours; on rapidly heating it fuses at 98°, and is completely decomposed at 155° into formic acid, water, oxyd of carbon, and carbonic acid. It dissolves in 10 parts cold, and in equal parts boiling water; in $2\frac{1}{2}$ parts cold, and in 2 parts boiling alcohol; little in ether.

Heated with concentrated sulphuric acid, it separates into carbonic acid and oxyd of carbon. Most of the Oxalates are insoluble in water; few of them dissolve in a solution of Oxalic acid or of chloride ammonium; all dissolve in diluted nitric acid, though more sparingly than most of the salts of the other organic acids, which are by themselves insoluble in water ; the soluble Oxalates precipitate the salts of lime, including the sulphate, from their solutions, except in the presence of an excess of any strong mineral acid.

The qualitative and quantitative determination of Oxalic acid is based upon its behaviour towards lime, with which it yields a white, pulverulent precipitate, insoluble in acetic acid, and convertible into carbonate of lime at a low red-heat without turning black. As the Oxalic acid generally occurs as Oxalate of lime, it is generally contained in the extract prepared by diluted hydrochloric acid. See No. vi., Div. iii., Pt. ii.

**Oxyacanthin**$=C_{32}$ $H_{23}$ $NO_{11}$. Alkaloïd, besides berberin in the root bark of Berberis vulgaris, and also discovered in another species of Berberis from Mexico, and likely to occur in many other species. Dilute the mother-ley obtained in the preparation of the chloride of berberin with water ; precipitate with carbonate of soda, wash the precipitate with water, treat with hydrochloric acid, filter, precipitate with ammonia, wash and dry the precipitate, draw out with ether, evaporate the solution, treat again with hydrochloric acid, precipitate with ammonia and dry.—Snow-white, amorphous, highly electric powder, turning yellow at the sunlight; is converted into fine needles by pouring on it a little ether or alcohol ; has a pure bitter taste and an alkaline reaction ; loses at 100° 3·13% ; fuses at 139°, and becomes later decomposed; nearly insoluble in water, in 30 parts cold, and in one part boiling alcohol of 90%, in 125 parts cold, and in 4 parts boiling ether, most readily in chloroform, also in oils, in concentrated sulphuric acid with brown-red colour ; liberates iodine from iodic acid; combines with acids to mostly crystallisable salts, soluble in water and in alcohol.

**Oxypinotannic Acid**$=C_{14}$ $H_8$ $O_9$. Occurring towards mid-winter in the leaves of Pinus sylvestris and allied species. Is obtained in the preparation of pinopicrin (see this) as Oxypinotannate of lead. Treat this with diluted acetic acid, throw down the filtrate with sub-acetate of lead, wash the precipitate, decompose with sulphuret of hydrogen, and evaporate the filtered liquid over the water-bath.—A grey or brownish powder, inodorous, of a very astringent taste ; dissolves readily in water and in alcohol, precipitates neither glue nor tartarated antimony, imparts a green colour to salts of oxyd of iron. By boiling with diluted sulphuric acid a red powder is formed, but no sugar.

[**Oxynarcotin** $= C_{44} H_{23} NO_{16}$. Alkaloïd, discovered by Beckett and Wright in opium. When partly purified Narceïn is boiled with water, a crystalline mass remains, which is dissolved in hot, dilute sulphuric acid, filtered and exactly neutralised with carbonate of soda. The dense, crystalline deposit, after resting for several hours, is separated from the liquid and repeatedly boiled with small quantities of water. The remaining crystals of O. are freed from traces of Narceïn by hot alcohol, dissolved in hydrochloric acid, re-precipitated by a small excess of potash-ley, washed and dried.—It forms small, asbestos-like, sandy crystals, little soluble in water and alcohol, even at the boiling heat. The concentrated solutions of its salts are precipitated by pure alkalies and their carbonates, but only slowly when diluted.]

**Palmitic Acid** $= C_{32} H_{31} O_3 + HO$. As to occurrence see Palmitin. Saponify palm oil with soda-ley, decompose the soap with sulphuric acid, and recrystallise repeatedly in alcohol, until the fusing point remains constant.—Small, white scales, inodorous and tasteless, readily friable; fuse at 62°; have an acid reaction; distil almost unaltered; are insoluble in water, dissolve in alcohol of 0·820 at 40° in every proportion, readily in ether. The palmitates of alkalies dissolve in water and in alcohol.

**Palmitin** (Tri-palmitin) $= C_{102} H_{98} O_{12}$. As palmitate of glyceryl $= C_6 H_5 O_3 + 3 C_{32} H_{31} O_3$ widely distributed among the fats of the vegetable kingdom, and especially abundantly in palmoil (from Elais Guineensis), in the tallow of Excaecaria sebifera, in Japanese wax (from Rhus succedanea in the wax of Myrica cerifera). Press the palm-oil strongly between calico, boil the remnant repeatedly with alcohol, in order to remove free palmitic and oleic acids, and recrystallise the remnant repeatedly in ether.—Small, pearly crystals, fusing at 61°, slightly soluble in boiling absolute alcohol, readily so in ether.

**Palm-Oil.** Obtained from the pericarp of Elais Guineensis.— Yellow, of butter-consistence, smells of violets, has a mild taste, fuses at 27°, when older, only at 32° to 36°; turns easily rancid, bleaches at the sunlight. Contains olein and palmitin.

**Panaquilon** $= C_{24} H_{25} O_{18}$. Peculiar ingredient of the root of the American Ginseng (Panax quinquefolius). Macerate with cold water, boil the solution, filter, evaporate to a syrup-consistence; mix with a concentrated solution of sulphate of soda, which yields a dense, brown deposit, which has to be washed with the same salt solution, and treated with absolute alcohol, the latter dissolving the Panaquilon. Evaporate the alcoholic solution, and dissolve the remnant in water; decolourise by means of animal charcoal; filter, evaporate and purify by redissolving in absolute alcohol.—Amorphous, yellow powder, has a taste similar to glycyrrhizin, but more bitter; fuses under decomposition, dissolves readily in water, and in alcohol, not in ether; becomes brown with alkalies; is precipitable by tannic acid, but not by other acids, or the chlorides of mercury and of platinum. Concentrated acids convert it under evolution of carbonic acid into a new body: crystalline, tasteless and insoluble in water.

**Papaverin** $= C_{40} H_{21} NO_8$. In opium. Precipitate the aqueous extract with soda-ley, treat with alcohol the deposit, consisting in the main of morphin, evaporate the brown tincture, dissolve the remnant in diluted acid and mix with ammonia, producing, at first, a brown resinous deposit. Dissolve the latter in diluted hydrochloric acid, and add acetate of potash, which separates a dark resin. Wash this with water, and boil with ether, the Papaverin crystallising on cooling from the ethereous solution.—Forms white needles of slightly alkaline reaction; is insoluble in water; with difficulty soluble in cold alcohol and ether, readily when warm; becomes blue or purple with concentrated sulphuric acid; dissolves in nitric acid unaltered; but is decomposed on warming the solution. Its salts are mostly sparingly soluble in water.

**Paramaleic Acid** $=$ FUMARIC ACID.

**Paramenispermin.** Composed like menispermin, and obtained together with it. Forms quadrangular rhombic prisms or concentrically radiated masses, fuses at $250°$ and sublimates undecomposed, is insoluble in water, soluble in absolute alcohol and in diluted acids, scarcely in ether; forms no salts with acids.

**Paramorphin** $=$ THEBAIN.

**Pararhodeoretin** $=$ JALAPIN.

**Parellin** or **Parellic Acid** $= C_{18} H_6 O_8 + HO$. Is sometimes obtained in the preparation of lecanoric acid from Lecanora Parella. The two acids are distinguished by their behaviour towards a solution of baryta. The lecanoric acid forms with the latter a soluble, the Parellic acid an insoluble salt. The Parellate of baryta is decomposed by hydrochloric acid, and the subsiding Parellin is recrystallised in alcohol.—Forms colourless crystals, which lose their water at $100°$, bitter when masticated or in alcoholic solution, the latter being of an acid reaction; fuses and becomes decomposed by heat; dissolves scantily in cold water, better in alcohol, and is thrown down by water as a jelly; is also soluble in ether.

**Paricin** $= C_{46} H_{26} N_2 O_6$. Alkaloïd of the Quina Carabaya from Cinchona succirubra. Exhaust with alcohol of $80°/_0$, distil off the alcohol from the solutions, evaporate to dryness, draw out with water and some hydrochloric acid, precipitate the solution with carbonate of soda, wash and dry the deposit, dissolve in ether, evaporate, treat with water mixed with some hydrochloric acid, decolourise the solution with animal charcoal, filter, throw down with carbonate of soda and dry.—Yellow, amorphous, resinous mass of very bitter taste, dissolves little in water, readily in alcohol and in ether, assumes a beautiful dark-green colour with nitric acid, likewise with concentrated sulphuric acid, but is destroyed afterwards. Yields amorphous salts.

**Paridin**$= C_{16} H_{14} O_7 + 2 HO$. Acrid substance of Paris quadrifolia. Exhaust the dried herb with water and acetic acid, prepare from the residue an alcoholic extract, remove from the latter the fat and chlorophyll by means of ether, dissolve the remaining extract in alcohol of 0·920, decolourise the solution with animal charcoal, filter warm, distil off the alcohol, dissolve the remnant in 16 to 20 parts hot water and allow the solution to stand cold for a few hours and the Paridin to form in crystals, which have to be purified by recrystallising.—Forms fine, shining laminæ or needles, at first tasteless, then poignantly acrid not bitter, loses its water at 100°; dissolves little in cold, a little more in hot water, much more readily in alcohol, little in ether; becomes red with concentrated sulphuric acid, and with phosphoric acid.

**Parietic Acid** ⎱
**Parietin** ⎰ $=$ Chrysophanic Acid.

**Parillin**$=$Smilacin.

**Parsley-Stearopten**$= C_{24} H_{14} O_8$. In the herb and seed of Carum Petroselinum besides a liquid volatile oil. Distil with water, collect the crystals forming in the distillate, press and recrystallise from alcohol. The stearopten is mostly contained in the oil that passes over towards the end of the distillation.—Forms fine, white, sexangular needles, heavier than water; smells faintly of parsley, has a burning, camphor-like, afterwards rancid taste; neutral; fuses at 30°, boils at 300° under decomposition; dissolves slightly in cold, more readily in hot water, readily in alcohol, ether, volatile and fixed oils.

**Paviin**$=$Fraxin.

[**Paytin** $= C_{21} H_{12} NO + HO$. Alkaloïd, discovered by Hesse in Quina blanca de Payta. The coarsely pulverised bark is extracted with alcohol; the solution is evaporated, saturated with an excess of soda, and shaken with ether, which takes up the alkaloïd. The latter is converted into the sulphate, almost completely neutralised with ammonia, and mixed with excess of iodide of potassium. The precipitate is rendered alkaline with solution of soda and shaken with ether, which on evaporation leaves the P. in well defined colourless crystals.—The P. is of alkaline reaction, and bitter taste, dissolves easily in ether, benzol, chloroform, petroleum ether, and alcohol, little in water; fuses at 156°; does not completely neutralise acids.]

**Pectin.** A matter, the chemical constitution of which is not known. It contains a larger amount of oxygen than the carbo-hydrates, and cannot therefore be regarded as one of the series. It occurs especially in fleshy fruits and roots, and is recognisable by its property of coagulating like a jelly, when the juice of such vegetable parts is mixed with alcohol. By pressing this jelly, re-dissolving in hot water, precipitating with alcohol, and drying the deposit of Pectin, it is obtained as an almost colourless, gummous, half-transparent, foliated, tasteless substance, which dissolves in water to a slimy and very thick liquid; it yields with nitric acid oxalic and mucic acids, with diluted sulphuric acid no sugar but other kinds of metamorphoses, as likewise by boiling with water. It contains always one or more per cent. of lime-compounds. Pectin, gum, and mucilage cannot be separated from each other, as they are all prepared and isolated in the same manner; as a distinguishing character of Pectin from either of the two serves its property of setting into a jelly, as mentioned above.

**Pelargonic Acid** $= C_{18} H_{17} O_3 + HO$. In the volatile oil of Pelargonium. Distil the herb with water, saturate the distillate with baryta, distil off the neutral oil, and decompose the remaining Pelargonate of baryta with sulphuric acid. The acid separates, and has, after washing, to be desiccated by means of chloride of calcium.—Colourless, oily liquid, smells faintly of butyric acid; congeals when cold, and fuses at 10°; boils at 260°, and distils unaltered; dissolves little in water, readily in alcohol and in ether. The Pelargonates of the alkalies and of the alkaline earths are soluble in water.

**Pelosin** $= C_{36} H_{21} NO_6$. Alkaloïd of the root of Cissampelos Pareira. Boil with water containing sulphuric acid, saturate the united decotions with carbonate of soda, collect the deposit, wash, dry, re-dissolve in acid water, treat with animal charcoal, filter, precipitate again with carbonate of soda, dry, draw out with absolute ether, and evaporate the ethereous solution.—Yellowish, amorphous mass, friable to a white powder, of a disagreeable sweetish-bitter taste, of alkaline reaction; dissolves not or little in water, most readily in alcohol, slowly but abundantly in ether; with ether containing water, hydrated Pelosin (with 3 eq. water: a nearly white amorphous powder) is formed, which does not dis-solve in ether. It fuses readily when warmed, and is decomposed by more heat; becomes brown when exposed to damp air, especially in the presence of alkalies. Its salts are not crystallisable, only the chloride of P. forms warty concrescences. [According to Flueckiger and Hanbury, Pelosin is identical with Bebirin.]

**Peppermint-Stearopten**$= C_{20} H_{20} O_2$. Forms in the oil of peppermint when kept at a low temperature. Colourless, translucent, shining prisms, of the smell and taste of the oil, fusible at 25° to 36°, boiling at 208° to 213°, readily soluble in alcohol. Yields with anhydrous phosphoric acid the hydrocarbon Menthen$= C_{20} H_{18}$.

**Pereirin.** Alkaloïd of the Pereira-bark, which comes from Picramnia ciliata. Draw out with acidulated water, precipitate the extract with ammonia, withdraw the P. from the deposit with ether, evaporate and purify by dissolving in acid water and precipitating with ammonia.—White-yellow, amorphous powder, very bitter, of alkaline reaction, fuses without loss of weight and is afterwards decomposed, dissolves very little in water, imparting to it a bitter taste, also in alcohol and in ether, in concentrated sulphuric acid with a beautiful violet colour passing into brown, on dilution with water successively into olive-green and grass-green; readily soluble in diluted acids, forming neutral, amorphous salts, mostly soluble in water and in alcohol.

**Peru-Balsam** or **Black Peru-Balsam.** Exudation, aided by artificial heat, of the stem of Myroxylon Pereiræ (Flueckiger and Hanbury), according to former authors, also of M. peruiferum and M. pubescens. It is similar in colour and consistence to dark molasses, smells vanilla-like but somewhat empyreumatic, tastes a little bitter, sharp and burning, has a density of 1·15-1·16 and an acid reaction, dissolves completely in strong alcohol, partially in ether and oils.—It consists of about 23°/₀ resin (agreeing with the resins of tolu-balsam and benzoin), a little styracin, 7°/₀ cinnamomic acid and 69°/₀ volatile oil. The latter is not obtainable by simple distillation, but the Balsam has to be treated first with soda-solution, as directed under Oil of Balsam of Peru.

**Peucedanin** $= C_{24} H_{12} O_6$. Crystalline substance of the root of Peucedanum officinale and P. Ostruthium. By concentrating the alcoholic extract of the root, the P. crystallises and has to be recrystallised in alcohol.—It forms colourless, rhombic prisms without smell and taste, while the alcoholic solution has a burning, aromatic and long-lasting acrid taste; it fuses at 75° and is decomposed with more heat, is insoluble in water, little soluble in cold, readily in hot alcohol, in ether, in volatile and in fixed oils, likewise unaltered in concentrated but not in diluted sulphuric acid; decomposes with alcoholic potash into angelic acid and Oreoselon $= C_{14} H_5 O_3$, isomeric with benzoic acid. The latter is a yellowish-white, crystalline, inodorous and tasteless substance, fusible at 190°, becomes carbonised by more heat, is insoluble in water, slightly soluble in alcohol and in ether, readily in concentrated sulphuric acid and in concentrated potash-ley, but in both not unaltered.

**Phaeoretin.** *See* APORETIN.

**Phaseolin.** Ingredient of French beans (seeds of Phaseolus vulgaris), which, like amygdalin of almonds, produces a volatile oil by decomposition. The beans which have no smell in the dry state, evolve, after being moistened with water, a disagreeable peculiar odour, caused by the formation of a volatile oil. By extracting the pulverised beans with absolute alcohol, they lose, like bitter almonds, the power of forming this oil with water. The Ph. has as yet only been obtained in the amorphous state, viz., by extracting with alcohol, treating the extract with ether (to remove the sugar) and evaporating the ethereous solution.

**Phillyrin**=$C_{54}$ $H_{34}$ $O_{22}$ + 3 HO. Bitter glucosid of the bark of Phillyrea angustifolia and P. latifolia. Boil the bark with water, clarify the decoction with albumen, precipitate with milk of lime, press the black-green precipitate (compound of lime with an acid resin and Ph.) after keeping for a long period, treat repeatedly with alcohol, shake the tinctures with animal charcoal and evaporate.—Forms silvery scales, inodorous, tasteless at first, then bitter; loses the water over sulphuric acid, or at 50° to 60°; fuses at 160°, and is decomposed by more heat; dissolves in 1300 parts cold, and copiously in boiling water, in 40 parts cold, and more readily in boiling alcohol; not in ether, oils, in warm acetic acid; is not altered by alkalies, separates, on boiling with diluted acids, into sugar and another crystalline product (phillygenin); dissolves in concentrated sulphuric acid with a violet-red colour.

**Pholobaphen**=$C_{20}$ $H_8$ $O_8$. Red pigment of the bark of many trees (species of Betula, Pinus, Cinchona, Platanus, &c.), also in the Polyporus annosus, and probably many of its congeners. Exhaust with ether first, draw out with alcohol afterwards, evaporate the alcoholic solution, boil with water and dry what has remained undissolved.—Red powder, inodorous, not fusible, insoluble in water and in diluted acids, readily soluble in alkalies with deep brown-red colour, almost insoluble in alcohol when previously dried at 100°.

**Phlorrhizin**=$C_{42}$ $H_{24}$ $O_{20}$ + 4 HO. Bitter glucosid of the bark of the root of fruit trees (apple, pear, cherry, plum tree), in less quantity in the bark of the stem and of the branches, also in the leaves of the apple tree, and, to all appearance, in the bark of the root of Ribes rubrum; not in the bark of the apricot, peach, almond, and walnut tree. Best adapted for its preparation is the root-bark of the apple tree. Draw out with weak alcohol, distil and re-crystallise what will have formed in the remaining liquid, with aid of animal charcoal.—Forms white, silky, often concentrically united needles of bitter, then sweetish taste, and of neutral reaction; loses the water at 100°, fuses at 109° to a colourless resin, becomes hard again at 130°, fuses again at 158° to 160°, and becomes decomposed with more heat; dissolves in 833 parts cold water, most readily at 50°, and in every proportion in boiling water, readily in alcohol, wood spirit, acetic acid, very little in ether, in warm concentrated sulphuric acid under decomposition with a red colour; breaks up, on heating with diluted acids, into grape-sugar and phlorrhetin ($C_{30} H_{14} O_{10}$); becomes with ammonia successively orange-red, purple and dark blue.

**Physalin** = $C_{28} H_{16} O_{10}$. The bitter ingredient of the leaves of Physalis Alkekengi. Exhaust with cold water, add to 1 liter of the liquid 20 grammes of chloroform, shake with two changes of the latter, evaporate the chloroform, dissolve the remnant in hot alcohol, shake with animal charcoal, throw down the filtrate with water, wash the deposit and dry.—Voluminous, white, or slightly yellowish powder, becomes electric by friction, not crystalline; of at first slight, afterwards lasting bitter taste; softens at 180°, becomes plastic at 190°, decomposes afterwards, dissolves very little in cold, a little more in boiling water, readily in alcohol and in chloroform, little in ether, very little in diluted acids, readily in liquor of ammonia.

**Physodin** = $C_{24} H_{12} O_{16}$. Crystalline resin of Parmelia physodes. Draw out with ether, evaporate, wash the remaining white powder with alcohol, and recrystallise in boiling anhydrous alcohol.—White, loose mass, consisting of microscopic needles (or larger crystals by spontaneous evaporation), neutral, fuses at 125°, behaves to water like a resin; is insoluble in alcohol of 80%, dissolves in boiling absolute alcohol, not in ether, acetic acid, readily in the hydrates and carbonates of alkalies.

**Physostigmin** = $C_{30} H_{21} N_3 O_4$. Alkaloïd of the Calabar bean (Physostigma venenosum). Mix the newly prepared alcoholic extract with an excess of bicarbonate of soda, shake the solution with an adequate quantity of ether, treat the ethereous solution with a very diluted acid, separate the acid solution completely from the ether, drive off the ether which has dissolved in the aqueous solution, filter through wet paper, add an excess of bicarbonate of soda, shake again with ether and evaporate the latter.—Forms colourless, rhombic laminæ*; fuses at 45°, becomes slightly decomposed when kept for a longer time at a temperature of 100°; dissolves readily in alcohol, ether, benzol, sulphide of carbon, chloroform, less readily in cold water; is of strongly alkaline reaction, and neutralises the acids completely. When divided under water, and

* Vée describes the Ph. as rhombic leaflets, fusible at 69°.

impregnated with carbonic acid gas, it dissolves soon, and yields a tasteless solution of alkaline reaction, but becomes slowly decomposed, and more quickly when warmed, while assuming a red colour. The salts of the Ph. are like the pure base, tasteless. Alkalies precipitate the Ph. from the solution of the salts, but decompose them at the same time with a red colour. They are also precipitable by chloride of mercury, chloride of gold, tannic acid, iodide of potassio-mercury.

**Phytomelin** = RUTIN.

**Pichurim tallow** = LAUROSTEARIN.

**Picrolichenin**$=C_{24} H_{20} O_{12}$. Crystalline bitter substance of Pertusaria communis. Evaporate the alcoholic tincture of the lichen to a syrup consistence, and allow to stand cold; the crystals which will have formed after a few weeks have to be purified by washing with a weak solution of carbonate of potash, and by repeatedly recrystallising in alcohol.—Forms colourless, shining, rhombic pyramids, inodorous, very bitter; fuses above the fusing point of sulphur, is decomposed by more heat; is insoluble in cold, slightly soluble in hot water, readily in alcohol, ether, sulphide of carbon, volatile oils, acetic acid, and in caustic alkalies.

**Picrotoxin**$=C_{20} H_{12} O_8$. Indifferent bitter substance of Cocculus indicus, the seeds of Anamirta paniculata. Pulverise the seeds, remove most of the fixed oil by pressing, extract the presscake with alcohol, bring the tinctures to dryness with pulverised charcoal, grind the mass, draw out with ether, drive off the ether from the tincture after addition of water, remove the supernatant congealed fat, allow the Picrotoxin to crystallise from the liquid, and recrystallise in alcohol.—White crusts and shining needles, inodorous, extremely bitter, fusible by heat, and decomposed in higher temperatures; dissolves in 150 parts cold, and in 25 parts boiling water, in 10 parts cold, and in equal parts boiling alcohol, in $2\frac{1}{2}$ parts ether, also in diluted alkalies; the latter solution turns yellow on heating, and leaves a brick-red residue.

**Pimaric Acid.** *See* ABIETIC ACID.

**Pine-sugar**$=$PINIT.

**Pinic Acid.** *See* ABIETIC ACID.

**Pinit** $= C_{12} H_{12} O_{10}$. Peculiar kind of sugar in the sap of Pinus Lambertiana. Dissolve the crude, hardened sap in water, decolourise the solution with animal charcoal, allow to evaporate spontaneously and recrystallise the crystals which have formed.— Forms colourless, hard, radiated warts, nearly as sweet as cane-sugar; fuses above 150°; is of 1·52 density, and of neutral reaction; is carbonised on heating, with an odour of burnt sugar; dissolves most readily in water, little in aqueous, scarcely in absolute alcohol, not in chloroform; yields with nitric acid nitro-compounds and a little oxalic acid; is not altered by diluted sulphuric acid, even on boiling; becomes black with sulphuric acid poured rapidly on it, but dissolves when brought into contact with it gradually and with a moderate heat; does not ferment with yeast, is not altered when boiled with potash-ley, ammonia, baryta, alkaline tartarate of copper, or chloride of iron.

**Pinocorretin** $= C_{24} H_{19} O_5$ (isomeric with Quinovic acid). In the bark of Pinus sylvestris. Boil the bark freed from the outer layers, with alcohol of 40%, remove by filtering the ceropic acid which forms in the decoction on cooling; evaporate most of the alcohol, mix the remnant with water and precipitate the turbid liquid with acetate of lead, throwing down pinocortannic acid and Pinocorretin, while cortici-pinotannic acid and sugar remain dissolved. The washed precipitate of lead, when treated with very dilute acetic acid, separates into pinocortannic acid, which dissolves (and may be obtained from the solution by precipitating with sub-acetate of lead, decomposing the deposit with sulphuret of hydrogen, and evaporating the filtrate under exclusion of the air) into a glutinous, dark residue, which has to be dissolved in strong alcohol, and is to be treated with sulphuret of hydrogen and filtered off from the sulphide of lead. The filtrate, on evaporating, leaves Pinocorretin, to be purified by dissolving in alcohol, filtering and evaporating the liquid. The solution, containing cortici-pinotannic acid and sugar, is freed from the former by precipitating with sub-acetate of lead, which forms cortici-pinotannate of lead.—The Pinocorretin is a black-brown, glutinous mass, almost completely soluble in ammonia.

**Pinocortannic Acid** $= C_{32} H_{19} O_{23}$. In the bark of Pinus sylvestris. As to preparation, see Pinocorretin. — Red-brown powder, after drying very scantily soluble in water, turning green with chloride of iron.

**Pinopicrin** $= C_{44} H_{36} O_{22}$. In the leaves and bark of Pinus sylvestris and allied species, also in the green parts of Thuja occidentalis and congeneric plants. Draw out with alcohol of 40%, distil off the alcohol, mix the remnant with water, pour off from the subsiding resinous mass, mix the liquid with a few drops of a solution of acetate of lead, filter, precipitate from the filtrate the oxy-pinotannic acid with acetate of lead, and after removing the deposit throw down the pinotannic acid with sub-acetate of lead boiling hot, filter, treat the filtrate with sulphuret of hydrogen, evaporate the liquid, freed from the sulphide of lead, in a current of carbonic acid gas to honey consistence, and shake with anhydrous ether-alcohol, which dissolves the Pinopicrin and leaves the sugar. Precipitate from the solution a small quantity of foreign matter by sub-acetate of lead, treat the filtrate with sulphuret of hydrogen, remove the sulphide of lead and evaporate.—Vivid-yellow powder, hygroscopic, of very bitter taste; softens at 55°, completely liquified at 100°, is decomposed by more heat; dissolves readily in water, also in alcohol, ether-alcohol, and in aqueous not in pure ether; the aqueous solution evolves immediately on heating the odour of ericinol ($C_{20}$ $H_{16}$ $O_2$ ), while breaking up completely into this substance and into sugar.

**Pinotannic Acid**$=C_{14}$ $H_8$ $O_8$. Contained towards midwinter in the leaves of old trees of Pinus sylvestris and other true pines, also in the green parts of Thuja occidentalis and allied trees. Is obtained in the preparation of pinopicrin [see this] as Pinotannate of lead. Heat the liquid, from which the oxypinotannic acid has been precipitated by acetate of lead, to the boiling point, precipi-tate carefully with sub-acetate of lead, and let cool. Decompose the washed precipitate with sulphuret of hydrogen, warm the liquid together with the sulphide of lead, filter, and bring to dry-ness in an atmosphere of carbonic acid gas.—Yellow-red; if pre-pared from thuja, brownish-yellow powder; of slightly bitter and acerb taste, becomes soft and glutinous at 100°, dissolves readily in water, alcohol, and ether; does not precipitate glue, precipitates chloride of iron with brown-red colour. By heating with diluted acids, a red product is obtained.

**Piperin**$=C_{34}$ $H_{19}$ $NO_6$. Alkaloïd of the fruit of Piper nigrum and P. longum. Exhaust the powder with cold water, digest the remnant with alcohol of 80%, evaporate the tincture to honey consistence, wash with cold water, dissolve the residue in alcohol, add a little hydrate of lime, digest for one day, filter and allow to crystallise.—Colourless (if not quite pure yellow), glassy, flat, klinorhombic prisms, almost tasteless (when yellow of pepper-like taste), neutral, fusible at 110°, decomposed by more heat; insoluble in cold, little soluble in hot water, in 30 parts cold and in 1 part boiling alcohol, in 100 parts ether, readily in acetic acid.

**Pipitzahoic Acid**$=C_{30}$ $H_{20}$ $O_6$. Discovered by Rio de la Loza in the Raiz (root) del Pipitzahuac, which comes from Perezia Humboldtiana. Draw out with alcohol, evaporate, and purify by recrystallising.—Tufts of gold-yellow, foliated crystals; fusible at 100°, and sublimable in gold-yellow leaflets; dissolves scarcely in water, readily in alcohol and in ether; alkalies produce a purple colour in the solutions, and yield amorphous compounds, which dissolve readily in water, alcohol, and in ether.

**[Pittosporin.** Glucosid of the bark and fruits of Pittosporum undulatum. The pulverised bark is extracted with hot alcohol, filtered when cold, mixed with an equal bulk of ether, filtered again, and evaporated.—Whitish, loose powder of at first sweetish, afterwards bitter and acrid taste; dissolves in water and alcohol, not in ether; froths with water, gives precipitates with acetate and sub-acetate of lead. Separates, by boiling with diluted acids, into sugar and a white substance, insoluble in water.—Baron F. von Mueller and L. Rummel.]

**Pityxylonic Acid** $= C_{25} H_{20} O_8$. In the stem of Pinus sylvestris, probably also in Pinus Abies and other species. Boil the finely rasped wood with water, strain, evaporate the liquid mixed with carbonate of baryta to a small bulk, filter, evaporate again and treat the remnant with ether. Digest the exhausted mass with alcohol, which dissolves the P. acid, and leaves it after evaporation.—Brown-yellow, amorphous, hygroscopic mass, very bitter, of acid reaction; dissolves slowly in cold, readily in boiling water with pale-yellow colour, readily in alkaline water.

**Plumbagin.** Acrid, crystalline ingredient of Plumbago Europæa. Draw out the bark of the root with ether, evaporate the liquid, boil the extract with water and purify the Plumbagin, which will have subsided on cooling from the decoction, by recrystallising in ether or in ether-alcohol.—Crystallises in delicate, orange-yellow, tuftily-united needles, tastes at first sweetish, irritating, afterwards burning and acrid, fuses readily, volatilizes partly unaltered, is of neutral reaction, dissolves sparingly in cold, more in hot water, readily in alcohol and in ether. Sulphuric and nitric acids dissolve it with yellow colour, ammonia with red colour, and acids restore the yellow colour. The aqueous solution becomes likewise coloured with sub-acetate of lead, under formation of a carmine-red precipitate.

**Pollenin.** The main ingredient of the pollen remains, after the pollen has been treated successively with water, alcohol, ether, diluted acids and alkalies.—Subtle, light, tasteless powder, putrifies in the moist state under evolution of ammonia. Is most likely not quite pure in this state, and perhaps in the main cellulose.

**Polychroit** = Crocin.

**Polygalin** = Saponin.

[**Polygonic acid.** Prepared by Rademaker from the herb of Polygonum hydropiper, by exhausting with dilute alcohol, evaporating to one-third, filtering and precipitating the filtrate by acetate of lead. The precipitate is well-washed, decomposed by sulphuret of hydrogen and the filtrate treated with ether, which dissolves the P. acid, and leaves it behind on evaporation.—Green, deliquescent, crystals of an acrid, bitter taste, and strongly acid reaction, soluble in alcohol, ether and chloroform, less so in aqueous alcohol; neutralise the bases completely, forming well-defined salts.]

**Populin**$=C_{40} H_{22} O_{16} + {}_4HO$.  Besides salicin the other glucosid of the bark, the leaves and the roots of species of Populus. Boil with water, precitate with subacetate of lead, free the filtrate from lead by sulphuric acid, concentrate, boil with animal charcoal and allow the salicin to crystallise.  The mother-ley yields with carbonate of potash a deposit of Populin, which has to be recrystallised in hot water.—Forms white, silky, shining, very voluminous needles, of the appearance of starch or of magnesia, tastes irritating, sweet, similar to liquorice, looses at 100° the whole of its water, fuses at 180°, and decomposes with more heat, yielding benzoic acid.  Dissolves in 2000 parts cold, and in 70 parts boiling water, in 100 parts cold absolute alcohol, in boiling alcohol more readily than in boiling water, scarcely in ether, behaves towards concentrated sulphuric acid like salicin, forms, on boiling with dilute sulphuric acid, benzoic acid, grape-sugar and saliretin, on heating with chromate of potash and sulphuric acid much salicylous acid.  Is not precipitable by any metallic salt.

**Porphyroxin.**  In opium.  Exhaust with hot ether, warm the remnant with water and a little carbonate of potash, and treat again with hot ether, which dissolves codein, thebain, porphyroxin, and caoutchouc; all these substances remaining after the spontaneous evaporation of the ether.  Dissolve the remnant in diluted hydrochloric acid, filter and precipitate with ammonia, which throws down thebain and Porphyroxin, while codein remains dissolved.  Dissolve the deposit, after drying and triturating, in boiling ether, and leave to evaporate at the air, obtaining thereby crystals of thebain and resinous Porphyroxin, separable by alcohol, which dissolves the P. readily.—Fine, shining needles, neutral, insoluble in water, readily soluble in alcohol and in ether; assumes with concentrated sulphuric acid or with nitro-sulphuric acid, an olive-green colour; dissolves colourless in diluted sulphuric, hydrochloric, and nitric acids, the solutions turning purple-red on boiling, but not the solution in acetic acid.  Alkalies decolourise the red liquids, and produce a white precipitate, but all acids (also acetic acid) reproduce the red colour even when cold.

**Porphyroxin.**  This particular alkaloid, according to what is known, is not identical with Porphyroxin of opium.  It occurs in the root of Sanguinaria Canadensis, accompanied by chelerythrin and puccin.  Draw out with water mixed with some acetic acid; precipitate the chelerythrin from the liquid by means of ammonia, neutralise the filtrate with acetic acid exactly, precipitate with tannic acid, wash the precipitate, mix intimately with lime, dry, draw out with alcohol, impregnate with carbonic acid, distil off the alcohol, bring the remnant to dryness, treat with boiling water, evaporate the solution, dissolve the remnant in ether, evaporate the ether and recrystallise in alcohol.—Small, white

N

tabular crystals, inodorous and tasteless, very little soluble in water, better in alcohol and in ether; form with acids colourless, neutral, crystallisable, bitter salts.

**Primrose Stearopten.** Passes over with the water in the distillation of the root of Primula Auricula and subsides in the turbid distillate. Has a strong and peculiar, pleasant odour; its alcoholic solution imparts a deep-red colour to solutions of iron.

**Primulin.** Indifferent, crystalline substance of the root of Primula veris. Treat the aqueous and well-dried extract of the root repeatedly with alcohol of $90°/_o$, evaporate the spirituous liquids slowly, press the separated crystalline mass between blotting paper, redissolve in alcohol, digest with carbonate of lead, filter and let crystallise.—Forms colourless needles or lustreless grains, inodorous and tasteless, neutral, readily soluble in water, also in alcohol (but more in aqueous than in anhydrous), not in ether, fuses by heat and decomposes in higher temperatures. Is not precipitable by metallic salts.

[Dr. L. Mutschler states, that according to his researches Primulin is identical with Cyclamin, and appears to be widely distributed among the order of Primulaceæ. He also believes that Cyclamin and Primulin may probably be identical with Saponin.]

**Prophetin.** *See* Ecbalin.

**Propionic Acid** $= C_6 H_5 O_3 + HO$. Has been found as yet only occasionally in the aqueous distillates of a few vegetable parts (Flores Millefolii, &c.), but is probably more widely distributed. Its presence in such a distillate is recognisable to some extent by its odour, resembling butyric and pyro-acetic acids. By saturating this distillate with carbonate of soda, drying and mixing the salts with sulphuric acid, the said odour becomes very striking, and on heating the salt by itself, the odour of alkarsin is evolved. The acid, after being isolated from the concentrated solution of the Propionate of soda by means of sulphuric acid, floats upon the surface in the form of an oily liquid and disappears only on addition of more water. The Propionates are unctuous to the touch, and are all soluble in water. The Propionate of soda, dried at 100°, is anhydrous and contains 64·87 % acid.

**Protein Substances.** In the vegetable and more so in the animal kingdom there exist, either dissolved or as solids, a number of amorphous, not volatile, inodorous and tasteless, nitrogenised, indifferent substances, which exhibit a great analogy in their composition and in their general properties. Being originally formed in the vegetable organism, they from thence are introduced by the food into the animal body, and are found there again with little or no alteration. Amongst the Protein-substances are classed albumin, fibrin, casein, legumin, &c., and which bear the same relation to each other as cellulose, starch, gum, sugar, &c. The bodies belonging to each of these two series, have nearly the same chemical composition and are convertible into each other, while the one series comprises the nitrogenised compounds which preponderate in the animal kingdom, the other the carbohydrates predominating in the vegetable life. These latter may be considered to contain as their radical the cellulose $= C_{12} H_{10} O_{10}$; starch having the same composition, while gum and sugar contain in addition one equivalent of water.

Mulder tried to show that those nitrogenised substances contained also a common radical, which is combined in the various Protein substances with varying quantities of sulphur and partly also of phosphor. He named this radical Protein, and expressed its composition by the formula $C_{36} H_{25} N_4 O_{10}$, which demands in 100: 57·29C, 6·64H, 14·85N, and 21·22O. In the free state, and combined with two equivalents water, Mulder's Protein is obtained by dissolving any of the abovenamed matters in potash-ley, boiling the solution until, by addition of an acid, sulphuret of hydrogen is evolved, neutralising carefully with acetic acid, and washing the gelatinous deposit with water. When freshly precipitated it is a transparent, grey, flocky mass, which becomes hard and brittle after drying, inodorous and tasteless; fuses by heat, yields ammoniacal and other products, and leaves a slowly, but completely combustible coal; it sinks in water, swells up and resumes a gelatinous appearance; is insoluble in water, alcohol, ether, oils. By continued boiling with water it is partially dissolved, but at the same time altered in its properties. Acetic, tartaric, citric, malic and phosphoric acids dissolve it readily, also diluted mineral acids; concentrated acids throw down from the solution a combination of Protein and acid, insoluble in the liquid. From the acid solutions the Protein is precipitated by ferrocyanide and ferricyanide of potassium, tannic acid, most of the metallic salts and by neutralisation with an alkali. Diluted sulphuric acid colours Protein purple-red on boiling. Concentrated hydrochloric acid dissolves it with indigo-blue colour, the solution turning black on boiling. Concentrated nitric acid colours it yellow. Nitrate of mercury, containing nitrous acid, produces on warming a red tinge. With sugar and sulphuric acid it assumes a beautiful purple violet colour.

According to recent investigations the above Protein retains more than 1 per cent. of sulphur, from which it cannot be freed. It is on this account that the Protein-theory has been abandoned by many chemists, though, to my belief, without a just cause, as the proportions of the four elements C, H, N and O to each other are not altered by the presence or absence of sulphur. Besides

Protein is in the same condition as many other organic radicals, which though not isolated as yet, are nevertheless acknowledged or assumed to exist.

The Protein substances behave in general like Protein (contaminated with sulphur); in their composition they only differ from it sometimes by a larger amount of sulphur, sometimes by an additional amount of phosphor; but in all instances they contain certain inorganic salts, notably phosphate of lime, and therefore leave, when incinerated, an ash, the quantity of which often amounts to 10%.

**Prussic Acid**=HYDROCYANIC ACID.

**Pseudocurarin.** As to distribution see Oleandrin. Boil the solution of tannate of Ps. obtained in the preparation of oleandrin, with oxyd of lead, evaporate the filtrate almost to dryness, remove from it the oleandrin by ether, dissolve the remnant in alcohol and evaporate the filtrate. The residue is Pseudocurarin. —A yellowish gum-like amorphous mass without smell or taste; most readily soluble in water and in alcohol, not in ether; forms with acids salts which do not crystallise; is not precipitable by the chlorides of platinum or of mercury. Acts not poisonous.

**Pseudomorphin**=$C_{34} H_{19} NO_8 + 2 HO$. In Opium. Is best obtained by connecting its preparation with that of morphin after the well known method of Robertson-Gregory (see Codein). By adding to the purified mixture of the chlorides of morphin—codein, etc., in alcoholic solution, a small excess of ammonia, the Ps. remains dissolved, while only the morphin is thrown down. Saturate the solution, freed from the morphin, with a slight excess of hydrochloric acid, distil off the alcohol and strain the remaining solution through a coal filter. The solution, which is now completely clear, but mostly coloured, yields, on neutralising with diluted ammonia, a voluminous deposit, consisting chiefly of Ps., which has to be collected, washed and dissolved in acetic acid. Add to the filtered solution carefully as much diluted ammonia as enables the liquid, after the precipitation, to redden very slightly blue litmus-paper. By this operation the Ps. subsides, which by combining with hydrochloric acid, yields a well-crystallising salt, which is purified by re-crystallisation in water. Finally dissolve the purified salt in much hot water and decompose by ammonia.— Fine, crystalline deposit, suspended in a liquid of a vivid, silky lustre, insoluble in water, alcohol, ether, chloroform, sulphide of carbon, diluted sulphuric acid, and in solutions of carbonate of soda, readily soluble in potash and soda-ley, and also a little in milk of lime, slightly in liquor of ammonia, readily in alcoholic ammonia; is of neutral reaction, does not saturate the acids, is tasteless in its combinations; dissolves in concentrated sulphuric acid, with olive-green colour, in concentrated nitric acid with orange-red colour, passing soon into yellow as with morphin; becomes blue with chloride of iron.

**Pseudoveratrin**=HELONIN.

**Pteritannic Acid**=$C_{24} H_{15} O_8$. In the rhizoma of Aspidium Filix mas. The ethereous solution, obtained in the preparation of tannaspidic acid, leaves on evaporating a black-brown residue, which has to be digested with petroleum as long as the latter assumes a brown colour. Collect the undissolved powder, press, triturate, and boil with water, dissolve the remaining resinous mass in ether and evaporate.—Black-brown, amorphous, shining mass, friable to a drab-coloured electric powder, tasteless, of a faint smell, of acidulous reaction, fuses with a gentle heat, is insoluble in water, dissolves in strong, less in diluted alcohol, readily in ether, not in volatile and in fixed oils; precipitates glue.

**Puccin.** In the root of Sanguinaria Canadensis, associated with chelerythrin and porphyroxin. Draw out with water and sulphuric acid, precipitate the solution with ammonia, wash the deposit with water, dry, draw out with ether, digest the solution with animal charcoal, filter and add sulphuric acid, which produces a deposit of sulphate of chelerythrin of a pale-cinnabar colour, insoluble in ether, like all other salts of this alkaloïd. The ethereous solution leaves, after filtering and evaporating, a dark-red, amorphous residue, which has to be redissolved in ether and mixed with diluted sulphuric acid, in order to remove the rest of the chelerythrin. After filtering and evaporating to dryness, the dark-red mass is treated with alcohol and the Puccin is thrown down from the solution with water.—Appears, after drying, as a red, tasteless powder, insoluble in cold water, fusing to a resin in boiling water. The alcoholic solution becomes pale-yellow with animal charcoal, and leaves a pale-red residue, which turns to a deep-red with hydrochloric acid and forms pinkish needles.

**Pulsatilla camphor**=ANEMONIN.

**Purpurin**=$C_{18} H_6 O_6$. In the madder (from Rubia tinctorum). Allow the pulverised root to ferment with yeast and water, wash with water, and boil with a solution of alum. Let cool and add sulphuric acid, which yields reddish flocks of purpurin, which have to be freed from alumina by boiling with diluted hydrochloric acid, and are recrystallised in alcohol or in ether.— Forms red needles, anhydrous (crystallising from weak alcohol as orange-yellow needles with 1 equiv. of water); fuses by heat and sublimates at 225°, mostly leaving a little coal; dissolves more readily in water than alizarin with a reddish, in diluted acids with yellow, readily in alkalies with crimson colour, also readily in alcohol and of a deeper red than alizarin, most readily in ether;

unaltered in concentrated sulphuric acid and reprecipitable by water; yields a purple deposit with acetate of lead.

[**Pyrocatechin** has been found by E. v. Gorup-Besanez in the green leaves of Vitis quinquefolia. Wiesner also recognised it as ingredient of the Eucalyptus kino, and F. A. Flückiger likewise as constituent of the kinoes of Pterocarpus Marsupium and Butea frondosa.]

**Pyrrhopin** = CHELERYTHRIN.

**Quassiin** = $C_{20} H_{12} O_6$. In the wood and bark of Quassia amara, Picraena excelsa and Simaruba amara. Treat the alcoholic extract with water, bring the solution to honey-consistence, treat repeatedly with small quantities of absolute alcohol, evaporate the solutions to dryness, draw out with hot water, decolourise the pale-yellow solution with animal charcoal, and evaporate.—Fine, white, silky, shining needles, permanent at the air, inodorous, neutral, very bitter, fuse a little less readily than colophony, decompose in higher temperatures, dissolve in 222 parts cold, more readily in hot water, readily in alcohol, very little in ether; the alcoholic solution is precipitable by tannic acid in dense, white flocks.

**Quercetin** = $C_{46} H_{16} O_{20}$. In the grains of Avignon (the fruits of Rhamnus græcus, R. prunifolius, R. infectorius, R. saxatilis, R. amygdalinus, R. oleoides), readily formed; probably in many other plants, too, as in the yellow berries of Hippophaë rhamnoides, according to Filhol, also in the green leaves, and in the flowers. Gellatly's Rhamnetin is alleged to be Quercetin. Draw out the berries with ether containing alcohol, evaporate the gold-coloured solution, mix the remnant with water, dissolve in alcohol what has been separated and evaporate the solution under addition of water.—Very fine, small, bright-yellow needles or citron-yellow powder, tasteless, of a slightly salty and somewhat styptic taste (according to other statements, very bitter like quinin), when dissolved in water, fusible above 250° and sublimable partly undecomposed; scarcely soluble in cold water, little in boiling water, readily even in weak alcohol, much less so in ether, readily in alkaline liquids with gold-yellow colour; turning dark-green with chloride of iron, and dark-red when warmed.

**Quercin.** Bitter ingredient of the bark of Quercus Robur. Draw out with milk of lime, precipitate the filtrate with carbonate of potash, evaporate the filtrate to honey-consistence, treat with alcohol, evaporate the tincture and recrystallise.—Small, white, inodorous, bitter crystals, readily soluble in water and in aqueous alcohol, not in absolute alcohol and in ether, has a neutral reaction, turns orange-yellow with concentrated sulphuric acid, dissolves also in lime-water.

**Quercit** $= C_{12} H_{12} O_{10}$. Peculiar kind of sugar of the fruits of Quercus racemosa and sessiliflora. Throw down, hot, with lime the tannic acid from an aqueous extract of acorns, filter, destroy any fermentable sugar by fermentation with yeast, evaporate to a syrup-consistence, wash the crystals which will form with cold alcohol, and recrystallise in water or in weak alcohol.—Forms hard, permanent, klinorhombic crystals of sweet taste, unaltered at 215°, fusing at 235°, partly sublimating, soluble in 8 to 10 parts cold water, also in hot weak alcohol, does not ferment with yeast, yields with nitric acid, on heating, oxalic but no mucic acid, dissolves in concentrated sulphuric acid colourless, is not altered on boiling with alkalies, acetate of copper, or with alkaline tartarate of copper.

**Quercitrin** $= C_{70} H_{36} O_{40}$. Yellow glucosid of the bark of Quercus tinctoria. Formerly confounded with rutin. Boil the bark with water, leave the decoction to stand cold, collect the Quercitrin which has formed, triturate it with a little alcohol of 35° B. to a pulpy state, heat over the water-bath, collect on calico, remove impurities by pressing, dissolve the remnant in a larger quantity of boiling alcohol, filter hot, mix with boiling water until it becomes turbid, and allow to stand cold. Collect the crystals of Quercitrin, and purify by again submitting them to the same treatment.—Forms sulphur or chrome-yellow, microscopic, rhombic, tabular crystals, inodorous and tasteless, slightly bitter when dissolved; fuses after desiccation at 168°; yields in higher temperatures crystals of quercetin under decomposition; dissolves in 2485 parts cold and in 143 parts boiling water, the straw-yellow solution becoming colourless by acids; dissolves in 23 parts cold and in 4 parts boiling alcohol, little in ether, most readily in diluted alkalies, these solutions turning dark at the air; breaks up on boiling with diluted acids into sugar and quercetin.

**Quina-Red** $= C_{12} H_{7} O_{7}$. In the bark of the genus Cinchona, produced by the oxydation of tannic acid. Draw out with diluted liquor of ammonia, precipitate the red-brown solution with hydrochloric acid, wash and heat the deposit (Quinovin and Quina-red) with thin milk of lime, dissolving the quinovin and leaving Quina-red lime undissolved, wash the latter with hot water; decompose with diluted hydrochloric acid, wash the deposit of Quina-red, redissolve in ammonia, precipitate with hydrochloric . acid, wash the precipitate, dissolve in alcohol, filter and evaporate to dryness.—Red-brown, inodorous and tasteless, not fusible, insoluble in water and in diluted acids, readily soluble in alcohol, ether, alkalies with dark-red colour; the ammoniacal solution after a rather long time throws down the glue on addition of water.

**Quinic Acid** $= C_{14} H_{11} O_{11} + HO$. In the genuine quina-barks (from Cinchona), in the Quina Maracaibo, Quina nova

Surinamensis (from Cascarilla magnifolia), in the seeds and leaves of Coffea Arabica, in species of Galium (G. Mollugo, &c.); in species of Vaccinium (V. Myrtillus, &c.), and probably in the following plants, the leaves of which yield kinon (quinon) on heating with sulphuric acid and peroxyd of manganese, viz.:—Cyclopia latifolia, and others; Fraxinus excelsior, and others; Hedera Helix, Ilex Aquifolium, Ilex Paraguayensis, Ligustrum vulgare, and other species; Quercus Ilex, Quercus Robur, and other oaks; Ulmus campestris, and other species. It appears hereby that Quinic acid is not confined to the genus of Cinchona, as formerly stated, but that it belongs largely to the family of Rubiaceae, and to many other orders of plants widely distant in natural affinities.

Preparation from the quina bark: Evaporate the liquid obtained in the preparation of quinin, by precipitating the sulphuric acid extract with milk of lime to the consistence of a syrup, decant from the lime-sulphate, evaporate over the water-bath to honey consistence, boil with alcohol several times and dissolve the remnant in little water. The solution yields, after a few days, a crystalline mass, which has to be strongly pressed, and is purified by recrystallisation. From the bilberry herb: Boil the green herb, gathered in spring, with water and lime, evaporate the decoction, and throw down the Quinate of lime with alcohol. Dissolve the glutinous precipitate in water mixed with some acetic acid, free from dyeing matters by means of acetate of lead and evaporate the filtrate, after removing the lead, to the density of a syrup, when Quinate of lime will form in crystals after a few days. The Quinate of lime, obtained by any of these methods, is purified by repeatedly recrystallising or by precipitating with alcohol of 36° B., and by dissolving in alcohol of 18° B. In order to isolate the acid, the Quinate of lime is dissolved in water and decomposed by oxalic acid. The filtered liquid is freed from the excess of oxalic acid by means of acetate of lead, the excess of the latter by sulphuret of hydrogen, and the filtered liquid is then allowed to crystallise.

Forms large, hard, tabular, klino-rhombic crystals with a characteristic hemitropism on the right side of the horizontal axis; has a pure and strong acid taste; loses at 100° nothing of its weight, fuses at 161°, and decomposes by more heat; dissolves in $2\frac{1}{2}$ parts cold and in less hot water, more readily in aqueous than in strong alcohol, scarcely in ether. By heating with superoxyd of manganese and sulphuric acid an orange coloured, needle-shaped sublimate of Quinon ($C_{12} H_4 O_4$) is obtained. The Quinates are mostly crystallisable and of neutral reaction; with the exception of the basic Quinate of lead, soluble in water, not in strong alcohol; yield by the destructive distillation tannic acid and Quinon.

[**Quinamin.** Alkaloid found by Hesse in the bark of Cinchona succirubra.—Very delicate, long, asbestos-like, white prisms, of alkaline reaction, readily soluble in ether, alcohol and petroleum-ether, little soluble in dilute alcohol, insoluble in water, potash-ley and ammonia. The sulphate and chloride are easily soluble. Chloride of platinum forms only in concentrated solutions a yellow precipitate. Chloride of gold becomes reduced to the metal. Concentrated sulphuric acid dissolves the Quinamin colourless, yellow to brown on heating; with concentrated nitric acid it becomes first yellow, then orange, and at last colourless. The Quinamin fuses at 172°, and on cooling, presents a radiated, crystalline mass; in higher temperatures it becomes brown and amorphous. It has a bitter taste.]

**Quinidin** $= C_{18} H_{11} NO$. In species of Cinchona. The preparation is similar to that of cinchonidin.—Appears in colourless, hard, klinorhomboidal prisms of glass-lustre and of moderately bitter taste; fuses at 175° without loss of weight and is decomposed by more heat, dissolves at 17° in 2580 pts., at 100° in 1858 pts. water, in 143 pts. ether, and in 12 pts. alcohol of 0·835°. The solution in chlorine-water is not altered by ammonia. Most of its salts dissolve more readily than those of quinin.

**Quinin** $= C_{20} H_{12} NO_2 + 3 HO$. In all true quina-barks of the genus Cinchona, always accompanied by larger or smaller quantities of cinchonin, in some barks also by quinidin, cinchonidin, and by other bases. Draw out with water acidulated with hydrochloric acid, saturate the liquid with an excess of hydrate of lime, collect the deposit, wash, dry, treat with ether, evaporate the solution, dissolve the remnant in the least possible quantity of water and sulphuric acid, prepare the pure sulphate by evaporating and decompose the sulphate by soda-ley. — Loose, white, easily friable mass or silky tufts of needles (crystallised from alcohol), loses only a little hygroscopic water at 100° to 150°, fuses at 196° without loss of weight, and decomposes in a higher temperature; has a very bitter taste, dissolves in 364 parts cold water, in 6 parts cold and in 2 parts boiling alcohol of 0·820, in 21 parts ether, in 2-6 parts chloroform, in benzol, in 200 parts glycerin, in 62 parts fixed oils; in chlorine water colourless, the solution, when oversaturated with ammonia, assuming a grass-green colour and yielding a precipitate of the same colour; in concentrated sulphuric acid colourless, the solution turning yellow-brown on heating. The solutions of its salts become brown in the direct sunlight, and are precipitable by the hydrates and by the carbonates of alkalies.

**Quinotannic Acid** $= C_{14} H_8 O_9$. In the barks of the genus Cinchona. Boil with water, mix the decoction with a little burnt magnesia, precipitating thereby quina-red; throw down the filtrate with acetate of lead, decompose the deposit under water with sulphuret of hydrogen; precipitate the liquid after filtering off from the sulphide of lead (quinovin and a little quina-red), with sub-acetate of lead; dissolve the deposit, after filtering, in diluted acetic acid (quina-red remaining undissolved), and precipitate the acid filtrate with ammonia. Wash the light-yellow deposit, decompose by sulphuret of hydrogen, filter off from the sulphide of lead, and precipitate the filtrate again with alcoholic solution of acetate of lead. After the deposit, consisting of Quinotannate of lead, has been decomposed under water with sulphuret of hydrogen, and after the liquid, freed from the sulphide of lead by filtering, has been evaporated in a vacuum over sulphuric acid and of a mixture of sub-sulphate of iron and lime, the Quinotannic acid remains, though already a little altered.—It is light-yellow, friable, very electric and hygroscopic; of an acidulous and very acerb, not bitter taste; dissolves readily in water, alcohol, and ether; yields on heating no pyrocatechnic acid; is precipitable by glue, becomes green with salts of oxyd of iron.

**Quinova-Red**$=C_{12} H_6 O_5$. In Quina-nova bark (from Cinchona species). Precipitate the decoction of the bark with acetate of lead, decompose the precipitate, consisting almost entirely of the Red, under water by means of sulphuret of hydrogen, wash the resulting mixture of Quinova-red and sulphide of lead with water, and boil with alcohol, and precipitate the Quinova-red from the filtrate by means of much water.—Almost black, lustrous, resinous substance, friable to a dark-red powder; dissolves sparingly in water, readily in alcohol, ether, and alkalies; is precipitable from the alcoholic solution by an alcoholic solution of acetate of lead, not by tartarated antimony.

**Quinova-Tannic Acid**$=C_{14} H_9 O_8$. In the bark Quina nova or Surinamensis (from Cascarilla magnifolia). Precipitate the decoction with acetate of lead, remove the deposit, containing quinova-red, divide the filtered liquid into three parts, precipitate one of them completely by sub-acetate of lead, and mix with the remaining two parts. The deposit, which contains quinovin, the rest of the quinova-red, and a little Quinova-tannic acid (but which cannot be used with advantage for the preparation of the latter) is also removed, and the liquid precipitated by sub-acetate of lead. Decompose the washed precipitate under water by sulphuret of hydrogen, remove the sulphide of lead, drive away the sulphuret of hydrogen by heating, add acetate of lead, and mix the filtrate with a great quantity of strong alcohol, whereby flocks of Quinova-tannate of lead are obtained. From these isolate the acid by sulphuret of hydrogen, and evaporate in a current of carbonic acid gas.—Amber-yellow, translucid, friable substance, of acerb and slightly bitter taste; dissolves in water and alcohol, not in ether, does not precipitate glue, colours chloride of iron dark-green; is not precipitated by tartarated antimony and acetate of lead, but is so by sub-acetate of lead, and by a solution of acetate of lead in alcohol. The aqueous solution throws down a red powder (quinova-red) when allowed to stand at the air.

**Quinovic Acid** $= C_{24} H_{19} O_5$. Found in the leaves of Pinus sylvestris, and in the green parts of Thuja occidentalis and many other coniferæ. Boil with alcohol of 40%, remove from the extracts the alcohol by evaporating, mix the remnant with water, dissolve the green resinous deposit in alcohol of 40%, precipitate the solution with alcoholic acetate of lead, throw down the excess of lead in the filtered liquid with sulphuret of hydrogen, filter and distil the alcohol; dissolve the remaining resin in very diluted potash-ley; throw down the resins by chloride of calcium, saturate the filtered liquid with hydrochloric acid, which precipitates pale-yellow flocks of Quinovic acid. These have to be re-dissolved in very diluted potash-ley, purified by treating with animal charcoal and precipitated with hydrochloric acid. — White or slightly yellowish, brittle mass, friable to a highly electric powder.

**Racemic Acid** $= C_4 H_2 O_5 + HO + Aq$. Observed as yet only in cream of tartar, associated with tartaric acid. The isolation and purification of this acid is identical with that of tartaric acid. It is distinguished from the latter by the klinorhomboidal form of its crystals, by its tendency to effloresce at the air, and by losing one equivalent of water at 100°, and then having exactly the same composition as tartaric acid. Its aqueous solution becomes turbid by sulphate of lime and the deposit obtained by lime water is insoluble in chloride of ammonium.

**Rape-Oil.** Obtained by pressing the seeds of several varieties of Brassica oleracea; is brownish-yellow, originally mild, assumes by keeping a nauseous odour and taste, is of 0·912 to 0·920 density, thickens below, 0°. Not drying.

**Raspberry Camphor.** Obtained by distillation with water from the fruits of Rubus Idæus.—Small, white laminæ, either lighter or heavier than water, soluble in water, alcohol, ether, and alkalies.

**Ratanhia Tannic Acid** $= C_{13} H_8 O_7$. In the Ratanhia-root and in the bark of the root of Savanilla-Ratanhia (from Krameria triandra and K. Ixina). Exhaust with ether, treat the ethereous extract with alcohol and evaporate the solution.—Ruby-red, amorphous, permanent at the air, of a bitter and astringent taste, of acidulous reaction, fusible by heat, dissolves in water, alcohol, ether; colours and precipitates chloride of iron dark-green, precipitates glue, not tartarated antimony; becomes decomposed with diluted acids into a brown-red, hard resin (ratanhia-red $=$ $C_{12} H_6 O_3$, also existing ready formed in the root), and into a sweet body which reduces the solution of copper.

[**Ratanhin** $= C_{20} H_{13} NO_6$ (Peckolt's Angelin). Contained in the American Ratanhia extract (according to Ruge), and in the resin of Hillia spectabilis. In order to prepare it from the last-named substance, the pulverised resin is repeatedly digested with water, the residue dissolved in water and hydrochloric acid and evaporated with a gentle heat to form crystals. The crystals are freed from the mother-ley by pressing, again dissolved in acid water and recrystallised. After repeating this process six or seven times, a white crystalline mass is obtained, which is dissolved in boiling distilled water and set aside to crystallise.—The Ratanhin, purified in this way, presents delicate, flexible needles, of a pure white colour and a vivid satin lustre. It is nearly insoluble in cold, and only sparingly soluble in boiling water; it is still less soluble in alcohol, both cold and boiling, and almost insoluble in ether. It is tasteless, without odour, and of neutral reaction. R. dissolves in acid and in alkaline liquids, but is thrown down again on neutralising the solution. It is likewise precipitated by alcohol, alcohol-ether, and in acid solutions by phosphomolybdic acid and by Nessler's reagent, not by chloride of platinum. Heated above 150° it melts and volatilises in a higher temperature, while emitting a not unpleasant aromatic odour. Quickly heated, it becomes carbonised with a horny odour and under formation of inflammable gases. Ratanhin, when formed into a thin pulp with water and an adequate quantity of dilute nitric acid, and boiled for some time, becomes first of a rose colour, and then changes from blood-red to violet and blue, while exhibiting a splendid red fluorescence.]

**Red Pigment of Berries** is mostly anthocyan, reddened by acids; yet there are some exceptions. For instance, the red of strawberries behaves like cissotannic acid, likewise, the red pigment of the berries of Ligustrum vulgare behaves differently and is named ligulin.

**Red Pigment of Flowers** is mostly anthocyan, reddened by acids.

[**Regianin**, found by T. L. Phipson in the green pericarp of the walnut (Juglans regia). Crystallises in yellow, protracted octahedra or needles, little soluble in water, better in alcohol and benzol; becomes after a few hours transformed into black, amorphous Regianic acid; forms with alkalies soluble salts of a splendid purple colour, with oxyd of lead an insoluble, brown-violet salt. On boiling the aqueous or alcoholic solutions of R. with hydrochloric acid, Regianic acid separates as a dense, black precipitate.]

**Resins.** A very large and widely-diffused class of bodies, which seem to originate from volatile oils by the oxydising influence of the atmosphere. Combined with volatile oils, they either exude spontaneously, or by the aid of incisions, or are extracted by solvents. They exhibit the following characteristics: They are colourless or coloured, translucid or transparent, not brittle, mostly amorphous, seldom crystalline, assume negative electricity with friction, are of 0·93 to 1·20 density, inodorous or odoriferous from traces of volatile oil, tasteless, or bitter and acrid, fusible by heat, are decomposed by more heat under carbonisation, burn with a smoking flame; they are insoluble in water, soluble in alcohol (sometimes only in the strongest), mostly in ether and in oils, the solutions being mostly of acid reaction; they dissolve as a rule in alkalies, yielding soap-like compounds. They contain principally carbon and hydrogen, mostly oxygen, too, but no nitrogen.

**Rhamnin.** Yellow, crystalline ingredient of the unmatured berries of Rhamnus cathartica, associated with rhamnocathartin. Press the berries, boil the remnant repeatedly with water and allow the decoctions to stand cold. Purify the cauliflower-shaped crystals which are formed, by pressing, dissolving in boiling alcohol, washing the crystals that have formed with cold water and weak alcohol, and by recrystallising in boiling alcohol with aid of animal charcoal.—Forms pale-yellow, cauliflower-shaped, small grains, seldom tuftily united needles, of a slight, peculiar taste, fuses by heat and decomposes afterwards; not or scarcely soluble in cold water, swells up considerably in boiling water, dissolves little in cold, readily in boiling alcohol, not in ether; in cold concentrated sulphuric and also in hydrochloric acid with saffron-yellow colour and precipitable by water, also in hot diluted sulphuric acid, and crystallising from it on cooling; in the hydrates and carbonates of alkalies with saffron-yellow colour, and precipitable by acids.

**Rhamnocarthartin.** The uncrystallisable bitter substance of the berries of Rhamnus cathartica. Evaporate the juice of the ripe berries to honey-consistence, exhaust with hot alcohol, evaporate the tinctures and mix the remnant with water, which throws down yellow-green, pulverulent rhamnotannic acid; the filtrate, when shaken with animal charcoal until devoid of bitter taste, yields up the Rh. to the coal. Wash the latter with cold water, dry, treat with hot alcohol and evaporate the tincture.—Transparent, amorphous, yellowish, brittle mass, friable to a yellow powder, assumes on friction a peculiar smell, tastes most nauseously bitter and acrid, is of neutral reaction, tolerably permanent at the air, fuses by heat to a yellow oil and is afterwards decomposed, dissolves in water in every proportion, likewise in alcohol, not in ether; the aqueous solution assumes with alkalies or with subacetate of lead a brownish, gold-yellow colour without any deposit, and becomes colourless with acids; colours chloride of iron dark-brown-green.

**Rhamnotannic Acid.** Is obtained in the preparation of rhamnocathartin and is purified by washing, drying, dissolving in ether and evaporating.—Green-yellow, amorphous, easily-friable mass of a bitter and acerb taste, fusible, dissolves scarcely in cold, slightly in boiling water, readily in alcohol and in ether, colours and precipitates the salts of oxyd of iron olive-green, precipitates also slowly tartarated antimony, but not glue.

**Rhamnoxanthin**$=C_{12} H_6 O_6$ or $C_{40} H_{20} O_{20}$. Yellow, crystalline substance of the bark of the root and of the stem, also of the seeds of Rhamnus cathartica and Rh. Frangula and doubtless many other species. Cover the branchlets of Rh. Frangula with sulphide of carbon and keep for three to four days, evaporate the liquid to dryness, treat the remnant with alcohol, which leaves the fat undissolved, evaporate again and recrystallise in ether.—Citron-yellow, crystalline mass of a dull, silky lustre, without taste or smell; fuses at 226° under evolution of yellow fumes, and sublimates under partial decomposition in gold-yellow needles; is not soluble in water, dissolves in 160 parts warm alcohol of 80%, and separates from it on cooling almost completely; scarcely soluble in ether, soluble in sulphide of carbon, fixed and volatile oils, in concentrated sulphuric acid with dark-ruby-red colour and reprecipitable by water, in hot concentrated nitric acid unaltered; in alkalies with a splendid purple colour.

**Rhaponticin** } $=$ Chrysophanic Acid.
**Rheic Acid** }

[**Rhinanthin**$=C_{58} H_{52} O_{40} + {}_8HO$. Glucosid, discovered by Ludwig in the seed of Alectorolophus hirsutus. To prepare it treat the pulverised seeds with strong, boiling alcohol, evaporate the filtrate to dryness, remove the oil by means of ether, dissolve the residue in water, filter and evaporate to a syrupy consistence. The crystals, which will slowly form, are to be purified by recrystallisation.—The R. forms colourless crystals of a bitter-sweet taste; is readily soluble in water and in alcohol, not in ether; of a neutral reaction. It is not precipitated by subacetate of lead, reduces ammoniacal silver solution when warmed, and separates with acids into glucose and a dark-blue-green body, soluble in alcohol but not in water. Heated with sulphuric or hydrochloric acids, R. turns brown, while emitting an aromatic rye odour.]

**Rhodeoretin**$=$Convolvulin.

**Rhodotannic Acid**$=C_{14} H_6 O_7$. In the leaves of Rhododendron ferrugineum and probably many of the congeners. Distil the alcoholic extract; mix the remnant with water, filter, precipitate the filtrate with acetate of lead, treat the deposit with diluted acetic acid, filter, heat the filtrate to boiling point and precipitate with subacetate of lead. Decompose the deposit under water with sulphuret of hydrogen and evaporate the filtrate in a current of carbonic acid gas.—Amber-yellow powder of an acidulous, acerb taste, greens the salts of oxyd of iron, separates on heating with diluted acids a red-yellow powder (Rhodoxanthin$=2$ $C_{14}$ $H_7$ $O_3$ + HO).

**Rhoeadin**$=C_{42}$ $H_{21}$ $NO_{12}$. Peculiar alkaloïd occurring in all parts of Papaver Rhoeas and allied species. Treat the whole herb with warm water, concentrate the extract, oversaturate with carbonate of soda and shake repeatedly with ether. Transfer the Rhoeadin from the ethereous solution to an aqueous solution of bitartarate of soda by shaking, throw away the ether and precipitate the aqueous liquid with ammonia, wash the deposit, dry, and boil with alcohol, in order to remove dyeing matters and an alkaloïd which exists in the plant in small quantity only, and seems to be thebain. The Rhoeadin remains by this process for the greatest part undissolved. To purify it completely, dissolve the remnant in acetic acid, shake with animal charcoal and precipitate with ammonia.—Forms small, white prisms, tasteless either by itself or in solution; fuses at $232°$ without loss of weight, becomes brown and sublimates partly; is almost insoluble in ether (in 1280 parts), benzol, chloroform, alcohol, water, liquor of ammonia, soda-ley and lime-water; the alcoholic solution has a scarcely perceptible alkaline reaction. Dissolves in acids, without being able to neutralise them or even to remain in contact with them without alteration (to turn red), especially hydrochloric and sulphuric acids, yielding purple-red solutions. Alkalies restore the original colour. The red colouration is accompanied by the production of a colouring substance, and of a new alkaloïd of highly basic properties and of the same composition as Rhoeadin. Concentrated sulphuric and nitric acids dissolve the Rhoeadin under decomposition, the former acid with olive-green, the latter with yellow colour. The colourless solution of Rh. is precipitable by tannin, chloride of mercury, etc.

**Rhoitannic Acid** $=C_{18}$ $H_{14}$ $O_{13}$. In the leaves of Rhus Toxicodendron. Shake the ethereous extract with warm water, filter, allow to rest for two days, filter anew, throw down sulphuric and phosphoric acids by means of a little acetate of lead, filter and precipitate completely with acetate of lead, decompose the latter deposit under water with sulphuret of hydrogen; filter and evaporate.—Amorphous, yellowish-green, gum-like mass, of slightly bitter taste and acidulous reaction, colours and precipitates the salts of oxyd of iron dark green, colours tartarated antimony dark yellow without precipitation, turbidifies solutions of glue when concentrated.

**Ricinin.** In the seeds of Ricinus communis. Boil repeatedly with water, evaporate the liquids, after the fixed oil has been removed as completely as possible, to honey-consistence, boil with alcohol, filter, free the tincture from the resin after 24 hours, and distil off the alcohol. In the remnant, crystals of Ricinin are formed after some time, which have to be purified by recrystallising in alcohol with aid of animal charcoal.—Forms colourless, rectangular prisms and scaly laminæ of a slight taste of bitter almonds; fuses by heat, sublimates unaltered; dissolves in water and in alcohol, scarcely in ether, in benzol, in concentrated sulphuric acid without colour, the solution assuming a green tinge with chromate of potash; in nitric acid without decomposition.

**Ricinoleic Acid**$= C_{36} H_{33} O_5 + HO$. In the oil of the seeds of Ricinus communis. Saponify, salt out, decompose the soap with hydrochloric acid, and refrigerate the oily mixture containing the R. acid and a little solid acids, under addition of $\frac{1}{3}$ volume of alcohol to a temperature of—10° to —12°, when the solid acids will crystallise. After removing the latter, and driving off the alcohol, Ricinoleate of lead is obtained by digestion with oxyd of lead, and which has to be dissolved in ether, and is decomposed with water and hydrochloric acid. Purify the R. acid which remains after the evaporation of the ethereous liquid by dissolving in liquor of ammonia, precipitating with chloride of baryum, recrystallising the baryum-compound in alcohol, decomposing with tartaric acid and washing with water.—Light wine-yellow, syrup like liquid of 0·94 density, inodorous, of a strong and lasting, disagreeable, acrid taste, dissolved in alcohol of acid reaction, congeals at —6° to —10° (according to others at 0°) to a granular mass, dissolves in alcohol and in ether in every proportion. The Ricinoleates dissolve all in alcohol, some of them also in ether, are not liable to oxydise by keeping.

**Riozolic Acid**$=$Pipitzahoic Acid.

**Robinin**$= C_{50} H_{30} O_{32} + 11 HO$. Yellow, crystalline glucosid of the flowers of Robinia Pseudacacia. Boil with water, use the decoction six to eight times for boiling anew fresh flowers, evaporate to a syrup thickness, treat with hot alcohol, filter, distil the alcohol off and allow the remnant to form in crystals. Dissolve the latter, after freeing from the bulk of the mother-ley by pressing and washing with cold alcohol, in boiling water, and mix the solution with acetate of lead, which throws down foreign matters and leaves the Robinin dissolved. The R. is obtained by evaporating the filtrate, freed from the lead by sulphuret of hydrogen, and is then recrystallised in water.—Forms very fine, straw-yellow needles of a slight satin-lustre, neutral, tasteless, in aqueous solution of a slightly astringent taste; loses its water at 100°, fuses at 195°, and decomposes in a higher temperature, producing quercetin and an odour of burnt sugar; dissolves little in cold, readily in boiling water, the light-yellow solutions becoming colourless with acids; little soluble in cold, more so in hot alcohol, not in ether, readily in alkalies with gold-yellow colour; the ammoniacal solution becomes brown by keeping, but not the solutions of the fixed alkalies; breaks up with diluted acids into sugar and quercetin.

**Roccellic Acid** $= C_{34} H_{30} O_6 + 2 HO$. In Roccella fuciformis and in Lecanora tartarea. Treat with water and ammonia, precipitate the filtrate with chloride of calcium, decompose the deposit with hydrochloric acid and purify the crystals that have formed by dissolving in ether.—Forms delicate, white, silvery, quadrangular, tabular crystals, obtained in short needles from alcohol, inodorous and tasteless, of acid reaction in the alcoholic solution; fuses at 130° without loss of weight, evaporates partly below 200°, and is partly converted into the anhydrous acid and decomposed by a higher temperature; is quite insoluble in water, dissolves in 1·8 parts alcohol of 0·819, readily in ether; forms with alkalies half-acid soluble, with the other bases mostly insoluble salts.

**Roccellinin** $= C_{36} H_{16} O_{14}$. In Roccella tinctoria. Treat with water containing lime, filter, precipitate the filtrate with hydrochloric acid and boil the deposit, consisting of Roccellinin and lecanoric acid (named formerly βorsellic acid), repeatedly with water, leaving the R. undissolved, which has to be recrystallised in alcohol.—Fine, hair-shaped crystals of silky gloss, insoluble in water, slightly soluble in cold alcohol and in ether, a little more when hot, readily in alkalies and in alkaline earths; assumes a permanent green-yellow colour with solutions of chloride of lime.

**Rottlerin** (Kamalin) $= C_{22} H_{10} O_6$. In the kamala, the stellated hairs and glandules that cover the fruit of Mallotus Philipinensis. It crystallises from the ethereous tincture in yellow needles of a silky gloss, fuses by heat and becomes decomposed in higher temperatures; is insoluble in water; dissolves in alkalies with deep-red colour, little in cold, more in boiling alcohol, and readily in ether.

**Ruberythric Acid** $= C_{72} H_{40} O_{40}$ or $C_{56} H_{31} O_{31}$. In the root of Rubia tinctorum, according to Rochleder; Schunck believes the above acid to be a product of decomposition of rubian, and perhaps identical with a substance named by him rubianic acid. Precipitate with acetate of lead the aqueous decoction of madder, remove the deposit (which may be used for the preparation of alizarin and purpurin), and precipitate the filtrate with subacetate of lead, but not in excess, producing a dark flesh-coloured and almost brick-red deposit, which contains Ruberythric, rubichloric, a little citric and phosphoric acids. Decompose the deposit under water with sulphuret of hydrogen; separate the liquid, containing mostly rubi-

o

chloric acid, by filtering off the sulphide of lead. Wash the latter for a short time and withdraw from it the R. acid by means of boiling alcohol. Evaporate the alcoholic solution to $\frac{1}{3}$, add water and a little solution of baryta, thereby obtaining a white deposit, which has to be removed when, after addition of more solution of baryta, Ruberythrate of baryta will be precipitated in crimson-red flocks. Collect the latter, dissolve in diluted acetic acid, neutralise the solution with ammonia almost completely, and add subacetate of lead, which throws down the compound of lead with a cinnabar-red colour. This compound has to be washed with diluted alcohol, and is decomposed under alcohol with sulphuret of hydrogen. Heat the whole to the boiling point, filter hot and evaporate, obtaining thereby light-yellow crystals of R. acid, which have to be purified by pressing and by recrystallising in a little water.—The Ruberythric acid forms yellow needles of a silky gloss, slightly bitter; loses at $100^c$ nothing of its weight; dissolves slowly in cold, readily in hot water; the aqueous solutions become cloudy when heated with hydrochloric acid, and form on boiling a yellow jelly, which conglomerates to flakes of alizarin; dissolves in alcohol and in ether with gold-yellow colour (Schunck's rubianic acid is insoluble in ether); dissolves in aqueous alkalies with blood-red colour, the solutions assuming on boiling the purple-red hue of the alkaline solutions of alizarin and throwing down alizarin after the addition of acids.

**Rubiacin** = $C_{32} H_{11} O_{10}$. In the root of Rubia tinctorum, perhaps combined with lime. Proceed at first, as indicated under Rubian; mix the residue, remaining after the edulcoration of the rubian, with the remnant left after the evaporation of the alcohol, containing verantin and rubiretin, and treat the mixture with a boiling solution of chloride or of nitrate of iron. Rubiretin and Rubiacin dissolve (the latter partly as such, partly by oxydation as rubiacate of iron-oxyd) in the liquid, while verantin remains in combination with oxyd of iron. Filter the deep red-brown solution after short boiling, keep the remnant for the preparation of verantin, precipitate from the filtrate with hydrochloric acid Rubiacin, rubiacic acid, and rubiretin, as a yellow (after washing, brown) deposit, dissolve moist in boiling alcohol, dissolving Rubiacin and rubiretin and separating the former on cooling in small, citron-yellow crystals. By evaporating more of the alcohol, a mixture of Rubiacin and rubiretin is obtained as a dark brown-red residue, which forms on boiling with water dark-brown drops of rubiretin, while Rubiacin remains suspended as a light powder, and is easily decanted. After repeating the boiling with water several times, and pouring off the yellow powder which has formed, rubiretin remains as a dark, red-brown mass.—Forms splendid tabular crystals and needles, similar to iodide of lead, of a strong, reddish-green lustre; dissolves slightly in boiling water with reddish-yellow colour, little in cold, more in boiling alcohol, readily in ether, little in diluted sulphuric acid, in concentrated sulphuric acid with yellow colour and undecomposed even on boiling, in ammonia with brownish, in potash-ley with rose-red, in soda-ley with orange colour, turning blood-red on boiling.

**Rubian** = $C_{56} H_{34} O_{30}$. Bitter glucosid of the root of Rubia tinctorum. Boil with water and precipitate the decoction with diluted sulphuric or hydrochloric acids, to produce a dark-brown deposit which, freed by cold water from the acid, contains different substances, viz., Rubian, alizarin, rubiacin, rubiretin, verantin, pectic acid, and a dark-brown decomposition-product. (In the filtrate remains chlorogenin and sugar). Boil the deposit moist with several changes of alcohol as long as the latter assumes a yellow colour, pectic acid and the decomposition-product remaining undissolved, while the dark-brown decoction on cooling frequently throws down the verantin as a dark-brown resinous powder, removable by filtering. Heat the alcoholic solution to boiling, adding newly precipitated hydrate of alumina and allow to digest until the solution is nearly decolourised, precipitating thereby alizarin, Rubian, rubiacin, and portions of rubiretin and verantin. Collect the alumina-deposit and add to it, after washing with alcohol, a concentrated boiling solution of carbonate of potash, filter the deep-red solution, containing all the other substances, from the undissolved alizarin-alumina (convertible into pure alizarin as described above); precipitate the alkaline filtrate with hydrochloric acid, to throw down Rubian, rubiacin, rubiretin, and verantin, collect the deposit and wash thoroughly with water. As soon as the water passes off without acid reaction, the Rubian, being insoluble in acid water but soluble in pure water, begins to dissolve, imparting to the latter a yellow colour and a bitter taste. By dissolving it in this manner and evaporating the filtered liquid, it is obtained as a yellow extract. The latter is freed from pectic acid by dissolving in alcohol, but retains 5% to 8% inorganic substances, from which it cannot be separated.—Hard, dry, brittle, amorphous mass, similar to dried varnish or to gum-arabic, permanent at the air, dark yellow, intensely bitter, is decomposed at 130° under loss of water; gives out orange-coloured vapours, consisting mostly of alizarin, when heated to a higher temperature, and leaves much coal; fuses on platinum-foil, dissolves most readily in water, a little less so in alcohol, not in ether, in concentrated sulphuric acid with blood-red colour, and becomes carbonised by heat; is decomposed by boiling with diluted sulphuric or hydrochloric acid, into sugar and into opalescent, afterwards orange-coloured flocks, consisting of alizarin, rubiretin, rubiacin and verantin; dissolves in hot caustic potash-ley with blood-red, then purple-red colour under decomposition. The Rubian in aqueous solutions also becomes blood-red with caustic baryta or ammonia. The aqueous solution of Rubian is not precipitated by acids; is not altered on boiling with phosphoric, oxalic, acetic, or tartaric acids, is not precipitated by alum or the acetates of alumina, lead, copper, or zinc, the chloride of tin, mercury, gold, the sub-nitrate of mercury, or the nitrate of silver.

**Rubichloric Acid** $= C_{14} H_8 O_9$. In the root and herb of Rubia tinctorum, in the herb of Asperula odorata, Galium Aparine, G. Molugo and G. verum, and probably therefore largely present in that group of the Rubiaceæ to which these herbs belong. Occurs in traces in the deposit produced by acetate of lead from the aqueous decoctions of the above vegetable parts; in a little larger quantity in the deposit produced by sub-acetate of lead, in the filtrate, and in the largest quantity in the deposit, effected by ammonia, from the liquid remaining after the filtration of the two former precipitations. In operating with Asperula adorata, for instance, the third mentioned deposit has to be washed with alcohol, and is then, suspended in alcohol, decomposed with sulphuret of hydrogen. Free the liquid from the sulphide of lead and from the sulphuret of hydrogen, precipitate again with alcoholic solution of acetate of lead and a little ammonia, decompose the deposit in alcohol with sulphuret of hydrogen; filter and evaporate in a vacuum.—Colourless or slightly yellowish, amorphous mass, inodorous, of an insipid nauseous taste, dissolves readily in water and in alcohol, not in ether, becomes yellow with alkalies and is decolourised by acids; becomes on heating with hydrochloric acid blue, afterwards green, forming dark-green flocks (chlorrubin), under formation of formic acid.

**Rubiretin** $= C_{14} H_6 O_4$ (isomeric with hydrated benzoic acid). In the root of Rubia tinctorum. It takes its origin, according to Higgin, like verantin by the boiling with water or as decomposition-product of purpurin, according to Strecker and Wolff, under the influence of alkalies. As to preparation see Rubian and Rubiacin.— Dark reddish-brown brittle resin, soft at 65°, fuses at 100°, yields usually with more heat a slight sublimate of alizarin, dissolves little in boiling water, readily in alcohol, in alkalies with purple-red colour.

**Rubitannic Acid** $= C_{14} H_8 O_9$. In the leaves of Rubia tinctorum. Precipitate the aqueous extract by means of acetate of lead, treat the deposit with diluted acetic acid, filter, precipitate the liquid with ammonia, wash the deposit with alcohol and decompose under alcohol with sulphuret of hydrogen; mix the liquid, after it has been filtered and the alcohol is driven off, with water, precipitate with subacetate of lead, decompose the deposit under water with sulphuret of hydrogen and evaporate the filtered liquid.—Very hygroscopic, precipitates the salts of oxyd of iron with a beautiful green colour.

**Rumicin** = CHRYSOPHANIC ACID.

**Rutin** or **Rutic Acid** = $C_{50} H_{23} O_{30} + 4$ HO. Glucosid of Ruta graveolens, of the flower-buds of Capparis spinosa, of the Waifa (the flower-buds of Sophora japonica). Formerly confounded with quercitrin. The safflower-yellow is, according to Stein, uncrystallisable Rutin, as likewise the pigments of straw, of Æthalium, of Hippophaë, and of Fagopyrum [also, according to Mylius, of Sedum acre]. Boil the herb of cultivated rue with vinegar, press and let rest ; wash the slowly forming Rutin with cold water, boil with a mixture of 1 pt. acetic acid and 4 pts. water, filter, allow to crystallise, wash the crystals, dissolve in boiling alcohol, treat the solution with animal charcoal, filter and let crystallise.—Forms light-yellow, fine needles of a feeble, silky gloss, inodorous and tasteless, bitter in solution ; neutral ; loses its water at $160^{\circ}$, conglutinates at $190^{\circ}$, fuses and becomes carbonized with a smell of burnt sugar; dissolves scarcely in cold, in 185 pts. of boiling water, little in cold absolute alcohol, readily in boiling alcohol of 76 %, not in ether, readily in alkalies and in alkaline earths, and reprecipitable unaltered by acids; yields on heating with diluted acids sugar and quercetin.

**Sabadillic Acid.** Peculiar volatile fat-acid of the melanthaceous group of Liliaceæ, especially observed in the seeds of Schœnocaulon officinale and of Colchicum autumnale, and in the root of Veratrum album. Treat preferably the seeds of Schœnocaulon (or Sabadilla) with ether, evaporate the solution, saponify the fixed oil, which has separated, with potash-ley, decompose the soap with tartaric acid, distil the aqueous liquid, saturate the distillate with baryta and distil the desiccated Sabadillate of baryta with concentrated phosphoric acid.—The S. acid sublimates in white needles of mother-of-pearl lustre, fusible at $20^{\circ}$, of the odour of butyric acid, soluble in water, alcohol and ether.

**Sabadillin** = $C_{20} H_{13} NO_5$. Alkaloïd, associated with veratrin in the seeds of Schœnocaulon officinale. Is obtained by extracting with alcohol, distilling the tincture, dissolving the remnant in diluted sulphuric acid, digesting the solution with animal charcoal, and precipitating with caustic potash. The deposit consists of veratrin, Sabadillin, and sabadillin-hydrate, and contains besides two, not basic substances (only one of which, the belonin, has been closely investigated). To separate these substances, redissolve the deposit in diluted sulphuric acid, add nitric acid as long as a black, pitch-like deposit is produced, precipitate the filtered solution with potash-ley, wash the deposit, dry, dissolve in absolute alcohol, evaporate the solution and boil the remnant with water, veratrin and helonin remaining behind, while Sabadillin and sabadillin-hydrate are dissolved.—From the aqueous solution nearly the whole of the Sabadillin crystallises in slightly reddish, concentrically arranged, sexangular prisms, which become white by recrystallising. It has an extremely acrid taste, fuses at 200°, losing 9·53 % water, is decomposed in higher temperatures; dissolves little in cold, readily in boiling water, also in alcohol, but crystallises not from it; is insoluble in ether. It has a strongly alkaline reaction, and forms with acids mostly crystallisable salts.

**Sabadillin-Hydrate** $= C_{20} H_{14} NO_6$ ($= C_{20} H_{13} NO_5 + HO$). By evaporating the liquid from which the sabadillin has crystallised, oily drops are formed, congealing to a red-brown, resin-like, brittle substance.—Dissolves readily in water and in alcohol, not in ether, is of alkaline reaction, forms with acids amorphous salts.

**Sagapenum.** Gum-resinous exudation of Ferula persica and F. Scovitziana. Yellow, brown, or reddish conglutinated grains of garlic-odour and of acrid, bitter taste, softening with the warmth of the hand. Contains two resins, gum, bassorin and volatile oil.—One of the resins is red-yellow, pellucid, at first tough, smells faintly garlic-like, tastes mild, afterwards bitter, dissolves readily in alcohol and in ether, little in ammonia and in oils, partially in potash-ley. The other resin is brown-yellow, brittle, inodorous and tasteless, dissolves readily in alcohol and in warm potash-ley, not in ether, ammonia and oils.

**Salicin** $= C_{26} H_{18} O_{14}$. Bitter glucosid of the bark, the leaves, and other parts of species of Salix and Populus, probably also in some species of Spiræa, which yield salicylous acid when distilled with water. Boil the bark with water containing lime, clarify the decoctions with albumen, strain, evaporate to a syrup consistence, add pulverised charcoal, dry, extract with alcohol, distil the tincture and allow the remnant to crystallise. Recrystallise what has formed in water with aid of animal charcoal.—Forms small, white, shining needles and scales, is inodorous, of a very bitter taste similar to willow-bark; fuses at 198° without loss of weight, decomposes in a stronger heat; dissolves in 22 parts cold and in half part boiling water, in 30 parts cold and in three parts boiling alcohol of 80 %, not in ether; has a neutral reaction; dissolves in concentrated sulphuric acid with purple-red colour, and water precipitates a dark-red powder from the solution; yields, on boiling with diluted sulphuric acid, grape-sugar and a resinous substance (saliretin $= C_{14} H_6 O_2$); on heating with the superoxyds of lead and manganese, or with chromate of potash and sulphuric acid, formic and carbonic acids are produced, in the latter case associated with salicylous acid. Metallic salts yield no precipitates.

**Salicylate of Methyl**$=C_{16} H_8 O_6 = C_2 H_3 O + C_{14} H_5 O_5$.
In the herb of Gaultiera procumbens, and doubtless of other species; it forms in the main the oil of wintergreen, obtained from that plant by distillation with water, and has also been observed lately as constituent of Monotropa Hypopitys. In the rectification of the above oil a light oil of the composition of oil of turpentine passes over at 200°; the boiling point rises rapidly, and when it is as high as 222° Salicylate of Methyl begins to distil.—This is a colourless oil, of a pleasant and very penetrating odour, and of a sweet, aromatic, refreshing taste, of 1·18 density, boils at 222°, dissolves little in water, the solution becoming purple-violet with salts of oxyd of iron ; mixes with alcohol, ether, and oils in every proportion, splits up with aqueous alkalies into salicylic acid and wood-spirit, forms with bases compounds wherein 1 eq. H is sub-stituted by 1 eq. metal, the potassium-compound dissolving readily in water, not so readily that of sodium, still less those of baryum, zinc, lead, copper, and mercury.

**Salicylic Acid**$=C_{14} H_5 O_5 + HO$. In the flowers of Spiræa Ulmaria, combined with methyl in Gaultiera procumbens. May be obtained from the distillation of the crude salicylite of soda (see Salicylous acid), or by extracting the above flowers with ether, distilling the ether from the solution, dissolving the remnant in water, saturating the solution (containing salicylic and tannic acids) with carbonate of soda, evaporating and distilling with sulphuric acid. The aqueous distillate, by slow evaporation, yields the acid in colourless needles. From the oil of Gaultiera it is obtained by heating with strong potash-ley, until the whole of the wood-spirit is driven off, precipitating the remnant with hydro-chloric acid, washing the deposit with hot water and recrystallising from hot alcohol.—The Salicylic acid crystallises in colourless needles and in tolerably large quadrangular prisms, has a sweetish, acid, afterwards irritating taste, reddens litmus paper, fuses at 150°, sublimates unaltered at 200° without boiling, dissolves little in cold, abundantly in hot water, readily in wood-spirit, alcohol and ether. The aqueous solution, like salicylous acid, colours the salts of oxyd of iron purple-violet. The Salicylates of the alkalies, of the alkaline earths and of zinc dissolve readily in water, those of lead, copper and silver with difficulty, all being crystallisable. The aqueous solutions of the Salicylates of alkalies turn brown at the air. In the destructive distillation most of its salts yield carbolic acid and carbonates.

**Salicylous Acid**$=C_{14} H_5 O_3$. Found as yet in all parts, but especially in the flowers of Spiræa Ulmaria; also in other herba-ceous kinds of Spiræa, in the flowers of Crepis fœtida, but seems to be only formed under the concurrence of water. It is obtained by distilling with water. The acid distillate is saturated with

soda, evaporated to dryness, and the remaining salt distilled with
sulphuric or better phosphoric acid, when Salicylous acid passes over
first and salicylic acid sublimates afterwards in long needles. The
liquid distillate is desiccated by chloride of calcium, and rectified.—
Colourless, oily liquid, smells pleasantly aromatic, somewhat like
bitter almonds; has a burning aromatic taste, congeals at —20° to
a translucent crystalline mass; has a density of 1·173, boils
between 160° and 170°; reddens litmus-paper first, and bleaches
it afterwards; dissolves copiously in water, in every proportion in
alcohol and ether; in alkalies with yellow colour; the aqueous
solution, even when largely diluted, colours the salts of oxyd of
iron purple-violet. The Salicylites of the alkalies are yellow and
moderately soluble, and give the same reaction with oxyd of iron
compounds as the free acid. The Salicylites of the other metallic
bases are for the greater part insoluble in water.

[**Samaderin.** De Vry's Glucosid (?) of the bark of Samadera
indica. Obtained by treating the alcoholic extract with water,
digesting the aqueous liquid with charcoal, and exhausting the
latter by hot alcohol.—Is extremely bitter, and only obtained in
the amorphous state.]

### Sandal-Red = SANTALIN.

**Sandarac.** Resinous exudation of Callitris quadrivalvis. Pale-
yellow grains, similar to mastic, but not softening in the mouth,
readily soluble in alcohol of 80%, also in ether. By treating with
cold alcohol of 60%, one-third remains undissolved (sandaracin).
It contains three resins, one of which is precipitable from the
alcoholic solution by alcoholic potash, while the two other ones are
separable by alcohol of 60%.

### Sanguinarin = CHELERYTHRIN.

[**Santal** = $C_{16} H_6 O_6$. Obtained by Weidel from sandal-wood
(Pterocarpus santalinus) by exhausting with boiling water, containing a little potash; precipitating with hydrochloric acid; dissolving the precipitate in boiling alcohol, and allowing to crystallise.—Forms colourless crystals, devoid of taste or smell, not
soluble in water, benzol, chloroform, sulphide of carbon, and but
sparingly in ether; yields with potash a faintly yellow solution,
which soon turns red and green.]

**Santalin** or **Santalic Acid** = $C_{30} H_{14} O_{10}$. The red pigment of
the wood of Pterocarpus santalinus. Boil the ethereous or the
alcoholic extract with water, the S. remaining undissolved.—
Forms microscopic, beautifully red prisms, inodorous and tasteless, of an acid reaction; fuses at 104°, decomposes above the
fusing point; is insoluble in water, dissolves readily in alcohol
with blood-red, in ether with yellow colour, less in fixed and in
volatile oils, readily in acetic acid and precipitable therefrom by

water, likewise in concentrated sulphuric acid, in alkalies with a violet-red hue. Combines with bases to amorphous salts, the soluble ones (of the alkalies) possessing a slightly acerb taste.

**Santonin** or **Santonic Acid** $= C_{30} H_{18} O_6$. Bitter, resinous acid of wormseed (Artemisia Cina and A. Siberi). Digest with alcohol of 40% and with hydrate of lime, strain, distil the alcohol, filter the remnant, concentrate and acidify with acetic acid. Collect the Santonin which has formed, wash with cold alcohol and re-crystallise from boiling alcohol with aid of animal charcoal.— Forms colourless, klino-rhombic needles and tabular crystals, in-odorous, slightly bitter; more bitter when dissolved in alcohol; fuses at 169°, sublimates and becomes decomposed afterwards; turns yellow slowly in diffused, rapidly in direct sunlight; dis-solves in 5000 parts cold and in 250 parts boiling water, in 43 parts cold and in 3 parts boiling alcohol of 80%, in 75 parts cold and in 42 parts boiling ether, in 4·35 parts chloroform, the solu-tions being of neutral reaction; in diluted acids not more abundantly than in water; readily in alkalies and in alkaline earths, also in oils, undecomposed and colourless in concentrated sulphuric acid and reprecipitable by water. It behaves to-wards bases like a weak acid; the compounds of the alkalies and alkaline earths being soluble in water, not the other compounds.

**Sapan-Red** $=$ BRASILIN.

**Saponin** $= C_{36} H_{28} O_{24}$. (Named, also, according to the origin, Githagin, Monesin, Monninin, Polygalin, Quillajin, Senegin, Struthiin.) In plants of various orders, especially in Caryo-phylleæ, as in the root and herb of Saponaria officinalis, in the root of Gypsophila Struthium, in the root and seed of Lychnis Githago, in the root-bark of Acacia lophantha, and perhaps other species, in the root of Monninia polystachya and Polygala Senega, in the Monesia-bark (Lucuma glycyphlæa), in the root of Quillaja Saponaria, in the fruit of Sapindus Saponaria and Aesculus Hippocastanum; in the root of Polypodium vulgare, and many other ferns. Boil, preferably, the root of Gypsophila Struthium, with alcohol of 0·824; let the decoctions stand cold, collect the sediment of Saponin, wash with ether and alcohol, and dry at 100°.—White, not crystalline, powder, which produces sneezing, of at first sweetish, afterwards burning, pungent and lastingly acrid taste, of neutral reaction, readily soluble in water, yielding a dense froth, even in solutions containing 1-10th %; dissolves more readily in aqueous than in strong alcohol, in 400 parts absolute alcohol, yielding solutions devoid of the frothy property; insoluble in ether and in volatile oils; is decomposed by heat; breaks up on boiling with diluted sulphuric acid into a carbo-hydrate and other products; dissolves little in cold alkalies, more in warm ones; is precipitable by acetate and sub-acetate.

**Scillitin.** In the fleshy bulb of Urginia Scilla. Bruise the fresh bulbs, digest with water mixed with some sulphuric acid, filter, saturate the filtrate with lime, evaporate, allow to stand cold, remove the sulphate of lime, bring the liquid to dryness, treat with strong alcohol and evaporate the tincture.—Small, white, hard prisms, of a bitter, not acrid taste, insoluble in water and in oils, soluble in 120 parts alcohol.

**Scoparin** $= C_{42} H_{22} O_{20}$. Yellow, crystalline pigment of Cytisus scoparius. Boil the herb with water, evaporate the decoction to a small bulk, leave to stand cold for a day, collect the greenish-brown jelly on a cloth, wash with cold water, treat with boiling alcohol, filter and evaporate slowly.—Small, light-yellow crytals, inodorous and tasteless, neutral, slowly soluble in cold water and in alcohol, readily soluble in both when warm; also easily dissoluble in the hydrates and carbonates of alkalies, also in limewater and in solution of baryta, in concentrated acids. It becomes dark green with solution of chloride of lime, is converted into picric acid by nitric acid; is precipitable by acetate and subacetate of lead.

**Scrophularin.** Bitter ingredient of Scrophularia aquatica, S. nodosa, and allied species, obtained as yet only in the impure state.

**Sebacic Acid** = STEARIC ACID.

**Secalin** = TRIMETHYLAMIN.

**Senegin** = SAPONIN.

**Sericic Acid** = MYRISTIC ACID.

**Sericin** = MYRISTIN.

[**Sicopirin** $= C_{32} H_{12} O_{10}$. Glucosid, found by Peckolt in the root-bark of Bowdichia virgilioides. Exhaust the powdered bark with absolute ether, distil and treat the residue with cold alcohol of 32° B.; dissolve the remaining crystalline mass in boiling alcohol with a little animal charcoal, filter and allow to crystallise.—Conglomerated needles of bitter, slightly pungent taste, slightly alkaline, soluble in ether and boiling alcohol, not in water, fusible to a clear liquid, and burning away without residue.]

**Sinapoleic Acid** $= C_{38} H_{35} O_3 + HO$. The liquid fat-acid of the oils of black and white mustard and of rape; is prepared like oleic acid.

**Sheabutter.** Probably from Lucuma Parkii; is greenish-white, fuses at 43°, and consists of about 30% olein and of 70% stearin.

**Sinalbin** = SINAPIN-SULPHOCYANIDE.

**Sinapin Sulphocyanide** $= C_{32} H_{23} NO_{11} + C_2 NS_2 H$. Observed in the seeds of Brassica alba, B. nigra and Arabis perfoliata. Free the pulverised seeds completely from the fixed oil by means of ether, exhaust with absolute alcohol (which dissolves a little sinapin), boil the remnant with alcohol of 90%, press, repeat the operation twice, and distil the tinctures, when the Sinapin-Sulphocyanide will crystallise from the remaining liquid.—It appears in white, very voluminous, pearly, tuftily united needles, is inodorous, has a bitter and mustard-like taste; is of neutral reaction, fuses at 130°, decomposes by more heat, dissolves in water and in alcohol with yellow colour, more readily when warm, the solutions becoming colourless with even traces of an acid; is insoluble in ether, sulphide of carbon and oil of turpentine, reddens the salts of oxyd of iron. [According to Will, Sinapin-sulphocyanide, or more properly called Sinalbin, has the composition $C_{60} H_{44} N_2 S_4 O_{32}$. When placed into contact with water and myrosin, it breaks up into Sulphocyanate of Acrinyl, Sulphate of Sinapin and sugar.]

**Sinapisin.** According to Simon, a fat occurring in the black mustard-seeds, and not saponifiable. Treat the pulverised seeds with alcohol of 94%, evaporate the tincture to honey consistence, treat with ether, evaporate the ethereous liquid to honey consistence, remove sugar, oil, and resin by washing with small quantities of ether, dissolve the residue in alcohol of 90%, decolourise the solution by means of animal charcoal, filter and evaporate. Recrystallise the scaly crystalline mass in ether.—Forms snow-white scales, dissolves readily in alcohol, ether and oils, not in acids or in alkalies; may be sublimated.

**Sinigrin** = MYRONATE OF POTASSIUM.

**Sipirin.** As to occurrence and preparation see Bebirin. Dark-red-brown, glossy, resinous mass, dissolves very slightly in water, readily in alcohol, not in ether, neutralises the acids, forming olive-brown salts.—Is, according to Tilley, impure Bebirin.

**Smilacin** $= C_{42} H_{34} O_{14}$. In the sarsaparilla, in the quina-root, and in other species of the genus Smilax. Boil with water, precipitate the decoction with hydrochloric acid, wash the deposit and dissolve in diluted sulphuric acid, precipitate with ammonia, and purify, if necessary, by redissolving in alcohol and treating with animal charcoal. Or, draw out with alcohol, precipitate the tincture with water, wash the deposit with ether, dissolve in alcohol, and decolourise with animal charcoal.—White warty mass or loose powder, permanent at the air, inodorous, of a bitter and acrid, somewhat astringent and nauseous taste, of neutral reaction, fuses by heat, and decomposes in higher temperatures, dissolves scarcely in cold, more copiously in hot water, yielding a froth by shaking; little soluble in cold, most readily in boiling alcohol, to a frothy liquid, scarcely in ether; dissolves in volatile, less in fixed oils, in caustic alkalies, in cold concentrated sulphuric acid, and re-precipitable by water unaltered; also in concentrated hydrochloric acid.

[**Socaloin** = $C_{34}$ $H_{19}$ $O_{15}$ + 5 HO. Prepared by Histed from Zanzibar or Socotrine aloes by moistening the pulverised drogue with alcohol of 0·960 sp. gr., pressing strongly between calico, dissolving the crystalline yellow residue in warm, weak alcohol, and purifying the crystals by recrystallisation.—Forms tufted needle-shaped prisms of a sweetish, afterwards bitter taste; melts at 118-120°; dissolves in 30 parts alcohol, 9 parts acetic ether, 380 parts pure ether, 90 parts water, and abundantly in methyl-alcohol. Over concentrated sulphuric acid it loses 12% of its weight, and at 100° 14%.]

**Soft Resins.** Viscid at ordinary temperature; are mostly ob-tained from vegetable parts by extracting with alcohol or with ether, and are probably in most cases mixtures of resin and of volatile or fixed oils, or may be hydrates. They are distinguish-able from balsams by the absence of smell.

**Solanin** = $C_{80}$ $H_{70}$ $NO_{32}$. Specific alkaloïd of the genus Solanum, easily obtained from the twigs of S. Dulcamara, the berries of S. nigrum, the sprouts of Solanum tuberosum, in S. verbascifolium, and to be found in numerous other species of Sola-num. Best adapted for its preparation are the sprouts of potatoes. Bruise them fresh, draw out with water and acetic acid, precipitate the liquid with acetate of lead, add milk of lime to the strained liquid, treat the deposit obtained thereby with alcohol, evaporate the tincture and purify the remaining Solanin by re-peatedly dissolving in alcohol.—White, flat, quadrangular prisms of mother-of-pearl lustre, or a powder of similar appearance; in-odorous, of a disagreeable, somewhat bitter, long lasting, rancid, and acrid taste, of very slightly alkaline reaction; fuses, but not without decomposition; dissolves little in water, the solution yielding a froth on shaking; is almost devoid of alkaline reaction; becomes turbid with tannic acid; dissolves in alcohol slowly, with a slightly alkaline reaction; not soluble in ether; dissolves in con-centrated sulphuric acid with successively brown and violet hue; breaks up when heated with diluted sulphuric acid (also hydro-chloric or oxalic acid) into sugar, and another stronger base (Solanidin = $C_{50}$ $H_{40}$ $NO_2$).

**Sorbin** = $C_{12}$ $H_{12}$ $O_{12}$. Peculiar kind of sugar of the ripe fruits of Pyrus aucuparia. Forms in the juice, when the latter is kept for a long time, and is purified by recrystallising with aid of animal charcoal.—Forms colourless, rhombic crystals of the taste of cane-sugar, fuses on heating, and burns with the odour of burnt sugar; dissolves in half part cold water, not in cold, little in boiling alcohol, yields oxalic acid by heating with nitric acid, assumes a red-yellow colour with cold concentrated sulphuric acid, and turns black when heated ; is not altered on heating with diluted sulphuric acid; becomes brown on heating with potash-ley, lime, baryta, and oxyd of lead, while evolving the odour of burnt sugar; reduces alkaline tartarate of copper; is not able to ferment with yeast; is not precipitable by subacetate of lead, but is so by ammoniacal acetate of lead.

**Spartein**$=C_{30} H_{26}$ N. Volatile alkaloïd of Cytisus scoparius. Concentrate the acid mother-ley of the impure scoparin (see this), distil with excess of carbonate of soda, saturate the distillate with chloride of sodium and distil again, ammonia passing over at first, followed by a colourless heavy oil which has to be freed from ammonia by washing with cold water.—Colourless, oily, thick liquid, heavier than water, of a faint odour, somewhat similar to anilin, of a very bitter taste, boils at 288°, dissolves little in water, but dissolves a little water when left in contact with it, and becomes turbid; has a strongly alkaline reaction; saturates the acids completely.

**Spiræa Yellow**$=C_{15} H_8 O_7$. Yellow matter of the flowers of Spiræa Ulmaria. Treat the flowers with ether, distil the ether from the tincture, mix the remnant with warm water to throw down impure dyeing matter, while a green oil floats on the water; remove the latter, dissolve the dyeing matter in hot water, remove the fat, which forms on cooling, and evaporate to dryness.— Yellow powder, consisting of fine needles, insoluble in water, soluble in alcohol and in ether with dark-green, or, when diluted, yellow colour; soluble in alkalies, in concentrated sulphuric acid with deep-yellow colour and reprecipitable by water unaltered.

**Staphisagrin**$=C_{32} H_{23} NO_4$. Alkaloïd of the seeds of Delphinium Staphisagria, is obtained by the method indicated under Delphinin.—Yellow-brownish resin of acrid taste, fuses at 200°, is almost insoluble in water, readily soluble in alcohol, insoluble in ether, dissolves readily in acids, but is not able to neutralise them.

**Starch**$=C_{12} H_{10} O_{10}$. A substance widely distributed in many vegetable organisms and especially in roots, in subterraneous stems and in seeds, but is also frequently met with in stems and in unmatured fruits, and the presence of which is recognised best by its property of acquiring with iodine a violet or dark-blue colour. The iodine used for this purpose may be kept ready prepared by dissolving 3 parts iodine and 4 parts iodide of potassium in 93 parts water.—Easy as the recognition of starch is, it frequently causes much trouble to separate it completely from the vegetable tissues, and with small quantities all efforts are in vain; with large quantities the final result depends on the vegetable tissues being reduced to a proper state by bruising, cutting, &c., and even then a small percentage of starch is retained tenaciously by the cellular membranes. Dry substances must be pulverised as fine as possible, and are converted into a paste with water; fleshy or tough parts have to be treated on a grater; the mass, obtained in either way, is brought on a square or circular piece of silk gauze (so-called bolting silk-gauze, Nos. 10–13), the latter is made to assume the shape of a bag and tied so as to enclose the contents firmly. The whole bag is then put into a basin containing pure water, is held by one hand above the knot, and is kneaded by the forefinger and thumb of the other hand. After the water has become very milky, it is poured into a glass jar capable of holding at least four to six times more water. Now put the bag into fresh water and knead again for some time, pour the liquid into the glass, and repeat these operations as often as the water becomes milky. Afterwards allow the liquids, all mixed together in the glass jar, to subside, collect the sediment on a filter, wash with pure cold water until the filtrate is found to leave no residue on evaporating, dry at first at a temperature not exceeding 40°, afterwards at 100°, and determine the weight.

Usually, instead of silk-gauze, linen or calico is used for the same purpose, but the starch-grains cannot pass through linen with equal facility; more kneading is therefore required, causing other parts (fibres, membranes) to pervade the pores and to contaminate the product. Even with silk-gauze part of those impurities are liable to pass through, unless another covering of the bag of the same material be employed, in which case it is possible to obtain the starch as nearly pure as possible.

After the Starch has been obtained, the remnant has to be tested on its thorough exhaustion by throwing a small sample, while moist, into a porcelain dish containing a solution of iodine diluted to a gold-yellow colour. If no violet or blue colouration ensues, the remnant is free from Starch; but if the latter be indicated by a more or less blue tinge, its quantity is determined in another way, suitable also in all cases where the mechanical process would be impracticable. But before this can be done, the substance has to be exhausted first successively by ether, alcohol and cold water in order to free it from any traces of sugar, gum, &c., which might be present.

This indirect estimation of Starch is effected by converting it into grape-sugar and by submitting the latter to the agency of an alkaline solution of sulphate of copper, containing tartaric acid (see under "Reagents," in Part II.). For this purpose the substance is dried after having been treated to the above-named solvents, and is put into a glass-flask containing an equal weight of concentrated sulphuric acid, which is diluted with 50 times its weight of water; the contents of the flask are then heated to boil gently. From time to time a drop of the liquid is taken out with a glass-rod and put into a porcelain dish, adding a drop of solution of iodine. When neither a blue nor a violet or reddish colour is produced, the flask is left to cool, the acid liquid (containing the Starch as grape-sugar) is filtered and washed until the acid reaction has disappeared; the liquid is mixed with the water used for washing; saturated (cold) with soda-ley, and its volume ascertained by cubic-centimeters.

Now, measure from the blue alkaline solution of sulphate of copper, 10 cubic centimeters, pour them into a flask, holding about 100 cubic centimeters, add 40 cubic centimeters water, heat the mixture to a gentle boiling heat, and add of the above neutralised solution of sugar gradually and in intervals, until every trace of blue has disappeared, and in its stead a yellowish tinge is observable. To find out the exact moment of the change of colour, place the flask on a piece of white paper. This decolouration of the cupric solution takes place after 0·05 grammes of grape-sugar, corresponding to 0·045 grammes of starch, have been added. It is easy herefrom to calculate the concentration of the liquid in question, and its previous amount of Starch. The quantity of Starch found in this way has to be added, if necessary, to that obtained before by kneading the substance.

The sulphuric acid, used for converting the Starch into grape-sugar, being very diluted, neither vegetable fibrin nor pectin which might be present, affect by changes of theirs the calculation.

Though, under all circumstances, the Starch is characterised with certainty and precision by its behaviour to iodine; there are differences of form and size, which to determine demands a microscope magnifying at least 400 diameters. Should it not be possible to isolate the Starch, thin slices of the substance, wherein Starch has been indicated by iodine, are submitted to the microscopical examination.

**Stearic Acid** $= C_{36} H_{35} O_3 + HO$. Contained as tristearin in fats, especially solid ones. Saponify with soda-ley, decompose the soap with hydrochloric acid, dissolve the fat-acids in hot alcohol, allow to crystallise, press and recrystallise repeatedly, until the product fuses at 69·1° to 69·2°. Shea-butter is of all vegetable and animal fats the best adapted for preparing pure stearic acid, as it contains only the one solid fat-acid.—Forms pearly needles and leaflets, inodorous and tasteless, of perceptibly acid reaction, fuses at 69·1° to 69·2°, has at 9° to 11° a density of 1·000, boils and distils in a vacuum unaltered, is not soluble in water, dissolves in 40 parts cold absolute and in every proportion in boiling alcohol, in 8·3 parts cold and in every proportion of boiling ether, in 3½ parts sulphide of carbon, in 4½ parts of benzol, in 10 parts cold concentrated sulphuric acid colourless. The stearates have the consistency of hard soaps and plasters, and are insoluble in water, except the stearates of the alkalies.

**Stearin** (Tri-stearin) = $C_{114} H_{110} O_{12} = C_6 H_5 O_3 + 3 C_{36} H_{35} O_3$. Is found principally in solid fats. Press, for instance, the shea-butter, which contains as solid fat only Stearin, and re-crystallise in hot alcohol.—White, pearly, radiated warty mass, fine needles and leaflets, inodorous and tasteless, of 0·986 density at 15°, fuses at 62°, dissolves not perceptibly in cold, in 6-7 parts boiling absolute alcohol, in 15 parts boiling alcohol of 0·805, in 66 parts boiling alcohol of 0·822 (on cooling the Stearin separates almost entirely), in 225 parts cold ether, most abundantly in boiling, readily in volatile oils.

**Stearoptens.** *See* ESSENTIAL OILS.

**Stillistearic Acid.** In the fat of Excæcaria sebifera; coincides with palmitic acid.

**Storax, liquid.** Exudation of the stem of Liquidambar orientalis. Dark-brown or greenish-grey, partly ash-grey mass of turpentine consistence, has a very pleasant balsamic odour similar to solid storax, a pungent aromatic taste, and an acid reaction.—Consists of a resin, cinnamic acid, volatile oil (Styrol), and a neutral crystalline body (Styracin=Cinnamate of Cinnamyl.

**Storax, solid.** Exudation of the stem of Styrax officinale. Appears mostly as a brown, glossy, somewhat glutinous mass, readily softening by the warmth of the hand, of an extremely pleasant balsamic odour, tastes sweetish-aromatic, stimulating and slightly bitter, and dissolves entirely in alcohol. Consists of resin, volatile oil and benzoic acid.

**Stramonin.** Peculiar, indifferent body of the seeds of Datura Stramonium, and some other species. Is obtained from the oil that forms by treating the alcoholic tincture with hydrate of lime and acidifying the filtered liquid, and has to be purified by re-crystallising.—White, small, inodorous and tasteless crystals, fusible at 150°, sublimating unaltered on careful heating, insoluble in water, slowly soluble in alcohol, better in ether, also in fixed and in volatile oils and in kreosot; of neutral reaction, concen-trated sulphuric acid yields a blood-red solution, diluted acids or alkalies have no effect. Metallic salts produce no deposit.

[**Strophanthin.** Poisonous principle of the Kombi-arrow-poison of West Africa, obtained from the seeds of Strophanthus hispidus, or another species of Strophanthus. Fraser obtained the Str. from the alcoholic extract of the seeds.]

**Strychnin**$=C_{42} H_{22} N_2 O_4 + 2$ HO. In various species of the genus Strychnos, for instance in the seeds and bark of Str. Nux vomica, in the seeds of Str. Ignatia, in the wood of Str. colubrina, in the root of Str. Tieute, therefore also in the arrow-poison, prepared from such like plants, associated with brucin. For its preparation preferably the seeds of Str. Nux vomica are used. Digest them in the rasped state with three changes of alcohol of 40%; distil the alcohol from the liquids, purified by straining, pressing and subsiding, evaporate the remaining liquid, until its weight be equal to that of the seeds employed, precipitate with acetate of lead, filter, digest the filtrate cold with burnt magnesia for several days, collect the sediment, wash, dry, triturate and digest warm with alcohol of 80%. Distil the tincture to a small volume, let rest cold, collect the crystals of Strychnin, wash with weak alcohol and recrystallise in hot alcohol of 80% (the mother-ley separated from the previously formed Strychnin, may be used for the preparation of brucin).—The Strychnin crystallises in white, quadrangular, acuminated prisms, is inodorous, of an insufferably bitter taste, undergoes no alteration in a gentle heat, fuses with more heat to a pale-yellow liquid and decomposes afterwards; dissolves in about 6000 parts water, in 120 parts cold and in 10 parts boiling alcohol of 80%, to solutions of a slightly alkaline reaction; is not soluble in ether and in alkalies; dissolves incompletely in chlorine water, producing by addition of ammonia voluminous white flocks, which change soon to rose-red; dissolves readily in concentrated sulphuric acid without colouration, the solution assuming a purple-violet hue on addition of a few particles of chromate of potash or of ferricyanide of potassium; nitric acid, when cold, dissolves the Str. colourless, the solution turning greenish-yellow on heating, and assuming a milky appearance with subchloride of tin.

[Strychnin, when mixed with concentrated sulphuric acid, assumes on addition of oxyd of cerium a beautiful blue colour, slowly changing to purple, and lasting several days. A very delicate test.]

**Styracin**$=C_{36} H_{16} O_4$, or Cinnamate of Cinnamyl$=C_{18} H_9 O + C_{18} H_7 O_3$. Crystalline substance of liquid storax (from Liquidambar orientalis), and probably also of balsam of Peru (from Myroxylon Pereirae). Treat liquid storax with 5 to 6 parts diluted soda-ley at a temperature not above 30°, until the remnant has become colourless, wash, dry, dissolve in ether-alcohol, and allow to crystallise.—Crystallises in colourless prisms and needles, inodorous, tasteless, fuses at 44°, is not volatile, insoluble in water, little soluble in cold alcohol, in 3 parts ether, yields on heating with chromate of potash and sulphuric acid, oil of bitter almonds, benzoic acid, and a resin.

P

**Styrol**$= C_{16} H_8$. The volatile oil of liquid storax. Distil the latter with water containing some carbonate of soda (to retain cinnamic acid), shake the oil floating on the distillate with chloride of calcium and rectify.—It is colourless, thin, smells like liquid storax, has a burning taste and 0·924 density, boils at 145·75° under conversion into an isomeric solid body (metastyrol) which is devoid of taste and smell, but is reconverted on slowly heating into the former liquid state.

**Suberin,** a modified woody fibre, main ingredient of the outer bark of Quercus Suber. Remains after the treatment of the rasped cork with water, alcohol, ether, and hydrochloric acid as a reddish-grey, very light, soft, elastic mass of cellular structure.

**Succinic Acid**$= C_4 H_2 O_3 + HO$. It undoubtedly occurs very frequently in the vegetable kingdom, though its presence, as recorded in most statements, has not been proved sufficiently. It is said to exist in turpentine, in the herbs of Lactuca sativa and L. virosa, and in Artemisia Absinthium; sometimes it is probably a product of decomposition of vegetable extracts, and therefore no primary constituent of plants. Its occurrence in turpentine is rendered probable by the fact that amber, which contains a large amount of the acid, comes from an extinct coniferous tree. The Succinic Acid in plants is combined with a base, and is in this state readily dissolved by water. Acetate of lead throws it down, yielding a compound which becomes anhydrous at 130°, and contains 30·94% acid. From the Succinate of lead the acid can be obtained without loss only by means of sulphuret of hydrogen; the liquid, after being separated from the sulphide of lead, yields, on evaporation, the acid as a hydrate. With larger quantities and when a small loss is of no consequence, the Succinate of lead is digested warm, with $\frac{1}{3}$ its weight of concentrated sulphuric acid, and with the necessary quantity of water; it is then filtered, evaporated to form in crystals, and purified by recrystallisation.—The pure acid forms klinorhombic prisms, inodorous, of a moderately acid taste, fuses at 180°, boils at 235°, and volatilises undecomposed in white, acrid fumes; dissolves in 25 parts cold and in two parts boiling water, readily in alcohol and in ether. Most of the Succinates are soluble in water.

**Sugar.** *See* Fruit, Cane and Grape Sugar.

**Sulphosinapin**
**Sulphosinapic Acid** $\Big\}$ = Sinapin-Sulphocyanide.
**Sulphosinapisin**

**Surinamin.** In the bark of Geoffroya Surinamensis. Treat the alcoholic extract of the bark with water, filter, precipitate with subacetate of lead, remove the lead from the filtrate by means of sulphuret of hydrogen, filter and evaporate, whereby a part of the Surinamin is obtained. The rest is got by digesting the liquid with magnesia, filtering and evaporating again. Wash the Surinamin with cold water and recrystallise in hot water.—Forms white, very fine, voluminous, cotton-like needles of insipid taste; neutral; is partly carbonised and partly sublimates on heating; dissolves very little in cold, readily in boiling water, almost insoluble in cold, little soluble in boiling alcohol, readily in diluted acids, also in potash-ley.

**Sycoceryl-Alcohol** $= C_{36} H_{30} O_2$. As acetate of Sycoceryl in the resin of Ficus rubiginosa. Withdraw the sycoretin from the resin by means of cold alcohol, and boil the remnant with alcohol, the solution forming on cooling crystals of acetate of sycoceryl and afterwards a small quantity of another flocky substance. By cooling the solution to 40°, straining the crystals, recrystallising in boiling alcohol and treating at 30° with ether so as to leave a little of the substance undissolved, the acetate of Sycoceryl is obtained pure, while a neutral crystalline substance, insoluble in ether, remains. Decompose the acetate through boiling with a solution of caustic soda in alcohol, precipitate the Sycoceryl-alcohol which has formed with water and recrystallise in alcohol.—Forms wawellite-like, very thin crystals, similar to caffein; fuses at 90°, and is volatilised partly undecomposed by more heat; is insoluble in water and in alkalies, readily soluble in alcohol, ether, benzol, and chloroform.

**Sycoceryl-Acetate** $= C_{36} H_{29} O + C_4 H_3 O_3$. Ingredient of the resin of Ficus rubiginosa, and doubtless of other species, obtained from it according to the foregoing paragraph.—Appears in thin, mica-like leaflets or sexangular tabular crystals; neutral; fuses at 118° to 120°, distils unaltered, dissolves most readily in hot alcohol, in acetic acid, aceton, ether, benzol, oil of turpentine.

**Sycoretin.** In the resin of Ficus rubiginosa. This resin separates on treating with cold alcohol into about 73% soluble Sycoretin, 14 sycoceryl-acetate, soluble in hot alcohol, and 13 residue (caoutchouc, sand and fragments of bark). By mixing with water the neutral, light-brown solution in cold alcohol, the Sycoretin subsides and may be obtained colourless by repeatedly dissolving and precipitating.—Amorphous, white, neutral, very brittle, very electric; fuses in boiling water and floats on it like an oil, fuses by itself only at 300°, and decomposes afterwards; is insoluble in water, diluted acids and alkalies, readily soluble in alcohol, ether, chloroform, oil of turpentine; in concentrated sulphuric acid with a beautiful green colour without formation of sugar.

**Sylvic Acid.** *See* ABIETIC ACID.

**Synaptase.** This is the nitrogenised body which possesses the faculty of separating amygdalin into hydrocyanic acid and its other products, and which, in combination with albumen, constitutes emulsin. It is obtained by mixing the press-residue of sweet almonds with water, pressing after two hours, filtering the liquid, precipitating the albumen with acetic acid, filtering again, removing the gum by acetate of lead, precipitating excess of lead by sulphuret of hydrogen and mixing the filtrate with alcohol, which throws down the Synaptase; the latter has to be washed with alcohol and dried in a vacuum.—Yellowish-white, either brittle and glossy like gluten, or opaque and spongy like sarcocolla, very soluble in cold water, almost insoluble in alcohol; the aqueous solution becomes soon decomposed at the air, curdles at 60°, is not precipitable by acids or by acetate of lead, but considerably so by tannic acid; acts strongly upon amygdalin even at 80°. The Synaptase does not behave towards starch similarly to diastase.

**Syringin** $= C_{38} H_{28} O_{20} + 2 HO.$ Tasteless glucosid of the bark of Syringa vulgaris, and other species especially developed in early spring, found in the leaves and half-matured fruits, and only in traces in the leafbuds; it disappears during the progress of vegetation, and in its stead appears syringopicrin; is also contained in the bark of Ligustrum vulgare and other Privets. Precipitate the decoction of the bark with subacetate of lead, treat the filtrate with sulphuret of hydrogen, filter, evaporate to a syrup consistence, press the crystalline mass and recrystallise in hot water with aid of animal charcoal.—Forms long, colourless needles, tasteless, neutral; loses its water at 115°, fuses at 212°, and becomes decomposed with the odour of burnt sugar; dissolves slowly in cold, readily in hot water, in alcohol, not in ether; dissolves in concentrated sulphuric acid with dark-blue colour, and forms, on addition of water, grey-blue flocks; dissolves in concentrated nitric acid with deep-red colour; breaks up with diluted acids under formation of sugar; is not precipitable by metallic salts.

**Syringopicrin** $= C_{26} H_{24} O_{17}.$ Bitter glucosid of the bark of Syringa vulgaris. Remains after the preparation of syringin in the mother-ley and is absorbed by animal charcoal. Wash the coal with warm water and boil with alcohol, which dissolves the S. and leaves it, after evaporating, in the form of a brown syrup-like liquid. Purify by dissolving in alcohol, decolourising with animal charcoal, evaporating and treating the remnant with ether, which dissolves an acrid substance, leaving the S. undissolved.—Yellowish, pellucid mass, friable to a permanent white powder, of a very bitter taste and of acidulous reaction; fuses below 100° and is decomposed by more heat; dissolves readily in water and in alcohol, not in ether, is not altered or precipitated by alkalies, by chloride of iron or by sub-acetate of lead, seems to yield sugar on treating with diluted sulphuric acid, for the liquid, obtained by this treatment, reduces the alkaline tartarate of copper.

**Tacamahac.** Exudation of the stem of Bursera tomentosa and of Calophyllum Inophyllum. The resin of the first-named tree is light-brown, opaque, of pleasant odour and of lasting bitter taste; dissolves almost completely in alcohol. The resin of the other tree is yellowish, smells after lavender, has a slightly acidulous taste; dissolves readily in alcohol and in alkalies.

**Taiguic Acid.** The yellow pigment of the Taigu-wood of Paraguay, the origin of which is unknown. Obtained from the wood by means of cold alcohol, and purified by treating repeatedly with alcohol and with ether.—Forms beautiful yellow crystals, turning slowly brown at the air, tasteless; fuses at 135°, volatilises at 180° undecomposed; contains no nitrogen; dissolves in 1000 parts boiling water, in 86 parts alcohol of 0.840, in 19 parts ether, in 16 parts aceton, in 45 parts benzol, also in sulphide of carbon, petroleum, in alkalies with red colour.

**Tallow, Chinese.** The fatty covering of the seeds of Excaecaria sebifera (the kernels contain a liquid fat). It yields a greenish-white tallow-like fat, fusing at 44°, and a white one, fusing at 37°; both containing olein and palmitin.

**Tanghinin.** The poisonous ingredient of the seeds of Tanghinia venenifera. Is obtained by extracting with ether from the seeds which have been freed by pressing from most of the fixed oil, and by evaporating the tincture.—Colourless crystals of a very bitter and acrid taste, fusible with a gentle heat, not volatile, insoluble in water, soluble in alcohol and in ether, little affected by acids or by alkalies.

**Tannaspidic Acid** $= C_{26} H_{14} O_{11}$. In the rhizome of Aspidium Filixmas and some other species. Boil with alcohol of 75 to 80 % and mix the decoction with water, hydrochloric acid and pulverised sulphate of soda, to produce a precipitate, which contains Tannaspidic and pteritannic acids. Collect this deposit, wash with solution of sulphate of soda, press, triturate with water and digest at 60 to 80° with water containing hydrochloric acid, for half an hour, removing thereby ammonia and other bases. Wash the remnant with water, dry and exhaust with anhydrous ether, which dissolves the pteritannic acid. Filter, warm the residue with strong alcohol, add a few drops of solution of acetate of lead, throw down the latter with sulphuret of hydrogen (to make the liquid apt for filtering), filter and evaporate the filtrate in a current of hydrogen and at last in a vacuum over sulphuric acid; the Tannaspidic acid thus obtained is pure when it is quite insoluble in water and in ether, but completely soluble in alcohol.—Black-brown, amorphous, glossy mass, friable to a spaniol-coloured powder, inodorous, of slightly adstringent taste and of acidulous reaction; is insoluble in volatile and in fixed oils, its alcoholic solution precipitates the glue, not the tartarated antimony, greens chloride of iron; throws down a red powder when digested hot with diluted acids.

**Tannicorticipinic Acid** $= C_{28} H_{13} O_{12}$. Known in the bark of Pinus sylvestris from 20-25 years old trees. Boil with alcohol of 40%; distil the alcohol completely from the tincture, remove a viscid resin from the remnant by filtering, precipitate the filtrate with acetate of lead, wash the deposit and add subacetate of lead to the filtered liquid, obtaining thereby a deposit, which contains, like the first, Tannicorticipinate of lead. Treat the deposit, produced by acetate of lead, three times with acetic acid, but avoid to dissolve the whole; free from the undissolved portion containing resin, by filtration, precipitate the liquids with subacetate of lead, collect the deposit, wash and decompose under water with sulphuret of hydrogen. The liquid, freed by filtering from the sulphuret of lead and evaporated to half its volume in a current of carbonic acid gas, forms brown-red crusts of the T. acid. B.—Decompose the deposit, obtained by subacetate of lead, under water with sulphuret of hydrogen, evaporate the filtrate in a current of carbonic acid gas, dissolve the remnant in alcohol, precipitate with alcoholic acetate of lead, wash and decompose the deposit under water with sulphuret of hydrogen, evaporate the filtrate in a current of carbonic acid gas, and dry at 100°.—Reddish-brown powder of astringent taste, colours chloride of iron—at first dark-green, afterwards red-brown, and produces at last a black-green deposit; yields a red product by heating with diluted acids.

**Tannic Acids.** Derive their name from one of their generally known properties, viz., of tanning animal membrane, *i.e.* of converting it into a durable compound, called leather. They are also distinguished by their faculty of forming with glue compounds more or less soluble in water; by their acid reaction towards litmus-paper; by their astringent, but not acid, taste; by their amorphous condition; by their property of forming deep-blue or green, sometimes brown, compounds with iron oxyd or with iron oxyd-suboxyd, not with pure iron suboxyd; and by their disposition to decompose in aqueous solutions and under access of the air and more readily so in the presence of alkalies.

The number of Tannic acids is very large, but of their chemical constitution we know as yet little or nothing. This accounts for the confusion which prevails in regard to their classification. Provisionally they are generally divided into acids which blue, and into those which green the iron-salts; but the colours of the liquids or of the deposits which are produced by the mutual action of these bodies and of iron-salts vary for one and the same substance, according to the state of oxydation of the iron, and are also influenced by the nature of the acid constituent of the iron-salt, by the greater or less acidity, and by the concentration of the liquids, and all this to such an extent that it is possible to produce all the gradations of blue or of green, and sometimes even from blue to green with the same substance. Besides, as stated already, there are a few Tannic acids which, under the same conditions, produce neither of the two colours, but only a dirty-brown deposit. Even the property of precipitating glue does not seem to be general, as evidenced by a few of the iron-greening acids.

Another distinguishing feature of Tannins is their behaviour, when submitted to dry distillation. Those which produce green iron compounds, yield, under these conditions, Pyrocatechin (Pyrocatechuic acid), while the Tannic acids of the blue reaction with iron-salts yield sometimes Pyrogallic, sometimes Pyrocatechuic acid.

Tannic acids are widely diffused in plants, especially prerennial ones. They occur in all parts of their organism, but predominantly in roots, barks, and young woods, less frequently in the teguments of fruits and seeds, seldom in the leaves. If not present in too small quantity, they are easily detected by the taste and by the reaction of their solutions with glue and iron-salts. Some plants contain two different Tannic acids, i.e., one of them producing a green iron compound, the other a blue one under equal conditions; or the two acids may appear identical in regard to iron-salts, but yet differ in other respects. For instance, the Tannic acids of nut-galls, and of oak-bark, produce severally blue iron-precipitates, but the former substance, by dry distillation, yields Pyrogallic acid, while the latter does not.

To determine quantitatively the amount of a Tannic acid, of whatever kind: prepare an aqueous solution or extract of one to two grammes of the substance in question, add to the filtered liquid solution of acetate of baryta, as long as a precipitate ensues, filter and precipitate the filtrate with acetate of lead, collect the deposit on a filter, wash dry at 120°, and note down the weight of the dry precipitate of Tannate of lead, incinerate the latter in a porcelain crucible at a red heat, moisten with nitric acid, and heat again, ascertain the weight of the remaining lead-oxyd, and deduct this from the weight of the dried Tannate of lead—the rest represents the weight of the Tannic acid.

### Tanningenic Acid = CATECHUIC ACID.

**Tannopinic Acid** = $C_{25} H_{15} O_{13}$. Known to exist in the leaves of Pinus sylvestris, and during spring-time, to replace the oxypino-tannic acid, oxydises readily in warm, moist air, yields a red product on heating with diluted acids.

**Tartaric Acid**$=C_4 H_2 O_5 + HO$. Occurs rather frequently, either in the free state or half or completely saturated, especially in sour, unmatured, and in sweet berries (for instance, in grapes); in small quantity in roots, barks, woods, herbage, and abundantly in Lycopodium complanatum, and probably to be found in most allied plants. From the aqueous extracts of vegetable substances it always passes into the deposit produced by acetate of lead, and into the liquid that results, after the deposit has been decomposed by sulphuret of hydrogen, and from which it may be obtained in crystals by slow evaporation. The deposit of lead, should it contain no other acid, may be used for its quantitative determination by drying at 100° and weighing. In 100 parts, are contained 37.08 parts acid. The pure acid crystallises in colourless, klinorhombic prisms and pyramids, is inodorous, of a pure and strongly acid taste, fuses at 170° and becomes carbonised with the odour of burnt sugar, dissolves in 2 parts cold and in 1 part hot water, also readily in alcohol, in 36 parts ether. The aqueous solution yields with lime-water a deposit soluble in sal-ammoniac. With potash, it forms tartar. The Tartarates of the true alkalies are in the neutral state readily soluble in water, the acid ones only sparingly; the neutral Tartarates of most of the other bases are sparingly, or not, soluble in water, but soluble in tartaric and in hydrochloric or in nitric acids. All Tartarates dissolve in liquor of ammonia and in the hydrates of potash and of soda, with the exception of Tartarate of silver, which does not dissolve in the latter two liquids, and of Tartarate of mercury, which does not dissolve in any of the three.

**Taxin.** Alkaloïd of the leaves of Taxus baccata, prepared according to Stas' method of the forensic investigation of alkaloïds. —A white, loose, amorphous powder, very bitter, slowly soluble in water, readily in alcohol and in ether, fusible to a yellow resin by a gentle heat; also soluble in diluted acids, the solutions of which yield no crystals. Precipitable by caustic alkalies, tannic acid, tincture of iodine, not by chloride of platinum. Concentrated sulphuric acid effects a purple-red solution which becomes decolorised by water.

**Thallochlor.** The green pigment of Cetraria islandica. Exists in the ether, containing oil of rosemary, used for washing the cetraric acid (see this). Evaporate to dryness the solution, remaining after the cetraric acid has been removed by crystallisation, dissolve the remnant in boiling alcohol, dilute the alcohol with water until of a strength of about 42%, and filter boiling hot. Repeat this several times, in order to remove lichestearic acid, and draw out the dried remnant with petroleum. Thallochlor and fat are dissolved, while cetraric acid and brown substances remain behind. Submit the solution to distillation under addition of water, dry the remnant, until the whole of the petroleum is driven away, dissolve in alcohol and throw down the Th. either by digesting with hydrate of lime or better by alcoholic acetate of lead. The green flocks obtained are, after filtering and boiling with ether, separated from the lead-oxyd by means of acetic acid. —Brittle, green mass, insoluble in water, scarcely soluble in hydrochloric acid, soluble in strong alcohol, ether, and oils.

**Thebain** $= C_{38} H_{21} NO_6$. In opium, to all appearance in Papaver Rhoeas too, associated with rhoeadin. By heating, as indicated under morphin, the aqueous extract of opium with milk of lime, the morphin remains dissolved, while the Thebain is left in the lime-sediment. Wash the latter, dry, boil with alcohol, evaporate the extract and treat the remaining brown, granular mass with ether, which dissolves the Thebain, and leaves it on evaporating as a brown, crystalline mass. Purify by dissolving in acid, precipitating with ammonia and recrystallising in alcohol or in ether.— White, silvery, quadratic leaflets, also needles, grains and cauliflower-shaped masses, very electric on rubbing, of a more acrid and styptic than bitter taste, of alkaline reaction; fuses at 150° without loss of weight, and decomposes in a higher temperature; dissolves not, or slightly in water; readily in alcohol, ether, diluted acids, becomes deep-red with concentrated sulphuric acid, and dissolves with yellow colour; becomes blood-red with sulphuric acid, containing nitric acid. Its salts are not crystallisable from water, but are so from alcohol and from ether.

**Thein** = CAFFEIN.

**Theobromin** $= C_{14} H_8 N_4 O_4$. Alkaloïd of the seeds of Theobroma Cacao. Precipitate the aqueous extract of the prepared seeds with acetate of lead, filter, free the filtrate from lead by sulphuret of hydrogen, concentrate to honey-consistence, boil with alcohol, evaporate the tincture and recrystallise what has formed. —White, crystalline powder, inodorous, of a bitter taste, slightly similar to cacao, sublimates at 290° to 295°, fusing at the same time; dissolves in 1600 parts of cold and in 55 parts boiling water, in 1460 parts cold and 47 parts boiling alcohol of 80 %, in 17,000 parts cold and in 600 parts boiling ether; all these solutions having a neutral reaction; dissolves readily in alkalies and in diluted acids. The very diluted solution in nitric acid, yields with nitrate of silver a silvery-white, crystalline deposit; the ammoniacal solution yields with nitrate of silver a jelly-like deposit soluble in warm liquor of ammonia. On boiling this solution ammonia is evolved and a colourless, granular, crystalline deposit of Theobromin-silver $= C_{14} H_7 Ag N_4 O_4$ is obtained.

**Thujin** $= C_{40} H_{22} O_{24}$. Dissolve the deposit, obtained in the following paragraph by means of acetate of lead, in diluted acetic acid, filter, precipitate the filtrate with subacetate of lead, decompose the deposit under water with sulphuret of hydrogen, heat the whole mixture, filter hot and evaporate in a current of carbonic acid gas and in a vacuum. Dissolve the crystals which have formed in boiling water under addition of alcohol, and recrystallise.—Forms glossy, citron-yellow, microscopic, tabular crystals of astringent taste; is decomposed by heat, insoluble in water, readily soluble in alcohol, separates by diluted acids into sugar and thujetin ($C_{28} H_{14} O_{16}$), dissolves in alkalies under decomposition, colours chloride of iron dark-green.

**Thujogenin** = $C_{28} H_{12} O_{14}$. Known to occur in small quantity in the green parts of Thuja occidentalis, is also obtained on warming thujin with hydrochloric acid. Boil with alcohol, strain, let cool, separate from the wax, distil the alcohol from the filtrate, and mix the remnant with water, and with a few drops of dissolved acetate of lead, in order to facilitate filtering. Precipitate the filtrate completely with acetate of lead and reserve the yellow deposit, containing thujin and thujetin ($C_{28} H_{14} O_{16}$) for the preparation of these substances. The filtered liquid produces with subacetate of lead another deposit, containing the Thujogenin; divide in water, decompose with sulphuret of hydrogen, heat the liquid with the sulphide of lead, filter hot, and evaporate in a current of carbonic acid gas and in a vacuum, when the Th. will form in flocks.—Microscopic needles, very sparingly soluble in water, readily in alcohol; the latter solution assuming a splendid green-blue colour with ammonia.

**Thymen** = $C_{20} H_{16}$. Forms with cymen the more volatile part of the oil of thyme. Rectify repeatedly with caustic potash the portion of the raw oil which distils between 160° and 165°, and distil afterwards by itself, when the Thymen will pass over at 160° to 165° and the cymen at 175°.—It is colourless, of a pleasant odour of thyme, of 0·868 density.

**Thymol** = $C_{20} H_{14} O_2$. The solid ingredient of oil of thyme, also contained in the volatile oil of Monarda punctata and of the seeds of Carum Ajowan. Distil the oil of thyme by itself, when thymen and cymen pass over first and afterwards Thymol; press the latter, after it has solidified, and recrystallise from alcohol.--Thin, rhomboidal, tabular crystals of a pungent, aromatic taste, neutral, fusing at 44° to 50°, boiling at 220° to 230°, of a density = 1·028 when solid, in the liquid state lighter than water.

**Tobaccocamphor** = NICOTIANIN.

**Tolen** = $C_{20} H_{16}$. In the balsam of Tolu. The oil, obtained from the latter by distillation with water, is a mixture of cinnamein (according to Flueckiger and Hanbury, it contains no cinnamein), cinnamic acid and Tolen; the latter distils, on heating the oil for a rather long time, to 160°, and is obtained pure by repeatedly rectifying with hydrated potash and retaining the portion which distils first.—Colourless, thin, smells like elemi, has a pungent, acrid, pepper-like taste; of 0·858 density; boils at 154° to 160°.

**Tolubalsam.** Exudation of the stem of Myroxylon toluiferum. In the fresh state thickish, yellow, becomes slowly darker and solid, has a very pleasant smell. The dry balsam is also named Opobalsamum siccum. Consists of resin, volatile oil, and cinnamic acid. The resin dissolves readily in alkalies, and has the formula $C_{18} H_{10} O_5$. The volatile oil, obtained by distilling the balsam with water, contains a hydrocarbon (Tolen = $C_{20} H_{16}$).

**Toncacamphor** = CUMARIN.

**Tragacanth-substance** = BASSORIN.

**Trehalose** = $C_{12} H_{11} O_{11} + 2$ HO. Peculiar kind of sugar in the trehala-manna of Syria, an amylaceous substance, gathered by coleopterous insects, and converted by them into a cocoon, consisting of about 66% starch, 5 gum, and 29 Trehalose. The latter forms rectorhombic crystals, has a less sweet taste than cane-sugar, loses at 25° to 30° partly, at 100° completely, 2 equivalents water; fuses on rapidly heating to 100°; but is not fused even at 180°, if desiccated before. Dissolves readily in water, is almost insoluble in cold, soluble in boiling alcohol, not in ether, ferments with yeast very slowly and incompletely; is not altered through boiling with alkalies, alkaline earths and alkaline tartarate of copper; yields with nitric acid, oxalic, but no mucic acid, is carbonised on heating with concentrated sulphuric acid, becomes converted into grape-sugar by heating with diluted sulphuric acid, is precipitated by ammoniacal acetate of lead.

**Trimethylamin** = $C_6 H_9 N$ (isomeric with propylamin and formerly confounded with it). Volatile alkaloïd of the herb of Chenopodium olidum, of the flowers of various Pomaceæ (for instance, Cratæguscoccinea, C. monogyna, C. Oxyacantha, Pyrus communis, P. aucuparia), of the seeds of Fagus sylvatica, of the ergot, of the rust fungus of wheat. Is obtained by distilling with a fixed alkali and water, saturating the distillate with sulphuric acid, evaporating, shaking the salty mass with ether-alcohol, removing the sulphate of ammonia by filtration, evaporating the filtrate, shaking the residue with potash-ley and afterwards with ether, decanting the ether and evaporating the ethereous solution in a vacuum.—Colourless liquid of a nauseous, ammoniacal, herring-like odour; precipitable by tannic acid, bi-iodide of potassium, chloride of mercury and iodide of potassio-mercury, soluble in water, alcohol and ether in every proportion.

[**Triticin** = $C_{12} H_{11} O_{11}$. Contained in the juice of the roots of Triticum repens.—A tasteless, amorphous, gummy substance, easily transformed into laevulose (fruit-sugar), if its concentrated solution is kept for a short time at 110°. Treated with nitric acid, it yields oxalic acid. Deflects in solution polarised light to the left.—H. MUELLER.]

**Tulucunin**$= C_{20} H_{14} O_8$. Bitter ingredient of the bark of Carapa guianensis. Boil the bark with water, evaporate the decoctions to a thick syrup consistence, treat repeatedly with alcohol of 33° Baumé, warm the whole of the alcoholic solutions, add milk of lime, which precipitates almost completely the colouring matters, let subside, filter, add water to the filtrate, distil the alcohol, concentrate the residue to honey-consistence, treat with strong alcohol, evaporate the solution to a syrup consistence, shake the latter with chloroform and allow the chloroform solution to evaporate.—Pale-yellow, amorphous substance, of very bitter taste and of acidulous reaction, dissolves not in ether, readily in alcohol, chloroform, in 150 parts cold, and a little more in hot water. With concentrated sulphuric acid it becomes at first brown, then slowly blue; by adding immediately after the acid a few drops of water a splendid blue colouration is instantly produced which lasts more than twenty-four hours. A similar blue tinge is obtained with hydrochloric and phosphoric acids, also with warm citric, tartaric and oxalic, not with acetic or nitric acids.

**Turpentine.** This name was originally applied to the resinous exudation of the terebinth-tree (Pistacia Terebinthus), and which is obtained either spontaneously or by incisions made in the stem; but afterwards all the similar exudations of coniferous trees in general have been termed likewise. The Turpentines are, as a rule, yellowish-white, very viscid, transparent or translucid masses of honey-consistence and of acid reaction, of a peculiar, strong, mostly unpleasant odour, and generally of a burning, aromatic, bitter, disagreeable taste; consist chiefly of resin and volatile oil. They dissolve in alcohol more or less readily, in ether, in oils, also in potash-ley, the latter solution being precipitable by excess of potash. According to their origin they have different names.

**Turpethin** $= C_{68} H_{56} O_{32}$. (Isomeric with jalapin, but not identical with it.) Resinous glucosid of the root of Ipomæa Turpethum. Shake with ether (which dissolves about 5%) the crude resin obtained by alcohol, &c. dissolve the remnant in alcohol, precipitate with ether and dry.—It is not decolourised by animal charcoal, is brown-yellow; inodorous, has at first no perceptible, afterwards an acrid and bitter taste, fuses at 183°, dissolves readily in alcohol, not in ether, separates on boiling with diluted acids into sugar and Turpetholic acid$= C_{32} H_{32} O_8$, a white mass consisting of microscopic needles and tufts, soluble in alcohol, less so in ether.

**Tyrosin** $= C_{18} H_{11} NO_6$. Has as yet been found only in the South American ratanhia extract from Krameria triandra, K. Ixina and K. secundiflora. To prepare the Tyrosin, precipitate the aqueous solution of the extract first with glue, afterwards with subsulphate of iron and next with lime, evaporate the liquid, to deposit at first sulphate and carbonate of lime, afterwards Tyrosin, which has to be recrystallised in hot water.—White, loose, warty mass, composed of fine needles, or isolated silky needles, without smell or taste ; is decomposed by heat ; dissolves in 2788 parts cold and in 138 parts boiling water, not in alcohol and ether, readily in the hydrates and carbonates of alkalies and in alkaline earths, likewise in diluted acids, in concentrated sulphuric acid, and the latter solution, after it has been saturated with carbonate of lime, and freed from the sulphate of lime by filtering, assumes a violet hue with chloride of iron.

[**Umbelliferon** $= C_{18} H_6 O_6$, seems, according to Flueckiger and Hanbury, to pre-exist to a small extent in galbanum, asafœtida and ammoniacum, and is also obtained by the dry distillation of resins of umbelliferous plants in general and of Daphne Mezereum.—It forms colourless, acicular crystals, soluble in water, ether, and chloroform. Its solution in water exhibits, especially on addition of an alkali, a brilliant blue fluorescence, which is destroyed by an acid. It may be prepared from galbanum, by heating the latter for some time to 100° with hydrochloric acid, and treating the cold acid liquid with ether or chloroform, which takes up the Umbelliferon.]

**Urson** $= C_{40} H_{34} O_4$. Crystalline substance of the leaves of Arctostaphylos (also found by Tonner in the leaves of Epacris spp.). Exhaust the leaves with ether, wash with ether the crystalline sediment, obtained by evaporation, and recrystallise from alcohol. —Forms colourless, silky needles, without taste and smell, fusible at 198° to 200°, boils in higher temperatures, and sublimes unaltered ; is insoluble in water, diluted acids and alkalies, sparingly in alcohol and in ether.

**Usnic Acid** $= C_{36} H_{18} O_{14}$. In various kinds of Usnea, Cladonia, Evernia, Lecanora, Parmelia, Ramalina ; is best obtained from Cladonia rangiferina or from Usnea florida. Boil with thin milk of lime, throw down the Usnic acid from the dark yellow solution by means of hydrochloric acid, dry the deposit and purify by recrystallisation from alcohol with aid of animal charcoal.— Appears in sulphur-yellow, pellucid needles and leaflets, friable to an electric powder ; tasteless ; fuses at 203°, is decomposed in higher temperatures, and yields a sublimate of Beta-orcin ; is not moistened by and is insoluble in water, dissolves in alcohol and in ether, still more when warm, in concentrated sulphuric acid with yellow colour, and precipitable by water unaltered, readily in alkalies, but becomes decomposed on boiling. Forms salts with bases; those of the alkalies are colourless, crystallisable, become slowly coloured at the air; those of the other bases are obtained in amorphous flocks by precipitation.

**Valeren** = Borneen.

**Valerianic or Valeric Acid** = $C_{10}$ $H_9$ $O_3$ + HO + 2 Ag. In the root and herb of Valeriana officinalis, in the root of Archangelica officinalis, Peucedanum Oreoselinum, and of other Umbelliferæ, in the flowers of Anthemis nobilis and other Compositæ, in various parts of Sambucus nigra and allied species, and probably widely distributed besides. To obtain it, distil the respective vegetable substances with water, saturate the distillate with carbonate of soda, evaporate to dryness, distil the salt with sulphuric acid and a little water, when a concentrated aqueous solution is obtained, on the surface of which most of the acid floats as an oil. This oil is the tri-hydrate; by distilling it by itself a milky fluid passes over, followed by a clear liquid, which is the mono-hydrate. The latter is a colourless, thin, oily liquid, smells peculiarly and more disagreeably than valerian, and at the same time like putrid cheese, has a burning, acid, afterwards aromatic, sweet, and apple-like taste; is of 0.935 density; boils at 132°; dissolves in 30 parts water, mixes with alcohol and ether in every proportion. The Valerates smell like the free acid, after putrid cheese, those of the alkalies and alkaline earths are fatty to the touch, dissolve readily in water, and have a sweet taste; those of the other bases are partly readily, partly slowly, and partly not soluble. The Valerate of lead is readily soluble.

As regards the quantitative estimation of Valerianic Acid, the latter, after being separated from the respective vegetable substance by distillation, ought to be in such a state of concentration as partly to float on the aqueous liquid. Now, add slowly and gradually cold water, until the oily liquid has disappeared, and ascertain the weight of the whole. One hundred parts of this cold, concentrated, aqueous solution of Valerianic Acid contain, or are equal to, 2.941 parts anhydrous acid.

**Valeriantannic Acids** = $C_{14}$ $H_9$ $O_8$ and $C_{12}$ $H_8$ $O_9$. In the root of Valeriana officinalis. Draw out with absolute alcohol, precipitate the tincture with an alcoholic solution of acetate of lead; filter and precipitate again with ammonia. The first deposit contains an acid = $C_{14}$ $H_9$ $O_8$, which does not produce a green colouration with chloride of iron. The other deposit contains a tannic acid = $C_{12}$ $H_8$ $O_9$, which turns iron combinations green.

**Vanillin** = Vanillic Acid.

**Vanillic Acid** = $C_{34}$ $H_{22}$ $O_{20}$. In the fruit of Vanilla aromatica. Forms the crystals, which effloresce on the fruit by keeping. To prepare it, treat the fruit with alcohol, evaporate the tincture to an extract, dilute with water to syrup-consistence, shake with ether, evaporate the ethereous solution and recrystallise from hot water. —Forms colourless, hard, quadrangular needles, has a pure but faint vanilla-like odour, and a similar afterwards pungent taste; fuses at 82°, sublimes above 260°, but unaltered only when rapidly heated; dissolves in 198 parts cold and in 11 parts boiling water, in 6 parts cold alcohol of 93 $°/_o$, and in equal parts boiling alcohol, in $6\frac{1}{2}$ parts cold and in equal parts boiling ether, also in volatile and fixed oils, the solutions in alcohol and ether, having a slightly acid reaction; the aqueous solution assumes a splendid dark-violet tinge with chloride of iron, and yields a pale-yellow deposit with chloride of platinum, and a yellowish-white one with acetate of lead, but is not affected by the nitrates of the suboxyds of palladium and mercury, and of the oxyds of silver.

[To estimate the amount of V. acid in Vanilla, Tiemann and Haarmann exhaust the finely-cut pods with ether, agitate the ethereous tincture with a solution of bisulphite of soda, which combines with the V. acid, and is decomposed by sulphuric acid.]

**Variolarin.** Known to exist in Pertusaria communis. Exhaust this lichen with boiling alcohol, evaporate to honey-consistence; remove from it the orcin by means of water, treat with ether and allow the latter to evaporate, leaving a crystalline remnant, which has to be freed from a soft resin by washing with cold alcohol, and is then dissolved in boiling alcohol. On cooling, long white needles are formed, which dissolve readily in alcohol and in ether, fuse with a gentle heat, decomposing afterwards, and are not altered by alkalies or by acids.

**Vateria Tallow,** obtained from the fruit of Vateria Indica. White or yellow, fatty and waxlike to the touch, of a globularly radiated fracture; of a faint, pleasant odour; tasteless; of 0.926 density; fuses at $36\frac{1}{2}°$.

**Vegetable Mucilage** or simply **Mucilage** $= C_{12} H_{10} O_{10}.$ Under this, in a chemical sense, undefined name, certain matters are comprised, which differ from gum, and are obtained on treating various vegetable substances with cold water, under the form of thickish, turbid, ropy liquids. Prominently rich in it are various seeds (from the orders of Labiatae, Lineae, Plantagineae, Rosaceae, etc.); but also leaves, stalks, barks, roots, for instance, Asperifoliae, Malvaceae, Orchideae, Algae, etc. By mixing with alcohol a liquid prepared as above with cold water, and freed, if necessary, from albumen by boiling and straining, the Mucilage immediately separates, but charged with no inconsiderable quantities of lime-salts, which may be abstracted for the greatest part from the deposit by alcohol containing hydrochloric acid.—The Mucilage is, after drying, yellowish, not so translucid as gum, more tough than brittle; swells up considerably in water, and dissolves to turbid, ropy, neutral liquids, which in many cases are thrown down by acids and by many salts that do not affect the solution of gum, as for instance by alum, subchloride of tin, acetate of lead; but on the other hand they are not turbidified by silicate of potash, nor thickened by borax. Nitric acid produces oxalic and partly mucic acid. By the agency of diluted sulphuric acid, at first gum, then sugar is formed. The different behaviour of different Mucilages is probably caused by the greater or less amount of salts, though this assumption is contradicted by Frank (Chem. Jahresb. für 1865, 598).

The presence of Mucilage can be best detected by directly treating with cold water. For this purpose the roots, barks, leaves and stalks have to be reduced to a proper state by bruising, cutting, etc.; seeds are left whole, as they contain the mucus in the epidermis, and with them, bruising would be either superfluous or detrimental to the purity of the product, because the water would dissolve, or keep suspended, substances that could not be easily removed, as albuminous substances, oils, etc.—The quantitative estimation of Mucilage is also effected by direct treatment with cold water; after subsiding and straining, the liquid must be boiled a few moments. Now, strain off the flocky deposit of albumen, evaporate to a small bulk, precipitate with alcohol, wash the deposit with alcohol and dry at 110°. As salts of lime (generally the phosphate) are always present, their weight has to be determined by incinerating a weighed quantity of the dried Mucilage and weighing the ash. The weight of the latter has to be deducted from that of the Mucilage.

**Verantin**=$C_{14}$ $H_5$ $O_5$. In the root of Rubia tinctorum. Liberate the Verantin oxyd of iron (*see* Rubiacin) from the metallic base by boiling with hydrochloric acid, wash and dissolve in boiling alcohol, when the Verantin will separate on cooling as a brown powder.—Reddish-brown, amorphous powder similar in appearance to snuff or ground roasted coffee; fuses scarcely in boiling water and dissolves sparingly in it, dissolves readily in boiling alcohol and separates on cooling in a pulverulent form, the solution retaining an acid reaction; dissolves in alkalies with a brown colour; is decomposed by heat.

**Veratric Acid**=$C_{18}$ $H_9$ $O_7$ + HO. In the seeds of Schœnocaulon officinale. Exhaust the pulverised seeds with alcohol acidulated with sulphuric acid, add to the tincture hydrate of potash, filter, distil the alcohol, collect the veratrin thus deposited and saturate the mother-ley with excess of sulphuric acid, when the V. acid will crystallise slowly before or after evaporation.

Wash with cold water and recrystallise in alcohol.—Forms colourless needles; reddens moist litmus paper; loses water on heating and becomes opaque, fuses and sublimes completely; dissolves slowly in cold, more readily in boiling water, readily in alcohol, not in ether. The Veratrates of alkalies are crystallisable and soluble in water, those of silver and of lead dissolve sparingly.

**Veratrin** $= C_{64} H_{52} N_2 O_{16}$. Alkaloïd of the seeds of Schœnocaulon officinale and of some allied plants. Exhaust with alcohol of 70% to 80%, add a little water to the tinctures and distil the alcohol, evaporate the remaining liquid under addition of pulverised charcoal to dryness, triturate the mass, treat with water mixed with concentrated sulphuric acid, filter, precipitate the solution with carbonate of soda, wash the deposit, divide in water, and add slowly and gradually diluted sulphuric acid to redissolution, digest the solution with animal charcoal, filter, precipitate with ammonia and dry the deposit after washing. To free the Veratrin from resin, dissolve in alcohol of 60%, filter if necessary, and evaporate. A white crystalline powder is formed, mixed with a brown resinous matter, which is removable by washing with cold water. The crystalline Veratrin has to be recrystallised from strong alcohol.—A tolerably white, light, amorphous powder, or, in the purest state, small, colourless crystals of an extremely acrid burning taste, inodorous, the dust of which produces violent sneezing and irritation, fuses and decomposes in higher temperatures, is insoluble in water, dissolves in 3 parts alcohol of 80% at the ordinary temperature, much more when boiling hot, the solution being of strongly alkaline reaction; dissolves also readily in ether when the product is quite pure, but the amorphous V. only in 50 parts. Concentrated sulphuric acid colours it passingly yellow, then carmine-red; concentrated hydrochloric acid produces, especially on warming, a deep dark-violet solution, on the surface of which small drops of oil begin to form.

**Veratrin Resin** = HELONIN.

**Vinetin** = OXYACANTHIN.

**Violin.** Alkaloïd of all parts of Viola odorata, Free the alcoholic extract of the root from chlorophyll and fat by means of ether, boil the residue with diluted sulphuric acid, in order to volatilise acetic acid, add hydrated oxyd of lead in excess, dry, treat with alcohol and evaporate the solution.—Pale-yellow powder, bitter, fusible, inflammable like a resin in higher temperatures; dissolves in water more and in alcohol less than emetin, not in ether, combines with acids.

**Viridic Acid.** *See* CAFFEIC ACID.

Q

**Virola Tallow.** From Myristica sebifera, is similar to nutmeg balsam.

**Viscin** $= C_{40} H_{48} O_{16} = C_{40} H_{32} + 16 HO$. Exudes from the receptacle and floral envelope of Atractylis gummifera; is found in the leaves and branches of Ilex Aquifolium, Viscum album, and especially in the berries of the latter, and of various kinds of Loranthus; has also been observed as constituent of other plants. [It is very plentiful in the fruits of Pittosporum undulatum.] Bruise the berries of Viscum with water, wash with water, treat afterwards with alcohol, and lastly with ether, evaporate the ethereous solutions, knead the glutinous yellowish mass, left after evaporation, first with alcohol, then with water, and heat to 120°, until all the water is driven off.—Clear mass of honey consistence; may be drawn out into threads; almost inodorous and tasteless; of an oily appearance at 100°, begins to boil at 210°, a thin yellow oil of 0·856 density distilling at 235°; has the density of water, and leaves greasy spots on paper.

**Volatile Acids.** Such as pass over in the distillation of vegetables with water, but are quickly decomposed, and have as yet not been isolated in the pure form; have been observed in plants of the following genera : Aconitum, Arum, Clematis, Ranunculus, Daphne, Pimelea, Polygonum, and many Fungi.

**Vulpulin** or **Vulpic Acid** $= C_{38} H_{14} O_{10}$. In Cetraria vulpina and in Parmelia parietina, but in the latter only when it has been gathered, not from trees, but, in the undeveloped state, from sandstone rocks. From the Parmelia the Vulpulin can be withdrawn nearly pure by sulphide of carbon. The Cetraria has to be soaked first in lukewarm water, containing a little lime; strain after a few hours, repeat the same treatment, saturate the united solutions with excess of hydrochloric acid, wash the flocks which have separated with cold water, and recrystallise from hot water or from alcohol. —Forms sulphur-yellow, translucid, large, rhombic pyramids or needles, crystallized from sulphide of carbon of the colour of bichromate of potash; fuses at 110°, and sublimes in small yellow laminar scales or needles, with the odour of benzoin; tasteless by itself, dissolved in alcohol, very bitter; is, even in boiling water, nearly insoluble, readily soluble in sulphide of carbon, in 376 parts cold, and in 200 parts boiling alcohol of 80%, in 588 parts cold. and in 88 parts boiling alcohol of 90%; also sparingly in boiling absolute alcohol, more readily in ether, and most readily in chloroform, in concentrated sulphuric acid with brown-red colour, which turns pale-yellow on addition of water.

**Walnut Oil.** Obtained by pressing the seeds of Juglans regia. Originally greenish, becomes soon pale-yellow, inodorous, of mild taste and of 0.926 density, congeals at —18°, butter-like, becomes hard at —27°, dries better at the air than linseed oil, yields like this soft soaps.

**Xanthin,** according to Higgin. In the root of Rubia tinctorum. Precipitate the fresh, filtered infusion of madder with acetate of lead, wash the deposit, decompose with sulphuret of hydrogen and boil the sulphide of lead with water. The solutions, after they have been neutralised with ammonia and digested with a little hydrated alumina, which deposit alizarin and rubiacin, yield the Xanthin on evaporating the filtrate and on extracting the residue. (By this method are obtained, according to Schunck, rubian and products of decomposition of the latter).—Dark-brown, deliquescent, gummous body, bitter (neither acerb nor sweet), fuses and decomposes in higher temperatures, dissolves readily in water with a beautiful yellow colour, also readily in alcohol, little in ether, in alkalies with purple-red hue. The aqueous solution is precipitable by alum and by subacetate of lead, but not by acetate of lead.

**Xanthorhamnin** $= C_{46} H_{25} O_{28} + 10$ HO, according to Gellatly. In the matured grains of Avignon (from Rhamnus infertorius and some other congeners), or, according to Kane, a decomposition-product of the chrysorhamnin of the unripe berries. After boiling the unripe berries with water for a few minutes and drying, no chrysorhamnin is obtained, but in its stead Xanthorhamnin. Likewise Xanthorhamnin is obtained from chrysorhamnin by boiling the latter with water under access of air. Gellatly boils the unripe pulverised berries with alcohol, frees the not too much concentrated tincture from a slowly forming dark-brown resin by repeated decantations, allows to crystallise and purifies by recrystallising.—Forms pale-yellow, shining silky tufts of almost tasteless crystals; loses its water at 100°; does not fuse at 130°; dissolves readily in cold and in hot water, and in alcohol, not in ether; colours black the solutions of iron; decomposes with diluted acids into sugar and Rhamnetin $= C_{22} H_{10} O_{10}$, forms with alkalies brown solutions.

**Xanthotannic Acid** $= C_{18} H_{13} O_4$. The yellow matter of autumnal leaves. Exhaust, for instance, elm-leaves with alcohol, evaporate the tinctures, filter off the wax, precipitate the filtrate with water, filter again, precipitate with acetate of lead and decompose the deposit by sulphuret of hydrogen. The liquid separated from the sulphide of lead, has an astringent taste, acid reaction and precipitates glue.

**Xanthoxylen** $= C_{20} H_{16}$. Obtained by distilling the so-called Japanese pepper (seeds of Xanthoxylum piperitum) with water, desiccating the oil with chloride of calcium and rectifying by means of potassium. Colourless, of great light-refracting power, and of a pleasant aromatic odour; boils at 162°.

**Xanthoxylin** $= C_{40} H_{24} O_{16}$. Forms in the volatile oil (xanthoxylen), obtained by distillation with water, from Xanthoxylum piperitum, and remains, after the oil has been freed from xanthoxylen by distillation at 130°. May also be obtained by evaporating the alcoholic tincture, and by freeing the crystals from resin by washing with liquor of ammonia.—Large, colourless, silky, klinorhombic crystals, neutral, smells faintly like stearin; tastes aromatic, fuses at 80°, volatilises in higher temperatures undecomposed; dissolves not in water, readily in alcohol and in ether.

**Xylochloric Acid** $= C_{30} H_{26} O_{24}$. Green pigment of decaying wood, especially that of beech. Treat with diluted liquor of ammonia, precipitate the green solution with hydrochloric acid, wash and dry the deposit.—Dark-green, friable mass, tasteless, not fusible, insoluble in water, alcohol, ether, diluted acids, becomes carbonised with concentrated sulphuric acid; is readily soluble in liquor of ammonia, the solution being of deep olive-green colour, and of neutral reaction, after the excess of ammonia has been driven off, yielding by spontaneous evaporation a green residue soluble completely in water. This aqueous solution is precipitated by chloride of iron and by acetate of lead dirty green, by sulphate of copper, and by subnitrate of mercury olive-green.

**Xylostein.** Bitter substance obtained from the berries of Lonicera Xylosteum. Boil the berries with water, precipitate the decoction with acetate of lead, remove excess of lead from the filtrate by sulphuret of hydrogen, evaporate to a syrup consistence, shake repeatedly with ether and evaporate the latter.—Crystallises in long, colourless needles or prisms, is inodorous, neutral, non-nitrogenised, of a slightly bitter taste; fuses at 100°, gives out a crystalline sublimate in higher temperatures under carbonisation; dissolves sparingly in cold, readily in boiling water, most readily in alcohol and in ether; is precipitable by subacetate of lead, yields with diluted acids, sugar, and other products.

# ADDENDA.

[**Aricin** = $C_{23}$ $H_{26}$ $N_2$ $O_4$. Alkaloïd, prepared from the Quina de Cusco by O. Hesse.—Beautiful, white prisms, fusing at 188°, of very feebly alkaline reaction, and very slightly astringent, not bitter taste, dissolves very easily in chloroform, rather easily in ether (1:20), in 235 parts alcohol of 80%. It behaves towards nitric and sulphuric acids like cusconin, and is precipitable by the same re-agents as cusconin and by iodide of potassium and tannic acid. Acetic acid likewise precipitates the solution of A. in hydrochloric acid under the form of small, white granules, very difficultly soluble in cold, more readily in boiling water, and crystallising therefrom in crystals after cooling.]

[**Capsaicin.** The active principle of the fruits of Capsicum annuum, prepared by Tresh.—Colourless prisms of extremely pungent taste, fusible and volatile without decomposition by itself and by the steam of water, the vapours strongly irritating the respiratory organs. The C. dissolves little in cold water, better in boiling water, easily in strong alcohol and ether; also in potash ley, and reprecipitable from the latter solution by addition of a solution of ammonium chloride.]

[**Cusconin** = $C_{23}$ $H_{26}$ $N_2$ $O_4$ + 2 H O. Prepared by O. Hesse, from the Quina de Cusco.—White leaflets, or short prisms of dull lustre, fuses at 110°, of very slightly alkaline reaction in alcoholic solution, dissolves in 35 parts ether, better in alcohol and aceton, very easily in chloroform, little in benzol and petroleum-ether; becomes green with nitric acid, and dissolves with greenish-yellow colour, with the same colour in concentrated sulphuric acid, changing to dark-brown on warming. Its solution in hydrochloric acid is precipitable by ammonia, soda, the sulphocyanide, bi-iodide, ferro- and ferricyanide of potassium, the chlorides of gold and of platinum, iodide of potassio-mercury, phosphotungstic acid. Its salts, when neutral and dissolved in water, react acid, and taste at first rancid, then slightly bitter; the neutral sulphate at last cooling, burning similar to inferior peppermint oil.]

[**Erythrophlaein.** Alkaloïd, discovered by Gallois and Hardy in the bark of Erythrophlaeum guinense.—It forms a clear, amber-yellow mass, of crystalline structure under the microscope, dissolves in water, alcohol, amylalcohol, acetic ether, very little in ether, chloroform, benzol; forms salts with acids. Concentrated potashley, also ammonia, yield crystalline precipitates in the solution of the hydrochloride. Precipitates are likewise formed by chloride of platinum, picric acid, potassium bi-iodide, the iodides of potassio-mercury, potassio-bismuth and potassio-cadmium, bichromate of potash, the chlorides of mercury and gold, and the subchloride of palladium. With hypermanganate of potash and sulphuric acid a similar but feebler violet colour is formed, as with strychnin. It acts as a strong poison.]

[Gardenin $= C_{23} H_{15} O_{10}$. Crystalline resin of the Dikamali resin from Gardenia lucida. Crystallises from the hot, alcoholic solution in yellow needles, which melt at 155°.—FLUECKIGER.]

[Glycyphyllin. Glucosid of the leaves of Smilax glycyphylla.—Brownish-yellow, amorphous mass, or, by slow evaporation of the ethereous solution, concentrically united tufts of crystals of aromatic odour and bitter-sweet taste; dissolves better in hot than in cold water, easily in alcohol and in ether; breaks up on boiling with dilute sulphuric or hydrochloric acid into sugar and another product.—Baron F. VON MUELLER and L. RUMMEL.]

[Jurubebin. Alkaloïd of the fruits of Solanum paniculatum.—Amorphous mass of bitter taste and slightly aromatic odour, little soluble in water, easily in liquor of ammonia, alcohol and chloroform. The hydrochloride of J. is crystalline and yields in solution precipitates with most alkaloïd-reagents except chloride of platinum, picric and chromic acids.—F. V. GREENE.]

[Muscarin $= C_5 H_{13} NO_2$. Poisonous alkaloïd of the Fly-agaric (Agaricus muscarius), isolated from the extract by dilute hydrochloric acid, evaporating the liquid to crystals and pressing the latter between blotting paper, which absorbs the deliquescent Muscarin salt, while the chloride of a non-poisonous alkaloïd, Amanitin $= C_5 H_{13} NO$ remains behind.—KOPPE AND SCHMIEDEBERG.]

[Nucit $= C_{12} H_{12} O_{12} + 4 HO$. Peculiar kind of sugar, contained in the leaves of the walnut tree (Juglans regia.) It crystallises in klinorhombic prisms of 1·54 sp. gr., has a sweet taste, dissolves easily in water, alcohol, ether and chloroform, is non-rotating, does not reduce copper salts, is unable to ferment even after boiling with dilute sulphuric acid. Treated with nitric acid it forms neither mucic nor oxalic acids.—TANNET AND VILLIERS.]

[Ostruthin $= C_{14} H_{17} O_2$. Discovered by v. Gorup-Besanez in the root of Peucedanum Ostruthium.—Odourless and almost tasteless, klinorhombic crystals, fusing at 115°, resolidifying at 91°; sublimable under decomposition, burning at the air with a bright, smoky flame. The O. yields on trituration an electric powder, is of neutral reaction, is insoluble in cold water, whereas boiling water dissolves only traces, little soluble in benzol and petroleum-ether, easily in alcohol and ether. The alcoholic solution exhibits, especially after addition of a little water, a blue fluorescence. It dissolves also in caustic potash, soda and ammonia, and is therefrom reprecipitated by carbonic acid. It is not precipitable by metallic salts, and forms with hydrochloric acid gas in alcoholic solution crystals of hydrochloride of Ostruthin.]

[**Oxynarcotin**$=C_{22} H_{23} NO_8$. Alkaloïd prepared by Beckelt and Wright, from the mother-liquor remaining in the preparation and purification of narcein from opium. It forms small, mica-like, sandy crystals, little soluble in water and alcohol, even on boiling, similar in appearance to narcotin, but differing from it by its sparing solubility in hot water and alcohol, and its insolubility in benzol, ether and chloroform. Fixed alkalies and their carbonates precipitate the O. from its concentrated solutions, not or only partially after some time from diluted ones. Analogous to the hydrochlorides of narcotin and narcein, the hydrochloride of O. by hot water breaks up into free acid and a basic salt.]

[**Picroroccellin.** Nitrogenised, neutral substance, found by Stenhouse and Groves, in Roccella fuciformis.—It crystallises in long, colourless prisms, of considerable lustre, and of very bitter taste, insoluble in water, petroleum and sulphide of carbon, little soluble in ether and benzol, moderately soluble in boiling alcohol. It fuses at 192° to 194°, and decomposes afterwards. Dissolves in concentrated sulphuric acid, with yellow colour on warming, reprecipitable as yellow powder on addition of water. The same takes place with nitric acid. On heating P. for ten minutes at 220°, or on boiling with dilute sulphuric or hydrochloric acid, it is converted into Xanthoroccellin, a nitrogenised body, crystallising in long, slender, yellow needles.]

[**Thevetosin.** Poisonous glucosid, discovered by Herrera in the seeds of Thevetia yccotli.—White, quadrangular prisms, inodorous, of intensely acrid taste, not soluble in water, very little soluble in ether, sulphide of carbon, fixed and volatile oils, easily soluble in alcohol. Yields, with dilute sulphuric acid, sugar and a resin. The alcoholic solution is not changed by nitrate of silver, the chlorides of platinum, gold and iron, iodide and iodate of potassium, tannic acid, the hydrates and carbonates of alkalies, ferro- and ferricyanide of potassium.]

# MOLECULAR WEIGHTS OF ORGANIC COMPOUNDS.

*Those formulas marked with an asterisk, as they differ from the text, are the results of more recent investigations.*

Abietic acid $= C_{44} H_{64} O_5$

Absinthin $= C_{20} H_{28} O_4$

Acetic acid $= C_2 H_4 O_2$

Aconitic acid $= C_6 H_6 O_6$

Aconitin $= C_{30} H_{47} NO_7$

Adansonin $= C_{48} H_{72} O_{33}$

Aesculin $= C_{21} H_{24} O_{13}$

Aesculetin $= C_9 H_6 O_4$

*Alizarin $= C_{14} H_8 O_4$ (Graeb. et Lieb.)

*Aloin $= C_{17} H_{20} O_7$ (Fl. et Hanb.)

Amygdalin $= C_{20} H_{27} NO_{11}$ (anhydrous)

Anacahuit tannic acid $= C_8 H_{12} O_5$

Anacardic acid $\quad C_{44} H_{64} O_7$

Anchusin $= C_{18} H_{20} O_4$

Anemonic acid $= C_{15} H_{14} O_7$

Anemonin $= C_{15} H_{12} O_6$

Anethol $= C_{10} H_{12} O$

Angelic acid $= C_5 H_8 O_2$

Annatto red $= C_8 H_{13} O$

Antiarin $= C_{14} H_{20} O_5 + 2 H_2 O$

Apiin $= C_{24} H_{28} O_{13}$

Arachidic acid $= C_{20} H_{40} O_2$

Arbutin $= C_{24} H_{32} O_{14} + H_2 O$

Hydrokinon $= C_6 H_6 O_2$

Aribin $= C_{23} H_{20} N_4 + 8 H_2 O$

*Aricin $=$ Cinchonidin (Hesse)

Arnicin $= C_{20} H_{30} O_4$

Asaron $= C_{20} H_{28} O_5$

Asclepion $= C_{20} H_{34} O_3$

Asparagin $= C_4 H_8 N_2 O_3 + H_2 O$

Aspartic acid $= C_4 H_7 NO_4$

Asperula tannic acid $= C_7 H_8 O_4$

Athamantin $= C_{24} H_{30} O_7$

Atherospermin $= C_{30} H_{40} N_2 O$

Atherosperma tannic acid $= C_{10} H_{14} O_2$

Atropin $= C_{17} H_{23} NO_3$

Azulen $= C_{16} H_{26} O$

Basilicum stearopten $= C_{10} H_{22} O_3$

Bassorin $= C_6 H_{10} O_5$

Bay oil hydrocarbons $= C_{10} H_1$ and $C_{15} H_{24}$

Bebirin $= C_{19} H_{21} NO_3$

Benic acid $= C_{22} H_{44} O_2$

Benzoic acid $= C_7 H_6 O_2$

Berberin $= C_{20} H_{17} NO_4$ (anhydrous)

Betulin $= C_{25} H_{40} O_2$

Betuloretic acid $= C_{36} H_{66} O$

Boheic acid $= C_7 H_{10} O_6$

Borneen $= C_{10} H_{16}$

Borneol $= C_{10} H_{18} O$

Brasilin $= C_{22} H_{20} O_7$

Brucin $C_{23} H_{26} N_2 O_4 + 4 H_2 O$

Bryoniu $= C_{48} H_{80} O_{19}$

Butyric acid $= C_4 H_8 O_2$

Caffeic acid $= C_{14} H_{16} O_7$

Viridic acid $= C_{14} H_{12} O_7$

Caffein $= C_8 H_{10} N_4 O_2 + H_2 O$

Calluna tannic acid $= C_{14} H_{14} O_9$

Camphor $= C_{10} H_{16} O$

Camphor oil $= C_{20} H_{32} O$

Cane-sugar $= C_{12} H_{22} O_{11}$

Caoutchouc $= C_{20} H_{32}$

Capric acid    $C_{10} H_{20} O_2$
Caproic acid $= C_6 H_{12} O_2$
Caprylic acid $= C_8 H_{16} O_2$
Capsulaescic acid $= C_{13} H_{12} O_8$
Cardol $= C_{21} H_{30} O_2$
Carminic acid $= C_{14} H_{14} O_8$
Carotin $= C_{18} H_{24} O$
Carthamin $= C_{14} H_{16} O_7$
Carven $= C_{10} H_{16}$
Carvol $= C_{10} H_{14} O$
Caryophyllic acid $= C_{10} H_{12} O_2$
Caryophyllin $= C_{10} H_{16} O$
*Catechuic acid $= C_{13} H_{12} O_5$
Ceradia resin $= C_{20} H_{28} O_2$
Ceropinic acid $= C_{36} H_{68} O_5$
Cerosin $= C_{24} H_{50} O$
Ceroxylin $= C_{20} H_{32} O$
Cetraric acid $= C_{18} H_{16} O_8$
Chelerythrin $= C_{19} H_{17} NO_4$
*Chelidonic acid $= C_7 H_4 O_5$
Chelidonin $= C_{19} H_{17} N_3 O_3 +$ $H_2 O$
Chenopodin $= C_6 H_{13} NO_4$
    Metagummic acid $= C_{12} H_{22} O_{11}$
Chica red $= C_8 H_8 O_3$
Chiratin $= C_{26} H_{48} O_{15}$
Chiratogenin $= C_{13} H_{24} O_3$
*Cholestearin $= C_{26} H_{44} O$
Chrysophanic acid $= C_{10} H_8 O_3$
Chrysorhammin $= C_{23} H_{22} O_{11}$
Cicuten $= C_{10} H_{16}$
Cinchona alkaloid, Howard's $=$ $C_{20} H_{24} N_2 O_2$
*Cinchonidin $= C_{20} H_{24} N_2 O$
Cinchonin $= C_{20} H_{24} N_2 O$
Cinnamic acid $= C_9 H_8 O_2$
Cissotannic acid $= C_{10} H_{12} O_8$
Citric acid $= C_6 H_8 O_7$
Cnicin $= C_{42} H_{56} O_{15}$
Cocain $= C_{16} H_{19} NO_4$
Codein $= C_{18} H_{21} NO_3 + H_2 O$
Colchicin $= C_{17} H_{19} NO_5$
Colocynthin $= C_{56} H_{84} O_{23}$
Colombic acid $= C_{42} H_{46} O_{13}$
*    ,,    ,, $= C_{22} H_{24} O_7$ (Boedeker)

Colombin $= C_{42} H_{44} O_{14}$
Conessin $= C_{25} H_{44} N_2 O$
Conhydrin $= C_8 H_{17} NO$
Coniferin $= C_{16} H_{22} O_8 + 2 H_2 O$
Coniin $= C_8 H_{15} N$
Convallamarin $= C_{23} H_{44} O_{12}$
Convallarin $= C_{34} H_{62} O_{11}$
Convolvulin $= C_{31} H_{50} O_{16}$
Copaivic acid $= C_{20} H_{30} O_2$
Coriamyrtin $= C_{40} H_{48} O_{14}$
Corticipino tannic acid $= C_{16} H_{14} O_7$
Corydalin $= C_{46} H_{58} N_2 O_7$
Crocin $= C_{29} H_{42} O_{15}$
*    ,,    $= C_{48} H_{60} O_{18}, H_2 O$ (Rochleder)
Crotonol $= C_9 H_{14} O_2$
*Cubebin $= C_{33} H_{34} O_{10}$
Cumarin $= C_9 H_6 O_2$
Cuminol $= C_{10} H_{12} O$
Curarin $= C_{10} H_{15} N$
*Curcumin $= C_{10} H_{10} O_{13}$
Cyclamin $= C_{28} H_{32} O_{12}$
Cymen (cymol) $= C_{10} H_{14}$
Dambonit $= C_4 H_8 O_3$
    Dambose $= C_3 H_6 O_3$
Dammaryl $= C_{20} H_{32}$
Dammaryl-hydrate $= 4 C_{20} H_{32} + H_2 O$
Dammarylic acid $= C_{45} H_{72} O_3$
*Daphnin $= C_{31} H_{34} O_{19}$
Daphnetin $= C_{19} H_{14} O_9$
Datiscin $= C_{21} H_{22} O_{12}$
    Datiscetin $= C_{15} H_{10} O_6$
Delphinin $= C_{24} H_{35} NO_2$
Dextrin $= C_6 H_{10} O_5$
Digitaletin $= C_{22} H_{38} O_9$
Digitalic acid, volatile $= C_5 H_{10} O_2$
Digitalin $= C_{28} H_{48} O_{14}$
    ,, crystallised $= C_{25} H_{40} O_{15}$
Dragon's blood $= C_{20} H_{21} O_4$
Dulcamarin $= C_{65} H_{100} N_2 O_{29}$
    ,, (Geissler) $= C_{22} H_{34} O_{10}$
Dulcamaritin $= C_{16} H_{26} O$
Dulcit $= C_6 H_{14} O_6$
Elaterin $= C_{20} H_{28} O_5$

Elemi resin $= C_{20} H_{32} O_2$
Elemin $= C_{40} H_{66} O$
Ellagic acid $= C_{14} H_6 O_8 + 3$
$H_2 O$
Emetin $= C_{20} H_{30} N_2 O_5$
,, $= C_{30} H_{44} N_2 O_8$ (Lefort)
Emodin $= C_{40} H_{30} O_{13}$
Equisetic acid $=$ Aconitic acid
Ericinol $= C_{10} H_{16} O$
Ericolin $= C_{34} H_{56} O_{21}$
Erucic acid $= C_{22} H_{42} O_2$
Erythric acid $= C_{20} H_{22} O_{10}$
Eugenin $= C_{10} H_{12} O_2$
Euphorbin $= C_{20} H_{32} O$
Euphrasia tannic acid $= C_{32} H_{40}$
$O_{17}$
Evernic acid $= C_{17} H_{16} O_7$
Fiber (Cellulose) $= C_6 H_{10} O_5$
*Filicic acid $= C_{14} H_{18} O_5$
Formic acid $= CH_2 O_2$
*Fraxin $= C_{16} H_{18} O_{10}$
*Fraxetin $= C_{10} H_8 O_5$
Fruit-sugar $= C_6 H_{12} O_6$
Fumaric acid $= C_4 H_4 O_4$
Galbanum resin $= C_{26} H_{38} O_5$
Galium tannic acid $= C_7 H_8 O_5$
Gallic acid $= C_7 H_6 O_5$
Gallotannic acid $= C_{27} H_{22} O_{17}$
Gambogic acid $= C_{20} H_{24} O_4$
Gardenia tannic acid No. 1 $= C_{24}$
$H_{28} O_{13}$
,, ,, ,, No. 2 $= C_{23}$
$H_{28} O_{13}$
Gaultierylen $= C_{10} H_{16}$
Gentianin $= C_{14} H_{10} O_5$
Gentiopicrin $= C_{20} H_{30} O_{12} +$
$H_2 O$
Ginkgoic acid $= C_{24} H_{48} O_2$
Globularia resin $= C_{20} H_{36} O_8$
Globularia tannic acid $= C_8 H_{12}$
$O_7$
Globularin $= C_{30} H_{44} O_{14}$
Glycerin $= C_3 H_8 O_3$
Glycyrrhizin $= C_{24} H_{36} O_9$
Gratiolin $= C_{20} H_{34} O_7$
Gratiosolin $= C_{46} H_{84} O_{25}$
Guaiacic acid $= C_{12} H_{16} O_6$

Guaiaconic acid $= C_{33} H_{40} O_{10}$
Guaiacum Beta resin $= C_{20} H_{20}$
$O_6$
*Guaiaretic acid $= C_{20} H_{26} O_4$
Gum $= C_{12} H_{22} O_{11}$
Gurjunic acid $= C_{44} H_{68} O_8$
Gutta $= C_{20} H_{32}$
Haematoxylin $= C_{16} H_{14} O_6 + 3$
$H_2 O$
Harmalin $= C_{13} H_{14} N_2 O$
Harmin $= C_{13} H_{12} N_2 O$
Hederic acid $= C_{15} H_{26} O_4$
Helenin $= C_{16} H_{28} O_{10}$
,, $= C_6 H_8 O$ (Kallen)
Helianthic acid $= C_7 H_9 O_4$
Helleborein $= C_{26} H_{44} O_{15}$
Helleborin $= C_{36} H_{42} O_6$
Helonin $= C_{14} H_{18} N_2 O_3$
Huanokin $= C_{20} H_{24} N_2 O$
Hydrastin $= C_{22} H_{23} NO_6$
Hydrocarotin $= C_{18} H_{30} O$
Hydrocyanic acid $= C N H$
*Hyoscyamin $= C_{15} H_{23} NO_3$
Hypogaeic acid $= C_{16} H_{30} O_2$
Ilixanthin $= C_{17} H_{22} O_{11}$
Indican $= C_{26} H_{31} NO_{17}$
Indigo blue $= C_8 H_5 NO$
Inosit $= C_6 H_{12} O_6 + 2 H_2 O$
Inulin $= C_6 H_{10} O_5$
Ipecacuanhic acid $= C_{14} H_{18} O_7$
Iris stearopten $=$ Myristic Acid
Isocetic acid $= C_{15} H_{30} O_2$
Jalapin $=$ Convolvulin
*Jervin $= C_{30} H_{46} N_2 O_3$
*Koussin $= C_{31} H_{38} O_{10}$
Lactic acid $= C_3 H_6 O_3$
Lactucerin $= C_{16} H_{26} O$
*Lactucin $= C_{11} H_{12} O_3 + H_2 O$
Laricin $= C_7 H_{12} O_2$
Laserpitin $= C_{24} H_{36} O_7$
Lauric acid $= C_{12} H_{24} O_2$
Laurostearin $= C_{27} H_{50} O_4$
Lecanoric acid $= C_{16} H_{14} O_7$
Ledum tannic acid $= C_7 H_6 O_3$
Lichenin $= C_6 H_{10} O_5$
Lichenostearic acid $= C_{14} H_{24} O_3$
Limonin $= C_{44} H_{52} O_{14}$

Linoleic acid $= C_{16} H_{28} O_2$

Lupulic acid $= C_{32} H_{50} O_7$

Luteolin $= C_{12} H_8 O_5$

Lycopodium resin $= C_{18} H_{32} O_2$

Lycopodium stearon $= C_{15} H_{30} O_2$

Malic acid $= C_4 H_6 O_5$

Mangostan resin $= C_{18} H_{22} O_5$

Mangostin $= C_{20} H_{22} O_5$

Mannit $= C_6 H_{14} O_6$

Masopin $= C_{22} H_{36} O$

Maynas resin $= C_{14} H_{18} O_4$

Meconic acid $= C_7 H_4 O_7 + 3 H_2O$

Meconin $= C_{10} H_{10} O_4$

Melezitose $= C_{12} H_{22} O_{11}$

Melitose $= C_{12} H_{22} O_{11}$

Menispermin $= C_{18} H_{24} NO_2$

Menyanthin $= C_{30} H_{46} O_{14}$

Morin $= C_9 H_8 O_5$

Morindin $= C_{28} H_{30} O_{15}$

Morus tannic acid $= C_9 H_8 O_5$

Morphin $= C_{17} H_{19} NO_3 + H_2 O$

Mycose $= C_{12} H_{22} O_{11} + 2 H_2 O$

Myristic acid $= C_{14} H_{28} O_2$

Myristin $= C_{45} H_{86} O_6$

*Myronate of Potash $= C_{10} H_{18} KNS_2 O_{10}$

Myroxocarpin $= C_{24} H_{35} O_3$

Narcein $= C_{23} H_{29} NO_9$

*Narcotin $= C_{22} H_{23} NO_7$

Nataloin $= C_{25} H_{28} O_{11}$

Nicotianin $= C_{20} H_{32} N_2 O_3$

Nicotin $= C_5 H_7 N$

Oenolin $= C_{10} H_{10} O_5$

Oil of Allium sativum $= C_3 H_5 S$

,, Artemisia Absinthium $= C_{10} H_{16} O$

,, A. Cina $= C_{12} H_{20} O$

,, Balsam of Copaiva $= C_{10} H_{16}$

,, ,, Peru $= C_{16} H_{14} O_2$

,, Brassica nigra $= C_4 H_5 NS$

,, Bursera gummifera $= C_{10} H_{16}$

,, Carum Petroselinum $= C_{10} H_{16}$

,, Cinnamomum $= C_9 H_8 O$

,, Citrus medica $= C_{10} H_{16}$

Oil of Citrus Limetta $= C_{10} H_{16}$

,, Coriandrum sativum $= C_{10} H_{18} O$

,, Elemi $= C_{10} H_{16}$

,, Humulus Lupulus $= C_{10} H_{16}$ et $C_{10} H_{18} O$

,, Juniperus communis $= C_{10} H_{16}$

,, ,, Sabina $= C_{10} H_{16}$

,, Myrrh $= C_{10} H_{14} O$

,, Origanum vulgare $= C_{50} H_{80} O$

,, Osmitopsis ast. $= C_{10} H_{18} O$

,, Pinus $= C_{10} H_{16}$

,, Piper nigrum $= C_{10} H_{16}$

,, Sassafrass offic. $= C_9 H_{10} O$

,, Spiraea Ulmaria $C_{10} H_{18} O_2$

,, Zingiber officinale $= 2 C_{10} H_{16} + H_2 O$

Oleic acid $= C_{18} H_{34} O_2$

Olein $= C_{54} H_{104} O_6$

Olivil $= C_{14} H_{18} O_5 + H_2 O$

Onocerin $= C_{12} H_{20} O$

Ononid $= C_{18} H_{22} O_8$

Ononin $= C_{30} H_{34} O_{13}$

Ophelic acid $= C_{13} H_{20} O_{10}$

Opianin $= C_{66} H_{72} N_4 O_{21}$

Orcin $= C_7 H_8 O_2 + H_2 O$

Oxalic acid $= C_2 H_2 O_4$

Oxyacanthin $= C_{32} H_{46} N_2 O_{11}$

Oxypinotannic acid $C_{14} H_{16} O_9$

Palmitic acid $= C_{16} H_{32} O_2$

Palmitin $= C_{51} H_{98} O_6$

Panaquilon $= C_{12} H_{25} O_9$

Papaverin $= C_{20} H_{21} NO_4$

Parellin $= C_9 H_6 O_4$ (anhydrous)

Paricin $= C_{23} H_{25} N_2 O_3$

Paridin $= C_{16} H_{28} O_7 + 2 H_2 O$

Parsley stearopten $= C_{12} H_{14} O_4$

Paytin $= C_{21} H_{24} N_2 O + H_2 O$

Pelargonic acid $= C_9 H_{18} O_2$

Pelosin = Bebirin

Peppermint stearopten $= C_{10} H_{20} O$

Peucedanin $= C_{12} H_{12} O_3$

Phillyrin $= C_{27} H_{34} O_{11}$ (anhydrous)

Phlobaphen $= C_{10} H_8 O_4$

Phlorrhizin $= C_{21} H_{24} O_{10} + 2 H_2 O$

Physalin $= C_{14} H_{16} O_5$

Physodin $= C_{12} H_{12} O_8$

Physostigmin $= C_{15} H_{21} N_3 O_2$

Picrolichenin $= C_{12} H_{20} O_6$

Picrotoxin $= C_{10} H_{12} O_4$

Pinocorretin $= C_{24} H_{38} O_5$

Pinocortannic acid $= C_{32} H_{38} O_{23}$

Pinopicrin $= C_{22} H_{36} O_{11}$

Pinotannic acid $= C_7 H_8 O_4$

Piperin $= C_{17} H_{19} NO_3$

Pipitzahoic acid $= C_{15} H_{20} O_3$

Pityxylonic acid $= C_{25} H_{40} O_8$

Populin $= C_{20} H_{22} O_8 + 2H_2 O$

Primulin $=$ Cyclamin

Propionic acid $= C_3 H_6 O_2$

Pseudomorphin $= C_{17} H_{19} NO_4 + H_2 O$

Pteritannic acid $= C_{12} H_{15} O_4$

Purpurin $= C_9 H_6 O_3$

Pyrocatechin $= C_6 H_6 O_2$

Quassiin $= C_{10} H_{12} O_3$

Quercetin $= C_{23} H_{16} O_{10}$

Quercit $= C_6 H_{12} O_5$

Quercitrin $= C_{35} H_{36} O_{20}$

Quina red $= C_{12} H_{14} O_7$

Quinic acid $= C_7 H_{12} O_6$

*Quinidin $= C_{20} H_{24} N_2 O_2$

Quinin $= C_{20} H_{24} N_2 O_2$

Quinotannic acid $= C_{14} H_{16} O_9$

Quinova red $= C_{12} H_{12} O_5$

Quinova tannic acid $= C_7 H_9 O_4$

Quinovic acid $= C_{24} H_{38} O_5$

Racemic acid $= C_4 H_6 O_6$

Ratanhia tannic acid $= C_{18} H_{16} O_7$

Ratanhin $= C_{10} H_{13} NO_3$

Rhamnoxanthin $= C_{20} H_{20} O_{10}$

Rhinanthin $= C_{29} H_{52} O_{20} + 4 H_2 O$

Rhodotannic acid $= C_{14} H_{12} O_7$

Rhoeadin $= C_{21} H_{21} NO_6$

Rhoitannic acid $= C_{18} H_{28} O_{13}$

Ricinoleic acid $= C_{18} H_{34} O_3$

Robinin $= C_{50} H_{60} O_{32} + 11 H_2 O$

Roccellic acid $= C_{17} H_{32} O_4$

Roccellinin $= C_{18} H_{16} O_7$

Rottlerin $= C_{11} H_{10} O_3^{\cdot\cdot}$

Ruberythric acid $= C_{36} H_{40} O_{20}$

Rubiacin $= C_{16} H_{11} O_5$

Rubian $= C_{28} H_{34} O_{15}$

Rubichloric acid $= C_{14} H_{16} O_9$

Rubiretin $= C_7 H_6 O_2$

Rubitannic acid $= C_{14} H_{16} O_9$

Rutin $= C_{25} H_{28} O_{15} + 2 H_2 O$

Sabadillin $= C_{20} H_{26} N_2 O_5$

Sabadillin hydrate $= C_{20} H_{28} N_2 O_6$

Salacin $= C_{13} H_{18} O_7$

Salicylate of Methyl $= C_8 H_8 O_3$

Salicylic acid $= C_7 H_6 O_3$

*Salicylous acid $= C_7 H_6 O_2$

Santal $= C_8 H_6 O_3$

Santalin $= C_{15} H_{14} O_5$

Santonin $= C_{15} H_{18} O_3$

Saponin $= C_{18} H_{28} O_{12}$

Scoparin $= C_{21} H_{22} O_{10}$

Sicopirin $= C_{16} H_{12} O_5$

Sinapoleic acid $= C_{19} H_{36} O_2$

*Sinapin sulphocyanide (or Sinalbin) $= C_{30} H_{44} N_2 S_2 O_1$ (Will.)

Smilacin $= C_{21} H_{34} O_7$

Solanin $= C_{40} H_{70} NO_{16}$

Sorbin $= C_6 H_{12} O_6$

Spartein $= C_{15} H_{26} N$

Spiraea yellow $= C_{15} H_{16} O_7$

Staphisagrin $= C_{16} H_{23} NO_2$

*Starch $= C_{18} H_{30} O_{15}$ (Musculus)

Stearic acid $= C_{18} H_{36} O_2$

Stearin $= C_{57} H_{110} O_6$

Strychnin $= C_{21} H_{22} N_2 O_2 + H_2 O$

Styracin $= C_{18} H_{16} O_2$

Styrol $= C_8 H_8$

Succinic acid $= C_4 H_6 O_4$

Sycoceryl alcohol $= C_{18} H_{30} O$

237

Sycoceryl acetate $= C_{20} H_{32} O_2$
Syringin $= C_{19} H_{28} O_{10} + H_2 O$
Syringopicrin $= C_{26} H_{48} O_{17}$
Tannaspidic acid $= C_{26} H_{28} O_{11}$
Tannicorticipinic acid $= C_{14} H_{13} O_6$
Tannopinic acid $= C_{28} H_{30} O_{13}$
Tartaric acid $= C_4 H_6 O_6$
Thebain $= C_{19} H_{21} NO_3$
Theobromin $= C_7 H_3 N_4 O_2$
Thujin $= C_{20} H_{22} O_{12}$
Thujogenin $= C_{14} H_{12} O_7$
Thymen $= C_{10} H_{16}$
Thymol $= C_{10} H_{14} O$
Tolen $= C_{10} H_{16}$
Trehalose $= C_{12} H_{22} O_{11} + H_2 O$
Trimethylamin $= C_3 H_9 N$
Tulucunin $C_{20} H_{28} O_8$
Turpethin $C_{34} H_{56} O_{16}$

Tyrosin $= C_9 H_{11} NO_3$
Urson $= C_{20} H_{34} O_2$
Usnic acid $= C_{18} H_{18} O_7$
Valerianic acid $= C_5 H_{10} O_2$
Valerian tannic acid $= C_7 H_9 O_4$
and $C_{12} H_{16} O_9$
*Vanillin $= C_{16} H_8 O_6$ (P. Carles)
Vegetable mucilage $- C_6 H_{10} O_5$
Verantin $= C_{14} H_{10} O_5$
Veratric acid $= C_9 H_{10} O_4$
Veratrin $= C_{32} H_{52} N_2 O_8$
Viscin $= C_{20} H_{48} O_8$
Vulpulin $= C_{19} H_{14} O_5$
Xanthorhamnin $= C_{28} H_{28} O_{14} + 5H_2 O$
Xanthotannic acid $= C_9 H_{18} O_2$
Xanthoxylen $= C_{10} H_{16}$
Xanthoxylin $= C_{20} H_{24} O_8$
Xylochloric acid $C_{15} H_{26} O_{12}$

# DIVISION II.

## SYNOPSIS OF THOSE PLANTS, WHICH YIELD THE PROXIMATE CONSTITUENTS, DESCRIBED UNDER DIVISION I.

———

*Abrus precatorius*, L. Leguminosæ. Leaf: Glycyrrhizin or allied principle.

*Acacia Arabica*, W. *A. Seyal*, Del.; *A Verek*, G. et P.; *A. stenocarpa*, Hochst.; *A. horrida*, Willd; *A. decurrens*, Willd.; *A. pycnantha*, Benth.; *A. harpophylla*, F. v. M.; *A. homalophylla*. Cunn.; *A. microbotrya*, Benth. and several others. Leguminosæ. Exudation of stem and branches: gum.

*Acacia Catechu*, W.; *A. Suma*, Kurz; *A. decurrens*, W. and various other species. Wood and bark: Catechu-tannic and Catechuic acids; gum.

*Acacia lophantha*, W. Root-bark: Saponin.

*Achillea Millefolium*, L. Compositæ. Whole plant: Achillein; Achilleic acid; volatile oil.

*Achillea moschata*, Jacq. Whole plant: Ivain; Achillein; Moschatin; volatile oil.

*Achillea nobilis*, L. Whole plant: volatile oil.

*Achras Sapota*, L. Sapotaceæ. Exudation of stem: Masopin.

*Aconitum ferox*, Wall. Ranunculaceæ. Root: Aconitin.

*Aconitum Lycoctonum*, L. Root: Acolyctin; Lycoctonin.

*Aconitum Napellus*, L. Root: Aconitin; Acolyctin; Mannit: sugar.

*Aconitum*, several species. Whole plant: Aconitic acid.

*Acorus Calamus*, L. Aroideæ. Root: volatile oil.

*Adansonia digitata*, L.; *A. Gregorii*, F. v. M. Malvaceæ. Bark: Adansonin.

*Aeranthus fragrans*, Ldl. Orchideæ. Whole plant: Cumarin.

*Aesculus Hippocastanum*, L. Sapindaceæ. Fruit: Saponin: Aesculic acid; Aesculin and Fraxin in the bark.

*Aesculus Pavia*, L. Bark: Fraxin.

*Aethusa Cynapium*, L. Umbelliferæ. Whole plant: Cynapin.

*Agaricus campestris*, L.  Fungi.  Agaricin.

    ,,    *croceus*, Pers.  Inosit.

    ,,    *muscarius*, L.  Amanitin; Licheno-stearic acid; Muscarin.

    ,, ·   *piperitus*, L.  Inosit.

*Ailantus excelsa*, Roxb.  Simarubeæ.  Bark: Ailantic acid.

*Alchornea latifolia*, Sw.  Euphorbiaceæ.  Bark: Alchornin.

*Aleurites triloba*, Forst.  Euphorbiaceæ.  Seed: fat-oil.

*Alisma Plantago*, L.  Alismaceæ.  Root: Alismin.

*Alkanna tinctoria*, Tausch.  Asperifoliæ.  Root: Alkanna-red.

*Allium sativum*, L.  Liliaceæ.  Bulb: volatile oil.

*Aloe*, several species.  Liliaceae.  Leaf: Aloin = Barbaloin; Nataloin; Socaloin.

*Alpinia Galanga*, Sw.; *A. officinarum*, Hance.  Scitamineæ.  Root: Kaempferid; volatile oil.

*Alsidium Helminthochorton*, Kuetz.  Algae.  Lichenin.

*Alstonia constricta*, F. v. M.  Apocyneæ.  Bark: bitter resin; volatile oil; tannic acid.

*Alstonia scholaris*, R. Br.  Bark: Ditamin; Echicerin; Echitin; Echitein; Echiretin; Echicoutchin.

*Althaea officinalis*, L.  Malvaceæ.  Root: Asparagin; mucilage.

*Alyxia stellata*, R. et S.  Apocyneæ.  Bark: Alyxia-stearopten.

*Amelanchier vulgaris*, Moench.  Rosaceæ.  Flower: Amygdalin.

*Ammi Copticum*, L.  Umbelliferæ.  Fruit: Thymol; Cymol.

*Amomum Meleguta*, Roscoe.  Scitamineæ.  Seeds: volatile oil.

*Amyris Plumieri*, D. C.  Burseraceæ.  Exudation of stem: Elemi.

*Anacardium occidentale*, Rottb.  Anacardiaceæ.  Pericarp: Anacardic acid; Cardol.

*Anagallis arvensis*, L.  Primulaceæ.  Root; Cyclamin.

*Anamirta paniculata*, Colebr.  Menispermeæ.  Pericarp: Menispermin; Paramenispermin.  Seed: Pikrotoxin; Stearin.

*Andira anthelminthica*, Benth.  Leguminosæ.  Wood: Andirin.

*Andropogon Ivarancusa*, Roxb.; *A. Calamus*, Royle; *A. citratus*, D. C.; *A. Martini*, Roxb.; *A. Schœnanthus*, L.; *A. muricatus*, Retz.  Gramineæ.  Whole plant: volatile oil.

*Anemone nemorosa*, L.; *A. pratensis*, L.; *A. Pulsatilla*, L.  Ranunculaceæ.  Whole plant: Anemonic acid; Anemonin.

*Anthemis nobilis*, L.  Compositæ.  Flower: volatile oil; Angelic and Valerianic acids; Butyl and Amyl.

*Anthoxanthum odoratum*, L.  Gramineæ.  Whole plant: Cumarin.

*Antiaris toxicaria*, Lesch.  Urticaceæ.  Juice of stem: Antiarin.

*Apium graveolens*, L.  Umbelliferæ.  Leaf: volatile oil; Apiin.  Fruit: volatile oil.

*Arabis perfoliata*, Lam.  Cruciferæ.  Seed: Sinapin.

*Arachis hypogaea*, L.  Leguminosæ.  Seed: fat-oil, containing Arabic and Hypogaeic acids.

*Aralia quinquefolia*, Dec. et Pl.  Araliaceæ.  Root (Ginseng): Panaquilon.

*Archangelica officinalis*, Hoffm. Umbelliferæ. Root: volatile oil; Hydrocarotin; Angelic and Valerianic acids.

*Arctostaphylos Uva Ursi*, Spr. Ericaceæ. Leaf: Arbutin; Ericolin; Ericinol; Urson.

*Areca Catechu*, L. Palmaceæ. Fruit: Catechutannic and Catechuic acids.

*Argemone Mexicana*, L. Papaveraceæ. Seed: fat-oil.

*Aristolochia Serpentaria*, L. Aristolochieæ. Root: volatile oil.

*Arnica montana*, L. Compositæ. Whole plant: Arnicin.—Flower and root: volatile oil.—Root: Caproic and Caprylic acids.

*Artemisia Absinthium*, L. Compositæ. Whole plant: Absinthiin. —Leaf and flower: volatile oil.—Herb: Succinic acid.

*Artemisia Cina*, Berg. *A. Sieberi*, Besser. Flower: volatile oil; Santonin.

     „      *Dracunculus*, L. Herb: volatile oil with Anethol.

     „      *vulgaris*, L. Root: volatile oil.

*Arum*, several species. Aroideæ. Whole plant: volatile acrid principle.

*Asarum Europaeum*, L. Aristolochieæ. Root: volatile oil with Asaron.

*Asclepias Syriaca*, L. Asclepiadeæ. Milky juice: Asclepion.

*Asparagus officinalis*, L. Liliaceæ. Herb and green fruit: Inosit. —Young shoots: Asparagin.

*Asperula odorata*, L. Rubiaceæ. Whole plant: Asperula-tannic and Rubichloric acids; Cumarin.

*Aspidium Filix Mas.*, Sw. Filices. Root: fat-oil; Filixoleic, Filicic, Pteritannic and Tannaspidic acids.

*Astragalus*, several species. Leguminosæ. Exudation of stem: Bassorin.

*Atherosperma moschatum*, Lab. Monimieæ. Bark: Atherospermin; Atherosperma-tannic acid; volatile oil.

*Atractylis gummifera*, L. Compositæ. Exudation of flower: Viscin.

*Atropa Belladonna*, L. Solaneæ. Whole plant: Atropin; Belladonnin.—Seed: fat-oil.

*Attalea funifera*, Mart. Palmaceæ. Seed: Apyrin.

*Avena sativa*, L. Gramineæ. Fruit: Avenin.

*Balsamodendron Africanum*, Arn. et B. *B. Roxburghii*, Arnott. Burseraceæ. Exudation of stem: Gum resin (Bdellium).

*Balsamodendron Opobalsamum*, Knth. Exudation of stem: Balsam of Mecca with volatile oil.

*Balsamodendron Myrrha*, Ehrenb. Exudation of stem, Myrrh, consisting of gum, resin and volatile oil.

*Banksia* species. Proteaceæ. Bark: Catechu-tannic acid.

*Barosma serratifolia*, Willd.; *B. crenulata*, Hk.; *B. betulina*, Brt. Rutaceæ. Leaf: Diosmin; volatile oil; mucilage.

*Bassia butyracea*, Rxb.; *B. latifolia*, Rxb.; *B. longifolia*, L. Sapotaceæ. Seed: Galam-butter with Myristin.

*Bassia sericea*, Blume. Yields part of the Java gutta-percha.

*Berberis vulgaris*, L. and other species. Berberideæ. Flower and root-bark: Berberin.—Root-bark: Oxyacanthin.

*Beta vulgaris*, L. Salsolaceæ. Root: Cane-sugar; Asparagin.

*Betula alba*, L. Amentaceæ. Leaf: volatile oil.—Leaf and shoots: Betuloretic acid.—Bark: Betulin.—Bark and its spongy excrescences: Phlobaphen.

*Bignonia Chica*, Humb. et B. Bignoniaceæ. Leaf: Chica-red.

*Bixa Orellana*, L. Bixaceæ. Fruit pulp: Annatto-red.

*Blumea balsamifera*, DC. Compositæ. Whole herb: Ngai-camphor.

*Boswellia Carterii*, Brdw. Burseraceæ. Exudation of stem: Olibanum, consisting of gum resin and volatile oil.

*Bowdichia virgilioides*, H. B. K. Leguminosæ. Root-bark: Sicopirin.

*Brassica alba*, Vis. Cruciferæ. Seed: fat-oil with Erucic and Sinapoleic acids; Erucin; Sinapin; Myrosin.

*Brassica nigra*, Koch. Seed: fat-oil with Erucic and Sinapoleic acids; Sinapin; Sinapisin; Myronic Acid; Myrosin.

*Brassica oleracea*, L. Seed: fat-oil with Erucic and Sinapoleic acids.—Leaf: Inosit.

*Bryonia alba*, L. Cucurbitaceæ. Root: Bryonin.

*Buphthalmum salicifolium*, L. Compositæ. Flower: Buphthalmum stearopten.

*Bursera gummifera*, Jacq. Burseraceæ. Exudation of stem: Caranna-resin with volatile oil; Central American Anime.

*Bursera Icicariba*, Benth. Exudation of stem: Elemi-resin with volatile oil.

*Butea frondosa*, Roxb. Leguminosæ. Furnishes part of the East Indian Kino; also, some Catechu.

*Buxus sempervirens*, L. Euphorbiaceæ. Bark: Bebirin.

*Cæsalpinia Crista*, L.; *C. Sappan*, L. Leguminosæ. Wood: Brasilin.

*Calamus Rotang*, L. Palmaceæ. Inflorescence: Dragon's blood.

*Calendula officinalis*, L. Compositæ. Leaf and flower: Calendulin.

*Callitris quadrivalvis*, Vent. Coniferæ. Exudation of stem: Sandarac.

*Calluna vulgaris*, Sal. Ericaceæ. Whole plant: Ericolin; Ericinol; Calluna-tannic acid.

*Calophyllum Calaba*, Jacq.; *C. longifolium*, Willd. Guttiferæ. Exudation of stem: Maynas-resin.

*Calophyllum Inophyllum*, L. Exudation of stem: Tacamahac.

*Calotropis gigantea*, R. Br.; *C. procera*, R. Br. Asclepiadeæ. Root: Mudarin.

*Camelina sativa*, Crantz. Cruciferæ. Seed: fat-oil.

R

*Camellia Thea*, Link. Ternstrœmiaceæ. Leaf: Caffein; Boheic acid.

*Canarium commune*, L. and allied species. Burseraceæ. Exudation of stem: Arbolabrea-resin.

*Canella alba*, Murr. Canellaceæ. Bark: volatile oil with Caryophyllic acid; Mannit.

*Cannabis sativa*, L. Urticaceæ. Herb: volatile oil.—Seed: fat-oil.

*Capparis spinosa*, L. Capparideæ. Flower buds: Rutin.

*Capsicum annuum*, L.; *C. baccatum*, L.; *C. fastigiatum*, Bl. Solanaceæ. Fruit: Capsicin.

*Carapa Guianensis*, Aubl. Meliaceæ. Bark: Carapin; Tulucunin; volatile oil.

*Carbenia benedicta*, Benth. Compositæ. Herb: Cnicin.

*Carthamus tinctorius*, L. Compositæ. Flower: Carthamin and Carthamus yellow.

*Carum Ajowan*, Benth. Umbelliferæ. Fruit: volatile oil with Cymol and Thymol.

*Carum Carui*, L. Fruit: volatile oil with Carven and Carvol.

*Carum Petroselinum*, Benth. Fruit and herb: volatile oil with stearopten.

*Cascarilla hexandra*, Wedd. Rubiaceæ. Bark: Paricin.

*Cascarilla magnifolia*, Wedd. Bark: Quina-red; Quinic and Quinova-tannic acids.

*Cassia acutifolia*, Del.; *C. angustifolia*, Vahl. Leguminosæ. Leaf: Cathartic acid.

*Castilloa elastica*, Cerv.; *C. Markhamiana*, Collins. Urticaceæ. Ulé-Caoutchouc, nearly as good as that of Hevea Guianensis.

*Centaurea Calcitrapa*, L. Compositæ. Herb: Cnicin.

*Cephaëlis Ipecacuanha*, Rich. Rubiaceæ. Root: Emetin; Ipecacuanhic acid.

*Ceratonia Siliqua*, L. Leguminosæ. Fruit: Butyric acid.

*Ceratophorus Leerii*, Hassk. Sapotaceæ. Yields part of the Sumatra gutta-percha.

*Ceroxylon andicola*, H. et B. Palmaceæ. Exudation of stem: Ceroxylin.

*Cetraria*. Several species. Lichenes: Lichenin.

*Cetraria Islandica*, Achar. Thallochlor; Cetraric, Fumaric and Lichenostearic acids.

*Chelidonium majus*, L. Papaveraceæ. Whole plant: Chelerythrin; Chelidonin; Chelidoxanthin; Chelidonic acid.

*Chenopodium album*, L. Salsolaceæ. Whole plant: Chenopodin.

*Chenopodium ambrosioides*, L. Herb: volatile oil.

*Chenopodium olidum*, Curtis. Whole plant: Trimethylamin.

*Chiococca racemosa*, Jacq. Rubiaceæ. Root: Caffe-tannic acid.

*Chlorea vulpina*, Nyl. Lichenes. Vulpulin.

*Chondodendron tomentosum*, R. et P. Menispermeæ. Root: Bebirin.
*Chondrus crispus*, Stackh. Algæ. Gœmin.
*Chrysanthemum Parthenium*, Pres. Compositæ. Whole plant: volatile oil.
*Cicuta virosa*, L. Umbelliferæ. Whole plant: Cicutin.—Seed: volatile oil with Cuminol, Cymen and Cicuten.
*Cinchona* species. Rubiaceæ. Bark contains alkaloïds as below:
*Cinchona Calisaya*, Wedd. Quinin; Cinchonidin.
   „    *cordifolia*, R. et P. Quinin; Cinchonin.
   „    *glandulifera*, R. et P. Quinin.
   „    *lancifolia*, Mutis. Quinin; Cinchonidin.
   „    *micrantha*, R. et P. Cinchonin; Quinidin; Huanokin.
   „    *nitida*, R. et P. Cinchonin; Quinidin.
   „    *officinalis*, L. Quinin; Cinchonidin.
   „    *Pahudiana*, How. Cinchonin.
   „    *Peruviana*, How. Cinchonin; Quinidin.
   „    *pubescens*, Vahl. Aricin=Cinchonidin (Hesse).
   „    *scrobiculata*, H. et B. Quinin; Cinchonin.
   „    *succirubra*, Pav. Quinin; Cinchonidin; Quinidin; Paricin; Quinamin; Howard's Cinchona alkaloid.
*Cinchona*, several species. Bark: Quina-red; Phlobaphen; Quinic and Quino-tannic acids.
*Cinnamomum Burmanni*, Blume. Lauraceæ. Bark: volatile oil.
   „    *Camphora*, F. Nees. Wood: camphor and volatile oil.
*Cinnamomum Cassia*, Blume. Bark: volatile oil.
   „    *Zeylanicum*, C. G. Nees. Leaf, flower and bark: volatile oil.
*Cinnamomum Culilaban*, Blume. Bark: volatile oil.
*Cissampelos Pareira*, L. Menispermeæ. Root; Bebirin.
*Cistus Creticus*, L. *C. ladaniferus*, L.; *C. Ledon*, Lam.; *C. laurifolius*, L.; *C. Monspeliensis*, L. Cistaceæ. Exudation of branches: Labdanum.
*Citrus Aurantium*, L. Rutaceæ. Flower: volatile oil.—Unripe fruit and rind: Hesperidin.—Rind of ripe fruit: volatile oil.—Seed: Limonin.
*Citrus Aurantium*, var. *Bergamia*, Riss. Rind of fruit: volatile oil.
*Citrus Aurantium*, var. *dulcis*, Volk. Rind of fruit: volatile oil.
   „    *medica*, L. Rind of fruit: volatile oil.—Seed: Limonin.
   „    „    var. *Limetta*, Risso. Rind of fruit: volatile oil.
*Cladonia rangiferina*, Hoffm. Lichenes. Usnic acid.
   „    several species. Lichenin.
*Clematis*, several species. Ranunculaceæ. Whole plant: volatile acrid substance.
*Coccoloba uvifera*, Jacq. Polygoneæ. West Indian Kino.

*Cochlearia Armoracia*, L. Cruciferæ. Root: volatile oil.
„   *officinalis*, L.; *C. Anglica*, L.; *C. Danica*, L. Herb: volatile oil.
*Cocos nucifera*, L. Palmaceæ. Seed: Cocos-fat with Laurostearin, Myristin, Caproic, Caprylic and Capric acids.
*Coffea Arabica*, L. Rubiaceæ. Leaf and seed: Caffein; Caffetannic and Quinic acids.
*Colchicum autumnale*, L. Liliaceæ. Whole plant: Colchicin.— Seed: Sabadillic acid.
*Conium maculatum*, L. Umbelliferæ. Whole plant: Coniin.— Flower and fruit: Coniin and Conhydrin.
*Convallaria majalis*, L. Liliaceæ. Whole plant: Convallarin.— Root: Convallamarin.
*Convolvulus Scammonia*, L. Convolvulaceæ. Root: Jalapin.
„   *scoparius*, L. fil.; *C. floridus*, L. fil. Stem and root: volatile oil.
*Copaifera*, several species. Leguminosæ. Exudation of stem: Copaivic and Metacopaivic acids; volatile oil.
*Copernicia cerifera*, Mart. Palmaceæ. Covering of leaf: Canauba wax.
*Coptis Teeta*, Wall. Ranunculaceæ. Root: Berberin.
*Cordia Boissieri*, A. DC. Asperifoliæ. Wood: Anacahuit-tannic acid.
*Cordyceps purpurea*, Fr. Fungi. Mycelium (ergot): Ecbolin, Ergotin; Trimethylamin; Ergotinin; Ergotic acid; Mycose.
*Coriandrum sativum*, L. Umbelliferæ. Fruit: volatile oil.
*Coriaria myrtifolia*, L. Phytolacceæ. Leaf and fruit: Coriaria-myrtin.
*Cornus florida*, L. Cornaceæ. Root-bark: Cornin.
*Corydalis fabacea*, Pers.; *C. tuberosa*, Cand. Fumariaceæ. Root: Corydalin.
*Corylus Avellana*, L. Amentaceæ. Seed: fat-oil.
*Coscinium fenestratum*, Colebr. Menispermeæ. Wood: Berberin.
*Cotoneaster vulgaris*, Lindl. Rosaceæ. Flower: Amygdalin.
*Cratægus coccinea*, L. Rosaceæ. Flower: Trimethylamin.
*Cratægus Oxyacantha*, L. Flower: Trimethylamin: Amygdalin.
—Bark of branches: Cratægin.
*Crepis fœtida*, L. Compositæ. Flower: Salicylous acid.
*Crocus sativus*, L. Irideæ. Pistil: volatile oil; Crocin.
*Croton Eleutheria*, Benn.; *C. Sloanei*, Benn. Euphorbiaceæ. Bark: Cascarillin; volatile oil.
*Croton erythrinum*, Mart. Furnishes Brazilian Kino.
„   *niveum*, Jacq. Bark: Copalchin.
„   *Tiglium*, L. Seed: fat-oil with Crotonol and Laurostearin.
*Cucumis Colocynthis*, L. Cucurbitaceæ. Fruit: Colocynthin.
„   *Melo*, L. Root: Melonemetin; cane-sugar.
„   *Prophetarum*, L. Fruit: Prophetin.

*Cucurbita Pepo*, L. Cucurbitaceæ. Seed: drying fat-oil.
*Cuminum Cyminum*, L. Umbelliferæ. Fruit: volatile oil.
*Curcuma longa*, L. Scitamineæ. Root: Curcumin, volatile oil.
   ,,    *Zedoaria*, Roxb. Root: volatile oil.
*Cyclamen Europæum*, L. Primulaceæ. Root: Cyclamin.
*Cyclopia latifolia*, Cand. Leguminosæ. Leaf: Quinic acid.
*Cynometra Spruceana*, Benth. Leguminosæ. Exudation of stem: Brazilian Copal.
*Cyperus esculentus*, L. Cyperaceæ. Root: fat-oil.
*Cytisus Laburnum*, L. Leguminosæ. Ripe seed: Cytisin.—Unripe seed: Laburnin.
*Cytisus scoparius*, Link. Whole plant: Spartein; Scoparin.
*Dæmonorops Draco*, Mart. Palmaceæ. Exudation of stem: Dragon's blood.
*Dahlia purpurea*, Poir. Compositæ. Root: volatile oil.
*Dammara australis*, Lamb. Coniferæ. Exudation of stem: Dammar resin with Dammaryl and Dammarylic acid.
*Dammara orientalis*, Lamb. Exudation of stem: Dammar-resin.
*Daphne alpina*, L. Thymeleæ. Bark: Daphnin.
   ,,    *Gnidium*, L. Seed: Coccognic acid.
   ,,    *Mezereum*, L. Bark: Daphnin.—Fruit: Coccognin and drying fat-oil.
*Datisca cannabina*, L. Datiscaceæ. Root: Datiscin.
*Datura Stramonium*, L. Solanaceæ. Whole plant: Daturin, Stramonin.
*Daucus Carota*, L. Umbelliferæ. Root: volatile oil, Carotin, Hydrocarotin.
*Delesseria*, several species. Algæ. Lichenin.
*Delphinium Consolida*, L. Ranunculaceæ. Herb: Aconitic= Equisetic acid.
*Delphinium Staphisagria*, L. Seed: Delphinin; Staphisagrin.
*Dicypellium coryophyllatum*, Nees. Lauraceæ. Bark: volatile-oil.
*Digitalis lutea*, L. Scrophularinæ. Leaf: Digitaletin; Digitalin.
   ,,    *purpurea*, L. Leaf: Digitaletin; Digitalin; Digitoleic, volatile and non-volatile Digitalic acids; Inosit.
*Dipterocarpus turbinatus*, Gærtn.; *D. incanus*, Roxb.; *D. alatus*, Roxb.; *D. Zeylanicus*, Thw.; *D. hispidus*, Thw.; *D. trinervis*, Bl.; *D. gracilis*, Bl.; *D. littoralis*, Bl.; *D. retusus*, Bl. Dipterocarpeæ. Exudation of stem: Gurjunic acid; volatile oil.
*Dipteryx odorata*, Schreb. Leguminosæ. Seed: Cumarin.
*Dorema ammoniacum*, D. Don. Umbelliferæ. Exudation of stalk and root: Ammoniacum.
*Dracæna Draco*, L. Liliaceæ. Exudation of stem: Dragon's blood.
*Drosera*, many species. Droseraceæ. Root and herb: Alizarin, or an allied substance.
*Dryobalanops Camphora*, Colebr. Dipterocarpeæ. Wood: Borneol; Borneen.

*Ecballion Elaterium*, A. Rich. Cucurbitaceæ. Fruit: Elaterid, Elaterin; Hydroelaterin; Pectin.

*Elais Guineensis*, Jacq. Palmaceæ. Pericarp: Palm-oil.

*Elettaria Cardamomum*, White. Scitamineæ. Seed: volatile oil.

*Epacris* species. Epacrideæ. Leaf: Urson.

*Equisetum fluviatile*, L. Equisetaceæ. Whole plant: Flavequisetin.

*Equisetum*, several species. Whole plant: Aconitic ( = Equisetic) acid.

*Erica carnea*, L. Ericaceæ. Whole plant: Ericinol; Ericolin.

*Erythrophlaeum Guineense*, G. Don. Leguminosæ. Bark: Erythrophlacin.

*Erythroxylon Coca*, Lam. Lineæ. Leaf: Cocain: Hygrin; Cocatannic acid.

*Eschscholtzia Californica*, Cham. Papaveraceæ. Whole plant: Chelerythrin; an acrid and a bitter alkaloïd.

*Eucalyptus*, many species. Myrtaceæ. Leaf: volatile oil; Eucalyptin; Eucalyptic acid.—Bark: Kino-tannic acid; Pyrocatechin; Catechin.

*Eucalyptus resinifera*, Smith. Botany Bay Kino.

*Eucalyptus rostrata*, Schl. Port Phillip Kino.

„ *viminalis*, Labill. Eucalyptus-manna with Melitose.

*Eugenia caryophyllata*, Thunb. Myrtaceæ. Flower: volatile oil with Caryophyllic acid; Caryophyllin; Eugenin.

*Euonymus Europaeus*, L. Celastrineæ. Seed: fat-oil.

*Eupatorium cannabinum*, L. Compositæ. Leaf and flower: Eupatorin.

*Euphorbia resinifera*, Berg. Euphorbiaceæ. Exudation of stalk: Euphorbium.

*Euphrasia officinalis*, L. Scrophularinæ. Whole plant: Euphrasia-tannic acid.

*Euryangium Sumbul*, Kauffm. Umbelliferæ. Angelic acid.

*Evernia prunastri*, Ach. Lichenes. Evernic acid.

*Evernia*, several species. Lichenes. Lichenin; Lecanoric and Usnic acids.

*Excaecaria sebifera*, J. M. Euphorbiaceæ. Pericarp: vegetable tallow with Palmitin.

*Fabiana imbricata*, R. et P. Solanaceæ. Flower: Crocin.

*Fagopyrum cymosum*, Meissn.; *F. emarginatum*, Meissn.; *F. esculentum*, Mœnch; *F. Tartaricum*, Gærtn. Polygoneæ. Whole plant: variety of Indigo.

*Fagus silvatica*, L. Amentaceæ. Seed: fat-oil; Trimethylamin.

*Ferula Asa-fœtida*, L.; *F. alliacea*, Boiss; *F. Narthex*, Boiss. Umbelliferæ. Exudation of stalk and root: Asafœtida; volatile oil.

*Ferula erubescens*, Boiss. Exudation of stalk and root: Galbanum; volatile oil.

*Ferula Persica*, W. *F. Szovitsiana*, DC. Exudation of stalk and root· Sagapenum; volatile oil.

*Ficus elastica*, Roxb. Urticaceæ. Exudation of stem: Assam-Caoutchouc.

*Ficus rubiginosa*, Dsf. Exudation of stem: Sycoretin; Sycoceryl acetate.

*Ficus subracemosa*, Blume; *F. variegata*, Blume; Exudation of stem: Getah-Lahöe.

*Fœniculum officinale*, All. Umbelliferæ. Fruit: volatile oil with Anethol.

*Fragaria vesca*, L. Rosaceæ. Fruit: Cisso-tannic acid.

*Fraxinus excelsior*, L. Oleaceæ. Leaf: Quinic acid.—Bark: Fraxin.

*Fraxinus Ornus*, L. Exudation of stem: Mannit; Fraxin.

*Fumaria officinalis*, L. Fumariaceæ. Whole plant: Fumarin.

*Galipea Cusparia*, St. Hil. Rutaceæ. Bark: Angusturin; volatile oil.

*Galium Aparine*, L. Rubiaceæ. Whole plant: Galium-tannic and Rubichloric acids.

*Galium Mollugo*, L. Whole plant: Asperula-tannic, Quinic and Rubichloric acids.

*Galium verum*, L. Whole plant: Galium-tannic and Rubichloric acids.

*Garcinia Indica*, Chois. Guttiferæ. Seed: Brindonia-tallow.

„ *Mangostana*, L. Exudation of stem: Mangostan-resin. —Pericarp: Mangostin.

*Garcinia pictoria*, Roxb.; *G. Morella*, Desr.; *G. Cochinchinensis*, Chois. Exudation of stem: Gamboge.

*Gardenia grandiflora*, Lour.; *G. lucida*, Roxb. Rubiaceæ. Pods: Gardenin.

*Gaultiera procumbens*, L. Ericaceæ. Whole plant: volatile oil with Methyl-salicylate.

*Gelseminum nitidum*, Michaux. Loganiaceæ. Whole plant: Gelsemin, Æsculin.

*Gentiana lutea*, L. Gentianeæ. Root; Gentiano-picrin (=Gentiamarin); Gentianin.

*Geoffroya inermis*, Wright. Leguminosæ. Bark; Berberin.

„ *Surinamensis*, Bondt. Bark: Surinamin.

*Geranium*, many species. Geraniaceæ. Root: Geraniin.

*Geum urbanum*, L. Rosaceæ. Root: volatile oil.

*Ginkgo biloba*, Salisb. Coniferæ. Fruit: Butyric and Ginkgoic acids.

*Glaucium luteum*, Scop. Papaveraceæ. Herb: Glaucin; Fumaric acid.—Root: Chelerythrin; Glaucopicrin.

*Globularia Alypum*, L. Scrophularinæ. Leaf: Globularin; Globulariæ resin; Globularia-tannic acid.

*Glycyrrhiza echinata*, L.; *G. glabra*, L. Leguminosæ. Root: Glycyrrhizin; Asparagin.

*Gratiola officinalis*, L.; *G. Peruviana*, L. Scrophularinæ. Whole plant: Gratiolin; Gratiosolin.

*Guaiacum officinale*, L.; *G. sanctum*, L. Zygophylleæ. Exudation of stem: Guaiacum-resin with Guaiacum beta-resin; Guaiacum yellow; Guaiacic, Guaiaconic and Guaiaretic acids.

*Gypsophila Struthium*, L. Caryophylleæ. Root: Saponin.

*Hæmatoxylon Campechianum*, L. Leguminosæ. Wood: Hæmatoxylin.

*Hagenia Abyssinica*, Lamarck. Rosaceæ. Flower: Koussin; Hagenic acid.

*Hancornia speciosa*, Gomes. Apocyneæ. Pernambuco Caoutchouc.

*Hedera Helix*, L. Araliaceæ. Leaf: Quinic acid.—Seed: Hederatannic and Hederic acids.

*Hedwigia balsamifera*, Sw. Burseraceæ. Balsam and volatile oil.

*Helianthus annuus*, L. Compositæ. Seed: drying fat oil; Helianthic acid.

*Helleborus niger*, L.; *H. viridis*, L. Ranunculaceæ. Root: Helleborin; Helleborein.

*Heracleum Sphondylium*, L. Umbelliferæ. Fruit: Capryl-alcohol; Capryl-acetate.

*Heracleum villosum*, Fischer. Fruit: Capryl-acetate; Capryl-butyrate.

*Herniaria glabra*, L. Paronychieæ. Whole plant: Cumarin.

*Hesperis matronalis*, L. Cruciferæ. Seed: drying fat-oil.

*Hevea Guianensis*, Aubl. and other species. Euphorbiaceæ. Exudation of stem: best Para-Caoutchouc.

*Hierochloa borealis*, Roem. et Sch. Gramineæ. Whole plant: Cumarin.

*Hillia spectabilis*, Fr. Vand. Rubiaceæ. Exudation of stem: Ratanhin.

*Hippophaë rhamnoides*, L. Elæagneæ. Fruit: Quercetin.

*Hordeum vulgare*, L. Gramineæ. Fruit: Hordein.—Germinated fruit (malt): Asparagin; Aspartic acid; Diastase; Cholestearin.

*Humulus Lupulus*, L. Urticaceæ. Leaf and flower: Lupulin; Trimethylamin; volatile oil.

*Hura crepitans*, L. Euphorbiaceæ. Exudation of stem: Hurin.

*Hybanthus Ipecacuanha*, F. v. M. Violaceæ. Root: Emetin.

*Hydrastis Canadensis*, L. Ranunculaceæ. Root: Berberin; Hydrastin.

*Hymenæa Candolleana*, H. et B.; *H. confertiflora*, Mart.; *H. confertifolia*, Hayne; *H. Courbaril*, L.; *H. latifolia*, Hayne; *H. Martiana*, Hayne; *H. Oljersiana*, Hayne; *H. rotundata*, Hayne; *H. stilbocarpa*, Hayne; *H. Selloviana*, Hayne; *H. venosa*, Vahl. Leguminosæ. Exudation of stem: Copal.

*Hyoscyamus niger*. L. Solanaceæ. Whole plant: Hyoscyamin. —Seed: fat-oil.

*Hyssopus officinalis*, L. Labiatæ. Whole plant: volatile oil.

*Iateorrhiza palmata*, Miers. Menispermeæ. Root: Berberin; Colombin; Colombic acid.

*Iatropha Curcas*, L. Euphorbiaceæ. Seed : fat-oil (Glycerids of Ricinoleic and Isocetic acids).

*Icica heptaphylla*, Aubl. Burseraceæ. Anime-Tacamahac (Batka).

*Ilex Aquifolium*, L. Aquifoliaceæ. Leaf : Ilicin ; Ilixanthin; Ilicic and Quinic acids; Viscin.

*Ilex Cassine*, L. Leaf: Coffein.

  ,, *Paraguensis*, St. Hil. Leaf : Coffein ; Coffeic and Quinic acids.

*Illicium anisatum*, L. Magnoliaceæ. Fruit: volatile oil, with Anethol.

*Imbricaria coriacea*, A. DC. Sapotaceæ. Yields part of the Madagascar gutta-percha.

*Indigofera Anil*, L.; *I. argentea*, L.; *I. tinctoria*, L. Leguminosæ. Whole plant: Indigo.

*Inula Helenium*, L. Compositæ. Root: Helenin; Inulin.

*Ionidium Ipecacuanha*, Vent. Violaceæ. Root: Emetin.

*Ipomœa Orizabensis*, Steud. Convolvulaceæ. Root: Convolvulin.

  ,, *Purga*, Wender. Root: Convolvulin.

  ,, *Turpethum*, R. Br. Root: Turpethin.

*Iris Florentina*, L. Irideæ. Root: Myristic acid.

*Irvingia Barteri*, J. Hook. Simarubeæ. Seed (Dica-bread): Dica-fat, with Laurostearin and Myristin.

*Isatis indigofera*, Fort.; *I. tinctoria*, L.; and other species. Cruciferæ. Whole plant: Indican.

*Isonandra Gutta*, Hook. Sapotaceæ. Exudation of stem: gutta-percha.

*Jeffersonia diphylla*, Bart. Berberideæ. Root: Berberin.

*Juglans regia*, L. Juglandeæ. Green pericarp: Nucin; Nucitannin; Regianin.—Seed: fat-oil.

*Juniperus communis*, L. Coniferæ. Fruit: Juniperin; volatile oil.

  ,, *Sabina*, L. Branches: volatile oil.

  ,, *Virginiana*, L. Wood : volatile oil with Cedren and Cedar-stearopten.

*Khaya Senegalensis*, A. Juss. Meliaceæ. Bark: Cailcedrin.

*Krameria triandra*, R. et P.; *K. ixina*, L.; *K. secundiflora*, DC. Polygaleæ. Root: Ratanhia-tannic acid ; Tyrosin.

*Lactuca sativa*, L.; *L. virosa*, L. Compositæ. Exudation of stalk (Lactucarium): Lactucin; Lactucerin; Succinic acid.

*Ladenbergia*, several species. Rubiaceæ. Bark: Paytin.

*Laetia apetala*, Jacq. Bixaceæ. Exudation of stem : Laetia-resin.

*Laserpitium latifolium*, L. Umbelliferæ. Root: Laserpitin.

*Laurus nobilis*, L. Lauraceæ. Fruit : fat-oil, with Laurostearin and Lauric acid; volatile oil.

*Lavandula angustifolia*, Ehrh.; *L. latifolia*, Vill.; *L. Stoechas*, L. Labiatæ. Flower: volatile oil.—Whole plant: volatile oil.

*Lecanora*, several species. Lichenes. Lecanoric and Usnic acids; Orcin.

*Lecanora Parella*, Achar. Parellin.

„ *Tartarea*, Achar. Erythric and Roccellic acids.

*Ledum palustre*, L. Ericaceæ. Whole plant: volatile oil with Ericinol; Ericolin; Ledum-tannic acid.

*Leontice thalictroides*, L. Berberideæ. Root: Berberin.

*Lepidium*, several species. Cruciferæ. Whole plant: Lepidin; volatile oil.—Seed: Drying oil.

*Liatris odoratissima*, Wlld. Compositæ. Whole plant: Cumarin.

*Ligustrum vulgare*, L. Oleaceæ. Leaf: Quinic acid.—Bark: Ligustrin.

*Limosella aquatica*, L. Scrophularinæ. Root: Cyclamin.

*Linum usitatissimum*, L. Lineæ. Seed: drying oil with Linoleic acid; mucilage.

*Liquidambar orientalis*, Mill. Saxifrageæ. Exudation of stem: Styrol; Styracin; Cinnamic acid.

*Liriodendron tulipifera*, L. Magnoliaceæ. Root-bark: Liriodendrin.

*Lobelia inflata*, L. Campanulaceæ. Whole plant: Lobeliin; Lobelacrin.

*Lonicera xylosteum*, L. Caprifoliaceæ. Fruit: Xylostein.

*Lucuma glycyphlaea*, Mart. et Eichl. Sapotaceæ. Bark: Glycyrrhizin; Saponin.

*Lucuma Parkii*, R. Br. Seeds: Shea-butter.

*Lunanea Bichi*, DC. Sterculiaceæ. Fruit: Coffein.

*Lupinus albus*, L.; *L. luteus*, L. and other species. Leguminoæ. Seed: Lupinin.

*Lychnis Githago*, Lam. Caryophylleæ. Seed: Agrostemmin; Saponin.

*Lycopodium Chamaecyparissus*, Al. Br. Lycopodiaceæ. Whole plant: Lycopodiamarin; Lycopodium-stearon; Lycopodium-resin.

*Lycopodium complanatum*, L. Whole plant: Tartaric acid.

*Lycopus Europaeus*, L. Labiatæ. Herb: volatile oil.

*Machura tinctoria*, D. Don. Urticaceæ. Wood: Morin; Morus-tannic acid.

*Madia sativa*, Mol. Compositæ. Seed: drying oil.

*Magnolia umbrella*, Lam. Magnoliaceæ. Fruit: Magnolin.

*Mallotus Philippinensis*, J. M. Euphorbiaceæ. Fruit hairs and glands (Kamala): Rottlerin.

*Marrubium vulgare*, L. Labiatæ. Whole plant: Marrubiin.

*Matricaria Chamomilla*, L. Compositæ. Flower: volatile oil.

*Melaleuca Leucadendron*, L. and many other species. Myrtaceæ. Foliage: volatile oil.

*Melampyrum nemorosum*, L. Scrophularinæ. Whole plant: Dulcit.

*Melilotus officinalis*, Desr. Leguminosæ. Whole plant: Cumarin.

*Melissa officinalis*, L. Labiatæ. Whole plant: volatile oil.

*Mentha australis*, R. Br.; *M. gracilis*, R. Br.; *M. laxiflora*, Bth.; *M. piperita*, L.; *M. Pulegium*, L.; *M. viridis*, L.; and other species. Labiatæ. Whole plant: volatile oil.

*Menyanthes trifoliata*, L. Gentianeæ. Whole plant: Menyanthin.

*Mercurialis annua*, L. Euphorbiaceæ. Whole plant: volatile oil; Mercurialin.

*Mikania Guaco*, H. et B. Compositæ. Leaf: Guacin.

*Mimusops Elengi*, L. Sapotaceæ. Yields part of the Indian gutta-percha.

*Mimusops Manilkara*, G. Don. Yields part of the China and Manilla gutta-percha.

*Monarda didyma*, L. Labiatæ. Flower: Carmic acid, or an allied one.

*Monarda punctata*, L. Whole plant: volatile oil with Thymol.

*Monnina polystachya*, R. et P. Polygaleæ. Root: Saponin.

*Monotropa Hypopitys*, L. Ericaceæ. Whole plant: volatile oil with Methyl-salicylate.

*Morinda citrifolia*, L. Rubiaceæ. Root: Morindin.

*Moringa oleifera*, Lam.; *M. aptera*, Gærtn. Moringaceæ. Seed: fat-oil, with Benic and Moringic acids.

*Myrica cerifera*, L.; *M. cordifolia*, L.; *M. quercifolia*, L.; *M. serrata*, Lam. Amentaceæ. Fruit: Myrica-wax with Palmitin and Myristin.

*Myrica Gale*, L. Leaf: volatile oil.

*Myristica Bicahyba*, Schott. Myristiceæ. Seed: Bicahyba-fat with volatile oil.

*Myristica fragrans*, Houtt. Pericarp (mace): fat; volatile oil.— Seed (nutmeg): fat with Myristin and volatile oil.

*Myristica Otaba*, H. et B. Seed: Otaba-fat with volatile oil.

      „      *sebifera*, Sw. Seed: Virola-tallow with volatile oil.

*Myroxylon Pereirae*, Kl. Leguminosæ. Exudation of stem (black balsam of Peru): Resin; Styracin; Benzyl-cinnamate; Cinnamic acid.—Fruit-juice (white balsam of Peru): Myroxocarpin.

*Myroxylon toluiferum*, L. fil. Fruit: Cumarin.—Exudation of stem (balsam of Tolu): Resin; Tolen; Cinnamic acid.

*Myrtus communis*, L. Myrtaceæ. Foliage, flower and fruit: volatile oil.

*Narcissus Jonquilla*, L. Amaryllideæ. Flower: volatile oil.

*Narthecium ossifragum*, Huds. Liliaceæ. Herb: Narthecin; Narthecic acid.

*Nasturtium officinale*, L. Cruciferæ. Herb: volatile oil.

*Nectandra Rodiei*, Schomb. Lauraceæ. Bark: Bebirin; Sipirin: Bebiric acid.

*Nerium Oleander*, L. Apocyneæ. Flower: Salicylous acid.—
Foliage and branches: Oleandrin; Pseudocurarin.
*Nicotiana*, many species. Solanaceæ. Whole plant: Nicotin;
Nicotianin.
*Nicotiana Tabacum*, L. Seed: fat-oil.
*Nigella sativa*, L. Ranunculaceæ. Seed: Nigellin; volatile oil.
*Nigritella angustifolia*, Rich. Orchideæ. Whole plant: Cumarin.
*Ocimum Basilicum*, L. Labiatæ. Whole plant: Basilicum-
stearopten.
*Ocotea Pichurim*, Kunth. Lauraceæ. Fruit: fat-oil with Lauro-
stearin and Lauric acid.
*Oenanthe Phellandrium*, Lam. Umbelliferæ. Seed: volatile oil.
*Olea Europaea*, L. Oleaceæ. Whole plant: Olivamarin.—Fruit;
fat-oil.—Exudation of stem: Olivil.
*Ononis spinosa*, L. Leguminosæ. Root: Ononid; Onocerin.
*Ophelia Chirata*, Griseb. Gentianeæ. Whole plant: Chiratin;
Ophelic acid.
*Opopanax Chironium*, Koch. Umbelliferæ. Exudation of stalk
and root: Opopanax.
*Orchis*, many species. Orchideæ. Tuber: Mucilage.
   „ *hircina*, Sw. Flower: Caproic acid.
   „ *purpurea*, Huds. Whole plant: Cumarin.
*Origanum Majorana*, L. Labiatæ. Whole plant: volatile oil.
   „ *vulgare*, L. Whole plant: volatile oil.
*Osmitopsis asteriscoides*, Cass. Compositæ. Flower: volatile oil.
*Othonna furcata*, Benth. Compositæ. Exudation of stem:
Ceradia-resin.
*Papaver Rhœas*, L. Papaveraceæ. Whole plant: Morphin;
Rhœadin and probably also Thebain.
*Papaver somniferum*, L. Seed: drying oil.—Exudation of peri-
carp: Codein; Metamorphin; Morphin; Narcein; Narcotin;
Opianin; Papaverin; Porphyroxin; Pseudo-morphin; Thebain;
Lanthopin; Meconidin; Codamin; Laudanin; Meconin;
Meconic acid.
*Paris quadrifolia*, L. Liliaceæ. Whole plant: Paridin.
*Parmelia parietina*, Ach. Lichenes. Chrysophanic acid.
   „ many species. Lichenin; Usnic acid.
   „ *physodes*, Ach. Physodin; Ceratophyllin.
*Pastinaca sativa*, L. Umbelliferæ. Fruit: Capryl-butyrate.
*Paullinia sorbilis*, Mart. Sapindaceæ. Fruit: Coffein.
*Payena macrophylla*, Benth. Sapotaceæ. Yields part of the
Java gutta-percha.
*Peganum Harmala*, L. Rutaceæ. Seed: Harmalin; Harmin.
*Pelargonium odoratissimum*, Ait.; *P. Radula*, Ait.; *P. capitatum*,
Ait. Geraniaceæ. Leaf and flower: Pelargonic acid; volatile
oil.
*Peltophorum Linnæi*, Bth. Leguminosæ. Wood: Brasilin.

*Perezia Humboldti*, A. Gr.  Compositæ.  Root: Pipitzahoic acid.
*Pertusaria communis*, Fries.  Lichenes.  Picrolichenin; Vario-
larin.
*Peucedanum galbaniferum*, Benth.  Umbelliferæ.  Exudation of
stalk and root: Galbanum with volatile oil.
*Peucedanum officinale*, L.  Root: Peucedanin.
,,  *Oreoselinum*, Cuss.  Foliage: volatile oil.—Root:
Athamantin; Valerianic acid.
*Peucedanum Ostruthium*, Koch.  Root: volatile oil; Ostruthin;
Peucedanin.
*Pharbitis Nil*, Choisy.  Convolvulaceæ.  Seed: Convolvulin.
*Phaseolus vulgaris*, L.  Leguminosæ.  Unripe fruit: Inosit.—Seed:
Phaseolin.
*Philadelphus coronarius*, L.  Saxifrageæ.  Flower: volatile oil.
*Phillyrea latifolia*, L.; *Ph. angustifolia*, L.  Oleaceæ.  Bark:
Phillyrin.
*Physalis Alkekengi*, L.  Solanaceæ.  Leaf: Physalin.
*Physostigma venenosum*, Balf.  Leguminosæ.  Seed: Physo-
stigmin.
*Picræna excelsa*, Lindl.  Simarubeæ.  Wood and bark: Quassiin.
*Picramnia ciliata*, Mart.  Simarubeæ.  Bark: Pereirin.
*Pimenta officinalis*, Lindl.  Myrtaceæ.  Fruit: volatile oil.
*Pimpinella Anisum*, L.  Umbelliferæ.  Fruit: volatile oil, with
Anethol.
*Pimpinella nigra*, L.  Root: volatile oil.
,,  *Saxifraga*, L.  Root: volatile oil.
*Pinckneya pubens*, Rich.  Rubiaceæ.  Bark: Aribin.
*Pinus Abies*, Du Roi.  Coniferæ.  Wood and bark: Formic acid.—
Seed: fat-oil.
*Pinus balsamea*, L.; *P. Fraseri*, Pursch.  Canada-balsam.
,,  *Lambertiana*, Dougl.  Exudation of stem: Pinit.
,,  *Larix*, L.  Exudation of stem: Melezitose; Venetian
turpentine.
*Pinus*, many species.  Whole plant and exudation of stem:
Abietic acid; volatile oil; Coniferin.
*Pinus Picea*, Du Roi.  Seed: drying oil.
,,  *Sabiniana*, Dougl.  Exudation of stem: Abieten; resin.
,,  *silvestris*, L.  Wood: Pityxylonic acid.—Foliage; Pinopicrin;
Quinovic acid; Oxypino-tannic, Pino-tannic and Tanno-pinic
acids.—Bark: Pino-picrin; Pino-corretin; Phlobaphen; Cor-
ticipino-tannic, Pinocortic-tannic and Tannocortici-pinic acids.—
Seed: drying oil.
*Piper angustifolium*, R. et P.  Piperaceæ.  Leaf: volatile oil.
,,  *Cubeba*, L. fil.; Fruit: volatile oil; Cubelin; stearopten.
,,  *longum*, L.; *P. officinarum*, C. DC.  Fruit: Piperin.
,,  *methysticum*, Forst.  Root: Methysticin.
,,  *nigrum*, L.  Fruit: Piperin; volatile oil.

*Pistacia Lentiscus*, L.    Anacardiaceæ.    Exudation of stem: Mastich.

*Pistacia Terebinthus*, L.    Exudation of stem: turpentine of Chios.

*Pisum sativum*, L.    Leguminosæ.    Unripe fruit.: Inosit; Cholestearin.

*Pittosporum undulatum*, Vent.    Pittosporeæ.    Flower: volatile oil.—Bark: Pittosporin; Viscin.

*Plantago decumbens*, Forsk.    Plantaginæ.    Seed: mucilage.

*Platanus orientalis*, L.    Plataneæ.    Bark: Phlobaphen.

*Plumbago Europæa*, L.    Plumbagineæ.    Root: Plumbagin.

*Podophyllum peltatum*, L.    Berberideæ.    Root: Berberin; Saponin.

*Polygala Senega*, L.    Polygaleæ.    Root: Saponin.

*Polygonum aviculare*, L.; *P. barbatum*, L.; *P. Chinense*, L.; *P. tinctorium*, Lour.    Polygoneæ.    Whole plant: Indigo.

*Polygonum Hydropiper*, L.    Whole plant: Polygonic acid; volatile acrid principle.

*Polylophium Galbanum*, F. v. M.    Umbelliferæ.    Exudation of stem and root: Galbanum with volatile oil.

*Polypodium vulgare*, L.    Filices.    Root: Glycyrrhizin and Saponin or allied substances.

*Polyporus officinalis*, Fries.    Fungi.    Laricin; Agaric acid.

*Populus*, several species.    Amentaceæ.    Whole plant: Populin; Salicin.—Leaf buds: volatile oil.

*Potentilla Tormentilla*, Sibth.    Rosaceæ.    Root: Ellagic acid.

*Primula auricula*, L.    Primulaceæ.    Root: Auricula stearopten.

„        *veris*, L.    Root: Primulin=Cyclamin.

*Prostanthera lasianthos*, Labill.; *P. rotundifolia*, R. Br.    Labiatæ.    Leaf: volatile oil.

*Prunus Amygdalus*, J. Hk.    Rosaceæ.    Seed: Amygdalin; Emulsin; Synaptase; fat-oil.

*Prunus Cerasus*, L.    Seed: Amygdalin.—Exudation of stem: gum.

„        *domestica*, L.    Seed: Amygdalin.—Exudation of stem: gum.

„        *Laurocerasus*, L.    Leaf and seed: Amygdalin.

„        *Mahaleb*, L.    Bark: Cumarin.

„        *Padus*, L.    Leaf, flower, bark, seed: Amygdalin.

„        *Persica*, J. Hook.    Leaf and seed: Amygdalin.    Exudation of stem: gum.

*Prunus spinosa*, L.    Flower, seed: Amygdalin.

„        *Virginiana*, L.    Bark: Amygdalin.

*Psychotria emetica*, Mutis.    Rubiaceæ.    Root: Emetin.

*Pterocarpus Draco*, L.    Leguminosæ.    Exudation of stem: Dragon's blood.

*Pterocarpus erinaceus*, Poiret.    Exudation of stem: West African Kino.

*Pterocarpus Marsupium*, Rxb.    Exudation of stem (Malabar and Amboina-Kino): Catechuic and Kino-tannic acids.

*Pterocarpus santalinus*, L. fil.    Wood: Santalin; Santal.

*Punica Granatum*, L. Lythraceæ. Fruit and root: Tannic acid; Mannit.

*Pyrola umbellata*, L. Ericaceæ. Whole plant: Chimaphilin.

*Pyrus Aria*, Ehrh.; *P. torminalis*, Ehrh. Rosaceæ. Flower: Amygdalin.

*Pyrus aucuparia*, Gaertn. Flower: Amygdalin; Trimethylamin. —Fruit: Sorbin.

*Pyrus communis*, L. Flower: Trimethylamin.

„ *Cydonia*, L. Seeds: mucilage.

„ *Malus*, L. Root-bark: Phlorrhizin.

*Quassia amara*, L. Simarubeæ. Wood and bark: Quassiin.

*Quercus coccinea*, Wang. Amentaceæ. Bark: Quercitrin.

„ *Ilex*, L. Leaf: Quinic acid or an allied acid.

„ *infectoria*, Oliv. Galls of branches: Ellagic and Gallo-tannic acids.

*Quercus Robur*, L. Fruit: Quercit; volatile oil.—Bark: Quercin; Tannic acid.

*Quercus Suber*, L. Bark: Cork substance.

*Quillaja Saponaria*, Mol. Rosaceæ. Root: Saponin.

*Ramalina*, several species. Lichenes. Lichenin; Usnic acid.

*Ranunculus bulbosus*, L.; *R. Flammula*, L.; *R. sceleratus*, L. Ranunculaceæ. Whole plant: Anemonic acid; Anemonin.

*Ranunculus*, many species. Whole plant: volatile acrid matter.

*Reseda Luteola*, L. Resedaceæ. Whole plant: Luteolin.—Seed: drying-oil.

*Reseda odorata*, L. Flower: volatile oil.

*Rhamnus catharticus*, L. Rhamnaceæ. Fruit: Rhamno-cathartin; Rhamnin; Rhamno-xanthin; Rhamno-tannic acid.—Bark of stem and root: Rhamno-xanthin.

*Rhamnus Frangula*, L. Fruit, bark of stem and root: Rhamno-xanthin.

*Rhamnus prunifolius*, Sbth.; *R. infectorius*, L.; *R. saxatilis*, L.; *R. amygdalinus*, Desf.; *R. oleoides*, L.; *R. Græcus*, Rent. Fruit: Chryso-rhamnin; Quercetin; Xanthorhamnin.

*Rheum australe*, D. Don. Polygoneæ. Root: Emodin.

*Rheum officinale*, Baill., and several other species. Root: Chrysophanic acid; Aporetin; Erythroretin; Phæoretin.

*Rhinanthus major*, Ehrh. Scrophularineæ. Seed: Rhinanthin.

*Rhododendron ferrugineum*, L. Ericaceæ. Leaf: Ericinol; Eri-colin; Rhodotannic acid.

*Rhus coriaria*, L. Anacardiaceæ. Bark: Gallotannic acid.

„ *semialata*, Marr. Galls of branches: Tannic acid.

„ *succedanea*, L. Fruit, leaf and branches: Japanese wax with Palmitin.

*Rhus toxicodendron*, L. Leaf: Rhoitannic acid.

*Ribes rubrum*, L. Saxifrageæ. Root-bark: Phlorrhizin.

*Richardsonia scabra*, St. Hil. Rubiaceæ. Root: Emetin.

*Ricinus communis*, L.   Euphorbiaceæ.   Seed : Ricinin ; fat-oil with Isocetic and Ricinoleic acids.

*Robinia Pseudacacia*, L.   Leguminosæ.   Flower : Robinin.—Unripe fruit : Inosit.

*Roccella fuciformis*, Ach.   Lichenes.   Roccellic and Erythric acids.
„    *tinctoria*, Cand.   Roccellinin.

*Roccella*, several species.   Lecanoric acid ; Orcin.

*Rosa centifolia*, L.; *R. Damascena*, Mill. ; *R. moschata*, Mill.; *R. Indica*, L.; *R. sempervirens*, L.   Rosaceæ.   Flower : volatile oil.

*Rosmarinus officinalis*, L.   Labiatæ.   Leaf and flower : volatile oil.

*Rubia tinctorum*, L.   Rubiaceæ.   Herb : Rubichloric and Rubitannic acids.—Root : Chlorogenin ; Xanthin ; Alizarin ; Purpurin ; Ruberythric acid ; Rubiacin ; Rubian ; Rubiretin ; Verantin ; Erythrozym ; Rubichloric acid.

*Rubus Idæus*, L.   Rosaceæ.   Fruit : Raspberry-stearopten.

*Rumex*, many species.   Polygoneæ.   Root : Chrysophanic acid.

*Ruta graveolens*, L.   Rutaceæ.   Whole plant : volatile oil ; Rutin.

*Saccharum officinarum*, L.   Gramineæ.   Stalk : Cane-sugar ; Cerosin.

*Salix*, many species.   Amentaceæ.   Whole plant : Salicin.

*Salvia officinalis*, L.   Labiatæ.   Whole plant : volatile oil.

*Samadera Indica*, Gærtn.   Simarubeæ.   Bark : Samaderin.

*Sambucus nigra*, L.   Caprifoliaceæ.   Flower : volatile oil.

*Sanguinaria Canadensis*, L.   Papaveraceæ.   Whole plant : Chelerythrin.—Root : Porphyroxin ; Puccin.

*Sapindus Saponaria*, L.   Sapindaceæ.   Fruit : Saponin ; Formic and Butyric acids.

*Saponaria officinalis*, L.   Caryophylleæ.   Herb and root : Saponin.

*Sassafras officinale*, Hayne.   Lauraceæ.   Root : volatile oil.

*Scabiosa succisa*, L.   Dipsaceæ.   Root : green acid.

*Schænocaulon officinale*, A. Gray.   Liliaceæ.   Seed : Sabadillin ; Veratrin ; Sabadillic and Veratric acids.

*Scrophularia aquatica*, L.; *S. nodosa*, L.   Scrophularineæ.   Whole plant : Scrophularin.

*Sedum acre*, L.   Saxifrageæ.   Whole plant : Rutin.

*Semecarpus Anacardium*, L. fil. and other species.   Anacardiaceæ.   Pericarp : Cardol.

*Sempervivum tectorum*, L.   Saxifrageæ.   Leaf : Formic acid.

*Sesamum Indicum*, L.   Pedalineæ.   Seed : fat-oil.

*Sideroxylon attenuatum*, A. DC.   Sapotaceæ.   Yields part of the Indian and also Philippine gutta-percha.

*Sideroxylon Muelleri*, J. Hook.   Exudation of stem : Batata.

*Simaba Cedron*, Planch.   Simarubeæ.   Fruit : Cedrin.

*Simaruba amara*, Aubl.   Simarubeæ.   Wood, bark : Quassiin.

*Sindora*, several species.   Leguminosæ.   Balsam of Sindor (also obtained from species of Dipterocarpus).

*Sisymbrium Alliaria*, Scop. Cruciferæ. Herb: volatile oil.—
Root: volatile oil.
*Smilax China*, L.; *S. cordato-ovata*, Rich.; *S. medica*, Cham. et.
Schl.; *S. officinalis*, Humb.; *S. papyracea*, Duh.; *S. Parhampui*,
Ruiz. Liliaceæ. Root: Smilacin.
*Smilax glycyphylla*, Smith. Leaves: Glycyphyllin.
*Solanum Dulcamara*, L. Solanaceæ. Stalk: Dulcamarin;
Solanin.
*Solanum nigrum*, L. Fruit: Solanin.
  „   *paniculatum*, L. Fruit: Jurubebin.
  „   *tuberosum*, L. Fruit: Solanin.—Germs: Solanin;
Inosit.—Tuber: Starch.
*Solanum verbascifolium*, L. Fruit: Solanin.
*Sophora Japonica*, L. Leguminosæ. Flower buds: Rutin.
*Spiræa Aruncus*, L.; *S. Japonica*, L. fil.; *S. sorbifolia*, L. Rosaceæ.
Leaf: Amygdalin.
*Spiræa Ulmaria*, L. Flower: volatile oil; Spiræa-yellow; Sali-
cylous and Salicylic acids.
*Squamaria elegans*, Hoffm. Lichenes. Chrysophanic acid.
*Sticta*, many species. Lichenes. Lichenin.
*Strophanthus hispidus*, DC. Apocyneæ. Seed: Strophanthin.
*Strychnos colubrina*, L. Loganiaceæ. Wood: Brucin; Strychnin.
  „   *Guianensis*, Mart. Curare: Curarin.
  „   *Ignatia*, Berg. Seed: Brucin; Strychnin.
  „   *Nux vomica*, L. Bark, seed: Brucin; Strychnin; Malic
acid.
*Strychnos Tieute*, Lesch. Root: Brucin; Strychnin.
*Styrax Benzoin*, Dryand. Styraceæ. Exudation of stem: Ben-
zoin with Benzoic acid.
*Styrax officinalis*, L. Exudation of stem: Storax.
*Syringa vulgaris*, L. Oleaceæ. Flower: volatile oil.—Leaf and
fruit: Syringin.—Bark: Syringin; Syringo-picrin.
*Tagetes glandulifera*, Schrank. Compositæ. Herb and flower:
volatile oil.
*Tamarindus Indica*, L. Leguminosæ. Fruit: Tartaric, Citric,
Acetic acids; sugar; Pectin.
*Tanacetum vulgare*, L. Compositæ. Herb and flower: volatile
oil.
*Tanghinia venenifera*, Poir. Apocyneæ. Seed: Tanghinin.
*Taraxacum officinale*, Web. Compositæ. Leaf: Inosit.
*Taxus baccata*, L. Coniferæ. Leaf: Taxin.
*Tetranthera calophylla*, Miq. Lauraceæ. Fruit: Fat with Lauro-
stearin.
*Teucrium Marum*, L. Labiatæ. Whole plant: Marum-stearopten.
*Theobroma Cacao*, L. Sterculiaceæ. Seed: Theobromin; Cacao-
fat.
*Thevetia Ycotli*, A. de Cand. Apocyneæ. Seeds: Thevetosin.

T

*Thlaspi arvense*, L. Cruciferæ. Herb and seed: volatile oil.
*Thuya occidentalis*, L. Coniferæ. Green parts: volatile oil.
*Thymus Serpillum*, L. Labiatæ. Whole plant: volatile oil.
„ *vulgaris*, L. Whole plant: volatile oil with Cymen, Thymen, Thymol.
*Tilia Europæa*, L. Tiliaceæ. Flower: volatile oil.
*Trachylobium Gærtnerianum*, Hayne; *T. Hornemanni*, Hayne. Leguminosæ. Exudation of stem: Copal.
*Triticum repens*, L. Gramineæ. Root: Triticin.
„ *vulgare*, Vill. Fruit: Starch; Gluten with Glutin, Mucin, &c.
*Tropæolum majus*, L. Geraniaceæ. Whole plant: Tropæolic acid.—Seed: volatile oil.
*Ulmus campestris*, L. Urticaceæ. Leaf: Quinic acid or one allied.
„ *fulva*, Mich. Bark: mucilage.
*Uncaria Gambir*, Roxb.; *U. acida*, Roxb. Rubiaceæ. Leaf: Catechu-tannic and Catechuic acids; Quercetin.
*Urceola elastica*, Roxb. Apocyneæ. Borneo and Sumatra caoutchouc.
*Urginia Scilla*, Steinh. Liliaceæ. Tuber: Scillitin; mucilage.
*Urtica urens*, L. Urticaceæ. Leaf: Formic acid.
*Usnea florida*, Hoffm. Lichenes. Usnic acid.
*Usnea*, several species. Lichenin.
*Vaccinium Myrtillus*, L. Ericaceæ. Whole plant: Quinic acid.
*Vahea gummifera*, Lam.; *V. Madagascariensis*, Bojer. Apocyneæ. Exudation of stem: Madagascar caoutchouc.
*Vahea Comorensis*, Bojer. Anjonan caoutchouc.
„ *florida (Landolphia florida*, Benth.) Niger caoutchouc, with other Vaheas of the section Landolphia.
„ *Heudelotii (Landolphia Heudelotii*, A. DC). Cazamanca caoutchouc.
„ *Owariensis (Landolphia Owariensis*, Beauv). West-African caoutchouc; Dambonit.
„ *Senegalensis*, A. DC. Senegambia caoutchouc.
*Valeriana officinalis*, L. Valerianeæ. Herb: Valerianic acid.— Root: volatile oil with Borneen and Borneol; Valerianic and Valeriana-tannic acids.
*Vanilla aromatica*, Sw.; *V. Guianensis*, Splitg.; *V. planifolia*, Andr.; *V. sativa*, Lindl. Orchideæ. Fruit: Vanillic acid.
*Variolaria*, many species. Lichenes. Lecanoric acid; Orcin.
*Vateria Indica*, L. Dipterocarpeæ. Fruit: Vateria tallow.— Copal.
*Veratrum album*, L.; *V. viride*, Aiton. Liliaceæ. Root: Jervin; Sabadillic acid.
*Vincetoxicum officinale*, Mœnch. Asclepiadeæ. Root: Asclepiadin.
*Viola odorata*, L. Violaceæ. Whole plant: Violin.
*Viscum album*, L. Loranthaceæ. Fruit, foliage and branches: Viscin.

*Vitex Agnus castus*, L. Verbenaceæ. Seed: Castin.

*Vitis quinquefolia*, Lam. Viniferæ. Red autumnal leaf: Cisso-tannic acid and Pyrocatechin.

*Vitis vinifera*, L. Fruit: Oenolin; Tartaric and Racemic acids.— Seed: fat-oil.

*Vouapa phaseolocarpa*, Hayne. Leguminosæ. Exudation of stem: Copal.

*Willoughbya edulis*, Roxb. Apocyneæ. Chittagong caoutchouc.

    „    *Martabanica*, Wall. And other species. Martaban caoutchouc.

*Wrightia antidysenterica*, R. Br. Apocyneæ. Bark: Conessin.

*Xanthorrhiza apiifolia*, L'Herit. Ranunculaceæ. Root: Ber-berin.

*Xanthorrhœa australis*, Br.; *X. arborea*, Br.; *X. Preissii*, Endl.; *X. quadrangulata*, F. v. M.; *X. semiplana*, F. v. M. Liliaceæ. Exudation of stem: Botany Bay resin with Benzoic acid; vola-tile oil; Bassorin.

*Xanthoxylon Caribæum*, Lam.; *X. fraxineum*, Willd. Rutaceæ. Bark: Berberin.

*Xanthoxylon piperitum*, Cand. Fruit: volatile oil with Xanthoxy-len and Xanthoxylin.

*Xylopia polycarpa*, J. Hook. et Th. Anonaceæ. Bark: Berberin.

*Zieria Smithii*, Andr. Rubiaceæ. Leaf: volatile oil.

*Zingiber officinale*, L. Scitamineæ. Root: volatile oil.

# DIVISION III.

---

## I. DICOTYLEDONEÆ.

### 1. DIALYPETALÆ.

#### *Ranunculaceæ.*

Aconitum ferox, Wall.
    „    Lycoctonum, L.
    „    Napellus, L.
    „    other species.
Anemone nemorosa, L.
    „    pratensis, L.
    „    Pulsatilla, L.
Clematis, several species.
Coptis Teeta, Wall.
Delphinium Consolida, L.
    „    Staphisagria, L.
Helleborus niger, L.
    „    viridis, L.
Hydrastis Canadensis, L.
Nigella sativa, L.
Ranunculus bulbosus, L.
    „    Flammula, L.
    „    sceleratus, L.
    „    many other species.
Xanthorrhiza apiifolia, L'Herit.

#### *Magnoliaceæ.*

Illicium anisatum, L.
Liriodendron tulipifera, L.
Magnolia umbrella, Lamarck.

#### *Anonaceæ.*

Xylopia polycarpa, J. Hook. et Th.

#### *Myristiceæ.*

Myristica Bicahyba, Schott.
    „    fragrans, Houtt.
    „    Otaba, H. et B.
    „    sebifera, Sw.

#### *Laurineæ.*

Cinnamomum Burmanni, Blume.
    „    Camphora, F. Nees.
    „    Cassia, Blume.
    „    Culilaban, Blume.
    „    Zeylanicum, C. G.
            Nees.

Dicypellium caryophyllatum, C.
G. Nees.
Laurus nobilis, L.
Nectandra Rodiei, Schomb.
Ocotea Pichurim, Kunth.
Sassafras officinalis, Hayne.
Tetranthera calophylla, Miq.

*Monimieæ.*

Atherosperma moschatum, Lab.

*Menispermaceæ.*

Anamirta paniculata, Colebr.
Chondodendron tomentosum, R.
et P.
Cissampelos Pariera, L.
Coscinium fenestratum, Colebr.
Jateorrhiza palmata, Miers.

*Berberideæ.*

Berberis vulgaris, L.
Jeffersonia diphylla, Bart.
Leontice thalictroides, L.
Podophyllum peltatum, L.

*Papaveraceæ.*

Argemone Mexicana, L.
Chelidonium majus, L.
Eschscholtzia Californica, Cham.
Glaucium luteum, Scop.
Papaver Rhoeas, L.
„ somniferum, L.
Sanguinaria Canadensis, L.

*Fumariaceæ.*

Corydalis fabacea, Pers.
„ tuberosa, Cand.
Fumaria officinalis, L.

*Cruciferæ.*

Arabis perfoliata, Lam.
Brassica alba, Vis.
„ nigra, Koch.
„ oleracea, L.
Camelina sativa, Crantz.

Cochlearia Armoracia, L.
„ Anglica, L.
„ Danica, L.
„ officinalis, L.
Hesperis matronalis, L.
Isatis indigofera, Fort.
„ tinctoria, L.
Lepidium, several species.
Nasturtium officinale, L.
Sisymbrium Alliaria, Scop.
Thlaspi arvense, L

*Capparideæ.*

Capparis spinosa, L.

*Droseraceæ.*

Drosera, many species.

*Violaceæ.*

Hybanthus (Ionidium) Ipecacuanha, F. v. M.
Viola odorata, L.

*Moringaceæ.*

Moringa oleifera, Lam.
„ aptera, Gaertn.

*Bixaceæ.*

Bixa Orellana, L.
Laetia apetala, Jacq.

*Cistaceæ.*

Cistus Creticus, L.
„ ladaniferus, L.
„ Ledon, Lam.
„ laurifolius, L.
„ Monspeliensis, L.

*Resedaceæ.*

Reseda luteola, L.
„ odorata, L.

*Datiscaceæ.*

Datisca cannabina, L.

*Canellaceæ.*

Canella alba, Murr.

*Pittosporeæ.*

Pittosporum undulatum, Vent.

*Polygalaceæ.*

Krameria Ixina, L.
„ secundiflora, DC.
„ triandra, R. et P.
Monnina polystachya, R. et P.
Polygala Senega, L.

*Guttiferæ.*

Calophyllum Calaba, Jacq.
„ longifolium, Willd.
„ Inophyllum, L.
Garcinia Indica, Chois.
„ Mangostana, L.
„ pictoria, Roxb.
„ Morella, Desr.
„ Cochinchinensis, Chois.

*Ternstrœmiaceæ.*

Camellia Thea, Link.

*Dipterocarpeæ.*

Dipterocarpus alatus, Roxb.
„ gracilis, Bl.
„ hispidus, Thw.
„ incanus, Roxb.
„ littoralis, Bl.
„ retusus, Bl.
„ trinervis, Bl.
„ turbinatus Gærtn.
„ Zeylanicus, Thw.
Dryobalanops Camphora, Colebr.
Vateria Indica, L.

*Lineæ.*

Erythroxylon Coca, Lam.
Linum usitatissimum, L.

*Geraniaceæ.*

Geranium, many species.
Pelargonium capitatum, Ait.
„ odoratissimum, Ait.
„ Radula, Ait.
Tropæolum majus, L.

*Malvaceæ.*

Adansonia digitata, L.
„ Gregorii, F. v. M.
Althaea officinalis, L.

*Sterculiaceæ.*

Lunanea Bichi, DC.
Theobroma Cacao, L.

*Tiliaceæ.*

Tilia Europæa, L.

*Rutaceæ.*

Barosma betulina, Bartl.
„ crenulata, Hk.
„ serratifolia, Willd.
Citrus Aurantium, L.
„ „ var. Bergamia, Riss.
„ „ „ dulcis, Volk.
„ medica, L.
„ „ var. Limetta, Riss.
Galipea Cusparia, St. Hil.
Peganum Harmala, L.
Ruta graveolens, L.
Xanthoxylon Caribaeum, Lam.
„ fraxineum, Willd.
„ piperitum, Cand.

*Zygophylleæ.*

Guaiacum officinale, L.
„ sanctum, L.

*Simarubeæ.*

Ailantus excelsa, Roxb.
Irvingia Barteri, J. Hook.
Picraena excelsa, Lindl.
Picramnia ciliata, Mart.

Quassia amara, L.
Samadera Indica, Gærtn.
Simaba Cedron, Planch.
Simaruba amara, Aubl.

### Burseraceæ.

Amyris Plumieri, DC.
Balsamodendron Africanum,
    Arn. et B.
  ,,   Myrrha, Ehrenb.
  ,,   Opobalsamum, Kunth.
Boswellia Carterii, Brdw.
Bursera gummifera, Jacq.
  ,,   Icicariba, Benth.
Canarium commune, L.
Hedwigia balsamifera, Sw.
Icica heptaphylla, Aubl.

### Anacardiaceæ.

Anacardium occidentale, Rottb.
Pistacia Lentiscus, L.
  ,,   Terebinthus, L.
Rhus coriaria, L.
  ,,   semialata, Murr.
  ,,   succedanea, L.
  ,,   Toxicodendron, L.
Semecarpus Anacardium, L. fil.

### Meliaceæ.

Carapa Guianensis, Aubl.
Khaya Senegalensis, A. Juss.

### Sapindaceæ.

Aesculus Hippocastanum, L.
  ,,   Pavia, L.
Paullinia sorbilis, Mart.
Sapindus Saponaria, L.

### Viniferæ.

Vitis quinquefolia, Lam.
  ,,   vinifera, L.

### Celastrineæ.

Euonymus Europæus, L.

### Rhamnaceæ.

Rhamnus catharticus, L.
  ,,     Frangula, L.
  ,,     prunifolius, Sibth.
  ,,     infectorius, L.
  ,,     saxatilis, L.
  ,,     amygdalinus, Desf.
  ,,     oleoides, L.
  ,,     Graecus, Reut.

### Aquifoliaceæ.

Ilex aquifolium, L.
 ,, Cassine, L.
 ,, Paraguensis, St. Hil.

### Plumbagineæ.

Plumbago Europæa, L.

### Caryophylleæ.

Gypsophila Struthium, L.
Lychnis Githago, Lam.
Saponaria officinalis, L.

### Salsolaceæ.

Beta vulgaris, L.
Chenopodium album, L.
    ,,     ambrosioides, L.
    ,,     olidum, Aubl.

### Paronychieæ.

Herniaria glabra, L.

### Polygoneæ.

Coccoloba uvifera, Jacq.
Fagopyrum cymosum, Meissn.
    ,,   emarginatum, Meissn.
    ,,   esculentum, Mœnch.
    ,,   Tartaricum, Gærtn.
Polygonum aviculare, L.
    ,,   barbatum, L.
    ,,   Chinense, L.
    ,,   tinctorium, Lour.
    ,,   Hydropiper, L.
Rheum australe, D. Don.
  ,, officinale, Baill.
Rumex, many species.

*Phytolacceae.*

Coriaria myrtifolia, L.

*Aristolochieæ.*

Aristolochia Serpentaria, L.
Asarum Europæum, L.

*Saxifrageæ.*

Liquidambar orientalis, Mill.
Philadelphus coronarius, L.
Ribes rubrum, L.
Sedum acre, L.
Sempervivum tectorum, L.

*Rosaceæ.*

Amelanchier vulgaris, Moench.
Cotoneaster vulgaris, Lindl.
Cratægus coccinea, L.
   „   Oxyacantha, L.
Fragaria vesca, L.
Geum urbanum, L.
Hagenia Abyssinica, Lam.
Potentilla Tormentilla, Sibth.
Prunus Amygdalus, J. Hook.
   „   Cerasus, L.
   „   domestica, L.
   „   Laurocerasus, L.
   „   Mahaleb, L.
   „   Padus, L.
   „   Persica, J. Hook.
   „   spinosa, L.
   „   Virginiana, L.
Pyrus Aria, Ehrh.
   „   torminalis, Ehrh.
   „   aucuparia, Gærtn.
   „   communis, L.
   „   Cydonia, L.
   „   Malus, L.
Quillaja Saponaria, Mol.
Rosa centifolia, L.
   „   Damascena, Mill.
   „   Indica, L.
   „   moschata, Mill.
   „   sempervirens, L.
Rubus Idæus, L.

Spiræa Aruncus, L.
   „   Japonica, L. fil.
   „   sorbifolia, L.
   „   Ulmaria, L.

*Leguminosæ.*

Abrus precatorius, L.
Acacia Arabica, W.
   „   Seyal, Del.
   „   Verek, G. et P.
   „   stenocarpa, Hochst.
   „   horrida, W.
   „   Catechu, W.
   „   Suma, Kurz.
   „   decurrens, W.
   „   lophantha, W.
Andira anthelminthica, Benth.
Arachis hypogæa, L.
Astragalus, several species.
Bowdichia virgilioides, H. B. K.
Butea frondosa, Roxb.
Cæsalpinia Crista, L.
   „   Sappan, L.
Cassia acutifolia, Del.
   „   angustifolia, Vahl.
Ceratonia Siliqua, L.'
Copaifera, several species.
Cyclopia latifolia, Caud.
Cynometra Spruceana, Benth.
Cytisus Laburnum, L.
   „   scoparius, Link.
Dipteryx odorata, Schreb.
Erythrophlæum guinense, G. Don
Geoffroya inermis, Wright.
   „   Surinamensis, Bondt.
Glycyrrhiza echinata, L.
   „   glabra, L.
Hæmatoxylon Campechianum, L.
Hymenæa Candolleana, H. et B.
   „   confertiflora, Mart.
   „   confertifolia, Hayne.
   „   Courbaril, L.
   „   latifolia, Hayne.
   „   Martiana, Hayne.
   „   Olfersiana, Hayne.
   „   rotundata, Hayne.
   „   stilbocarpa, Hayne.

Hymenæa Sellowiana, Hayne.
    „       venosa, Vahl.
Indigofera Anil, L.
    „       argentea, L.
    „       tinctoria, L.
Lupinus albus, L.
    „    luteus, L.
Melilotus officinalis, Desr.
Myroxylon Pereiræ, Kl.
         „    toluiferum, L. fil.
Ononis spinosa, L.
Peltophorum Linnæi, Benth.
Phaseolus vulgaris, L.
Physostigma venenosum, Balf.
Pisum sativum, L.
Pterocarpus Draco, L.
        „     erinaceus, Poiret.
        „     Marsupium, Roxb.
        „     santalinus, L. fil.
Robinia Pseudacacia, L.
Sindora Sumatrama, Miq.
Sophora Japonica, L.
Tamarindus Indica, L.
Trachylobium Gærtnerianum,
    Hayne.
Trachylobium Hornemanni,
    Hayne.
Vouapa phaseolocarpa, Hayne.

*Myrtaceæ.*

Eucalyptus resinifera, Smith.
        „    rostrata, Schl.
        „  .  viminalis, Labill.
        „    many other species.
Eugenia caryophyllata, Thunb.
Melaleuca Leucadendron, L.
        „    many species.
Myrtus communis, L.
Pimenta officinalis, Lindl.

*Lythraceæ.*

Punica Granatum, L.

*Piperaceæ.*

Piper angustifolium, R. et P.
    „    Cubeba, L. fil.

Piper longum, L.
    „    officinarum, C. DC.
    „    methysticum, Forst.
    „    nigrum, L.

*Euphorbiaceæ.*

Alchornea latifolia, Sw.
Aleurites triloba, Forst.
Buxus sempervirens, L.
Croton Eleutheria, Benn.
    „    Sloanei, Benn.
    „    erythrinum, Mart.
    „    niveum, Jacq.
    „    Tiglium, L.
Euphorbia resinifera, Berg.
Excæcaria sebifera, J. M.
Hevea Guianensis, Aubl.
Hura crepitans, L.
Iatropha Curcas, L.
Mallotus Philippinensis, J. M.
Mercurialis annua, L.
Ricinus communis, L.

*Urticaceæ.*

Antiaris toxicaria, Lesch.
Cannabis sativa, L.
Castilloa elastica, Cerv.
    „    Markhamiana, Collins.
Ficus elastica, Roxb.
    „    rubiginosa, Desf.
    „    subracemosa, Blum.
    „    variegata, Blum.
Humulus Lupulus, L.
Maclura tinctoria, D. Don.
Ulmus campestris, L.
    „    fulva, Mich.
Urtica urens, L.

*Juglandeæ.*

Juglans regia, L.

*Amentaceæ.*

Betula alba, L.
Corylus Avellana, L.
Fagus silvatica, L.
Myrica cerifera, L.

Myrica cordifolia, L.
„ quercifolia, L.
„ serrata, Lam.
„ Gale, L.
Platanus orientalis, L.
Populus, several species.
Quercus coccinea, Wang.
„ Ilex, L.
„ infectoria, Oliv.
„ Robur, L.
„ Suber, L.
Salix, many species.

*Cucurbitaceæ.*

Bryonia alba, L.
Cucumis Colocynthis, L.
„ Melo, L.
„ Prophetarum, L.
Cucurbita Pepo, L.
Ecballion Elaterium, A. Rich.

*Loranthaceæ.*

Viscum album, L.

*Proteaceæ.*

Banksia, several species.

*Thymeleæ.*

Daphne alpina, L.
„ Gnidium, L.
„ Mezereum, L.

*Elæagneæ.*

Hippophaë rhamnoides, L.

*Cornaceæ.*

Cornus florida, L.

*Araliaceæ.*

Aralia quinquefolia, Dec. et
Planch.
Hedera Helix, L.

*Umbelliferæ.*

Æthusa Cynapium, L.
Ammi Copticum, L.
Apium graveolens, L.
Archangelica officinalis, Hoffm.
Carum Ajowan, Benth.
„ Carui, L.
„ Petroselinum, Benth.
Cicuta virosa, L.
Conium maculatum, L.
Coriandrum sativum, L.
Cuminum Cyminum, L.
Daucus Carota, L.
Dorema Ammoniacum, D. Don.
Euryangium Sumbul, Kauffm.
Ferula Asa-fœtida, L.
„ alliacea, Boiss.
„ Narthex, Boiss.
„ erubescens, Boiss.
„ Persica, W.
Fœniculum officinale, All.
Heracleum villosum, Fischer.
„ Sphondylium, L.
Laserpitium latifolium, L.
Œnanthe Phellandrium, Lam.
Opopanax Chironium, Koch.
Pastinaca sativa, L.
Peucedanum galbaniferum,
Benth.
„ officinale, L.
„ Oreoselinum, Cuss.
„ Ostruthium, Koch.
Pimpinella Anisum, L.
„ nigra, L.
„ Saxifraga, L.
Polylophium Galbanum, F. v. M.

2. GAMOPETALÆ.

*Rubiaceæ.*

Asperula odorata, L.
Cascarilla hexandra, Wedd.
„ magnifolia, Wedd.
Cephaëlis Ipecacuanha, Rich.
Chiococca racemosa, Jacq.
Cinchona Calisaya, Wedd.
„ cordifolia, R. et P.

Cinchona glandulifera, R. et P.
   ,,    lancifolia, Mutis.
   ,,    micrantha, R. et P.
   ,,    nitida, R. et P.
   ,,    officinalis, Hook.
   ,,    Paludiana, How.
   ,,    Peruviana, How.
   ,,    pubescens, Vahl.
   ,,    scrobiculata, H. et B.
   ,,    succirubra, Pav.
Coffea Arabica, L.
Galium Aparine, L.
   ,,    Mollugo, L.
   ,,    verum, L.
Gardenia grandiflora, Lour.
   ,,    lucida, Roxb.
Hillia spectabilis, Fr. Vaud.
Ladenbergia, several species.
Morinda citrifolia, L.
Pinckneya pubens, Rich.
Psychotria emetica, Mutis.
Richardsonia scabra, St. Hil.
Rubia tinctorum, L.
Uncaria acida, Roxb.
   ,,    Gambir, Roxb.
Zieria Smithii, Andr.

*Caprifoliaceæ.*

Lonicera xylosteum, L.
Sambucus nigra, L.

*Valerianeæ.*

Valeriana officinalis, L.

*Dipsaceæ.*

Scabiosa succisa, L.

*Compositæ.*

Achillea Millefolium, L.
   ,,    moschata, Jacq.
   ,,    nobilis, L.
Anthemis nobilis, L.
Arnica montana, L.
Artemisia Absinthium, L.
   ,,    Cina, Berg.
   ,,    Sieberi, Besser.

Artemisia Dracunculus, L.
   ,,    vulgaris, L.
Atractylis gummifera, L.
Blumea balsamifera, DC.
Buphthalmum salicifolium, L.
Calendula officinalis, L.
Carbenia benedicta, Benth.
Carthamus tinctorius, L.
Centaurea Calcitrapa, L.
Chrysanthemum Parthenium,
      Persoon.
Crepis fœtida, L.
Dahlia purpurea, Poir.
Eupatorium cannabinum, L.
Helianthus annuus, L.
Inula Helenium, L.
Lactuca sativa, L.
   ,,    virosa, L.
Liatris odoratissima, Willd.
Madia sativa, Mol.
Matricaria Chamomilla, L.
Mikania Guaco, H. et B.
Osmitopsis asteriscoides, Cass.
Othonna furcata, Benth.
Perezia Humboldti, A. Gr.
Tagetes glandulifera, Schrank.
Tanacetum vulgare, L.
Taraxacum officinale, Web.

*Campanulaceæ.*

Lobelia inflata, L.

*Ericaceæ.*

Arctostaphylos Uva Ursi, Spr.
Calluna vulgaris, Sal.

Erica carnea, L.
Gaultiera procumbens, L.
Ledum palustre, L.
Monotropa Hypopitys, L.
Pyrola umbellata, L.
Rhododendron ferrugineum, L.
Vaccinium Myrtillus, L.

*Epacrideæ.*

Epacris, several species.

*Styraceæ.*

Styrax Benzoin, Dryand.
„   officinale, L.

*Sapotaceæ.*

Achras Sapota, L.
Bassia butyracea, Roxb.
„   latifolia, Roxb.
„   longifolia, L.
„   sericea, Blume.
Ceratophorus Leerii, Hassk.
Imbricaria coriacea, A. DC.
Isonandra Gutta, Hook.
Lucuma glycyphlæa, Mart. et Eichl.
„   Parkii, R. Br.
Mimusops Elengi, L.
„   Manilkara, G. Don.
Payena macrophylla, Benth.
Sideroxylon attenuatum, A. DC.
„   Muelleri, J. Hook.

*Oleaceæ.*

Fraxinus excelsior, L.
„   Ornus, L.
Ligustrum vulgare, L.
Olea Europæa, L.
Phillyrea latifolia, L.
„   angustifolia, L.
Syringa vulgaris, L.

*Asclepiadeæ.*

Asclepias Syriaca, L.
Calotropis gigantea, R. Br.
„   procera, R. Br.
Vincetoxicum officinale, Moench.

*Apocyneæ.*

Alstonia constricta, F. v. M.
„   scholaris, R. Br.
Alyxia stellata, R. et S.
Hancornia speciosa, Gomes.
Nerium Oleander, L.
Strophanthus hispidus, DC.
Tanghinia venenifera, Poir.

Thevetia Ycotli, A. de Cand.
Urceola elastica, Roxb.
Vahea Comorensis, Bojer.
„   gummifera, Lam.
„   Madagascariensis, Bojer.
„   Senegalensis, A. DC.
„   florida, F. v. M.
„   Heudelotii, F. v. M.
„   Owariensis, F. v. M.
Willoughbya edulis, Roxb.
„   Martabanica, Wall.
Wrightia antidysenterica, R. Br.

*Gentianeæ.*

Gentiana lutea, L.
Menyanthes trifoliata, L.
Ophelia Chirata, Griseb.

*Loganiaceæ.*

Gelsemium nitidum, Michaux.
Strychnos colubrina, L.
„   Guianensis, Mart.
„   Ignatia, Berg.
„   Nux vomica, L.
„   Tieute, Lesch.

*Plantagineæ.*

Plantago decumbens, Forsk.

*Primulaceæ.*

Anagallis arvensis, L.
Cyclamen Europæum, L.
Primula Auricula, L.
„   veris, L.

*Convolvulaceæ.*

Convolvulus floridus, L. fil.
„   Scammonia, L.
„   scoparius, L. fil.
Ipomæa Nil, Roth.
„   Orizabensis, Steud.
„   Purga, Wender.
„   Turpethum, R. Br.

*Solanaceæ.*

Atropa Belladonna, L.
Capsicum annuum, L.
Datura Stramonium, L.
Fabiana imbricata, R. et P.
Hyoscyamus niger, L.
Nicotiana Tabacum, L.
    ,,    many species.
Physalis Alkekengi, L.
Solanum Dulcamara, L.
Solanum nigrum, L.
    ,,    paniculatum, L.
    ,,    tuberosum, L.
    ,,    verbascifolium, L.

*Scrophularineæ.*

Digitalis lutea, L.
    ,,    purpurea, L.
Euphrasia officinalis, L.
Globularia Alypum, L.
Gratiola officinalis, L.
Limosella aquatica, L.
Melampyrum nemorosum, L.
Rhinanthus major, Ehrh.
Scrophularia aquatica, L.
    ,,    nodosa, L.

*Bignoniaceæ.*

Bignonia Chica, H. et B.

*Pedalineæ.*

Sesamum Indicum, L.

*Asperifoliæ.*

Alkanna tinctoria, Tausch.
Cordia Boissieri, A. DC.

*Labiatæ.*

Hyssopus officinalis, L.
Lavandula angustifolia, Ehrh.
    ,,    latifolia, Vill.
    ,,    Stœchas, L.

Lycopus Europæus, L.
Marrubium vulgare, L.
Melissa officinalis, L.
Mentha australis, R. Br.
    ,,    gracilis, R. Br.
    ,,    laxiflora, Benth.
    ,,    piperita, L.
    ,,    Pulegium, L.
    ,,    viridis, L.
Monarda didyma, L.
    ,,    punctata, L.
Ocimum Basilicum, L.
Origanum Majorana, L.
    ,,    vulgare, L.
Prostanthera lasianthos, Labill.
    ,,    rotundifolia, R. Br.
Rosmarinus officinalis, L.
Salvia officinalis, L.
Teucrium Marum, L.
Thymus Serpillum, L.
    ,,    vulgaris, L.

*Verbenaceæ.*

Vitex Agnus castus, L.

*Coniferæ.*

Callitris quadrivalvis, Vent.
Dammara australis, Lamb.
    ,,    orientalis, Lamb.
Ginkgo biloba, Salisb.
Juniperus communis, L.
    ,,    Sabina, L.
    ,,    Virginiana, L.
Pinus Abies, Du Roi.
    ,,    balsamea, L.
    ,,    Fraseri, Pursch.
    ,,    Lambertiana, Dougl.
    ,,    Larix, L.
    ,,    Picea, Du Roi.
    ,,    silvestris, L.
    ,,    Sabiniana, Dougl.
Taxus baccata, L.
Thuya occidentalis, L.

## II. MONOCOTYLEDONEÆ.

### Orchideæ.

Aeranthus fragrans, Lindl.
Nigritella angustifolia, Rich.
Orchis hircina, Sw.
„ purpurea, Huds.
„ many species.
Vanilla aromatica, Sw.
„ Guianensis, Splitg.
„ planifolia, Andr.
„ sativa, Lindl.

### Scitamineæ.

Alpinia Galanga, Sw.
„ officinarum, Hance.
Amomum Melegueta, Roscoe.
Curcuma longa, L.
„ Zedoaria, Roxb.
Elettaria Cardamomum, White.
Zingiber officinale, L.

### Irideæ.

Crocus sativus, L.
Iris Florentina, L.

### Amaryllideæ.

Narcissus Jonquilla, L.

### Liliaceæ.

Allium sativum, L.
Aloe, several species.
Asparagus officinalis, L.
Colchicum autumnale, L.
Convallaria majalis, L.
Dracæna Draco, L.
Narthecium ossifragum, Huds.
Paris quadrifolia, L.
Schœnocaulon officinale, A. Gray.
Smilax China, L.
„ cordato-ovata, Rich.
„ glycyphylla, Smith.
„ medica, Cham. et Schl.
„ officinalis, Humb.
„ papyracea, Duh.

Smilax Parhampui, Ruiz.
Urginia Scilla, Steinh.
Veratrum album, L.
„ viride, Aiton.
Xantorrhœa australis, Br.
„ arborea, Br.
„ Preissii, Endl.
„ quadrangulata, F. v. M.
„ semiplana, F. v. M.

### Aroideæ.

Acorus Calamus, L.
Arum, several species.

### Alismaceæ.

Alisma Plantago, L.

### Palmaceæ.

Areca Catechu, L.
Attalea funifera, Mart.
Calamus Rotang, L.
Ceroxylon andicola, H. et B.
Cocos nucifera, L.
Daemonorops Draco, Mart.
Elais Guineensis, Jacq.

### Cyperaceæ.

Cyperus esculentus, L.

### Gramineæ.

Andropogon Calamus, Royle.
„ citratus, DC.
„ Ivarancusa, Roxb.
„ Schœnanthus, L.
„ muricatus, Retz.
Anthoxanthum odoratum, L.
Avena sativa, L.
Hierochloa borealis, Roem. et Schult.
Hordeum vulgare, L.
Saccharum officinarum, L.
Triticum repens, L.
„ vulgare, Vill.

# III. ACOTYLEDONEÆ.

*Equisetaceæ.*

Equisetum fluviatile, L.
,,      several species.

*Lycopodiaceæ.*

Lycopodium Chamæcyparissus,
Al. Br.
,,      complanatum, L.

*Filices.*

Aspidium Filix mas, Sw.
Polypodium vulgare, L.

*Lichenes.*

Cetraria Islandica, Achar.
,,      several species.
Chlorea vulpina, Nyl.
Cladonia rangiferina, Hoffm.
,,      several species.
Evernia prunastri, Achar.
,,      several species.
Lecanora Parella, Achar.
,,      Tartarea, Achar.
,,      several species.
Parmelia parietina, Achar.

Parmelia physodes, Achar.
,,      many other species.
Pertusaria communis, Fries.
Ramalina, several species.
Roccella fuciformis, Achar.
,,      tinctoria, Cand.
,,      several species.
Squamaria elegans, Hoffm.
Sticta, many species.
Usnea florida, Hoffm.
,,      several species.
Variolaria, many species.

*Algæ.*

Alsidium    Helmintochorton,
Kuetz.
Chondrus crispus, Stackh.
Delesseria, several species.

*Fungi.*

Agaricus campestris, L.
,,      croceus, Pers.
,,      muscarius, L.
,,      piperitus, L.
Cordyceps purpurea, Fries.
Polyporus officinalis, Fries.

# PART II.

## DIVISION I.

### APPARATUS REQUIRED FOR PHYTO-CHEMICAL ANALYSES.

These will be described under the following heads :—

    A. Weighing apparatus.

    B. Drying apparatus.

    C. Comminuting apparatus.

    D. Extracting apparatus.

    E. Straining apparatus.

    F. Evaporating apparatus.

    G. Miscellaneous apparatus.

### A.—Weighing Apparatus.

Three different balances are required for use :—A so-called *weigh-bridge* to carry 50 kilograms, and to turn with 10 grams or less, when so loaded; a *common balance*, which indicates distinctly 1 decigram when carrying 1 kilogram in each pan; and a *chemical balance* to carry 50 grams, and to turn, thus loaded, with 1 milligram. For better security of dust, fumes, &c., the last-named balance is always to be kept in a glass case.

The *weights*, now in common use amongst analytical chemists, are made according to the French or metrical system, and consist of brass or German silver. Only the smaller weights, from 1 decigram downwards, are made of aluminium. To use platinum for this purpose is inconvenient, on account of the smallness of the pieces, and the risk of losing them. The weights, and especially those under 5 grams, are never touched with the bare fingers, but with *pincers* made of metal.

v

## B.—Drying Apparatus.

Freshly-gathered plants or parts of plants are freed from adhering impurities, as earth, sand, or dust, by beating, dusting, or quickly washing with water. Thick roots, stems, branches, or fleshy fruits, are first split or cut into pieces of a proper size, and afterwards spread on *wooden sieves*, previously covered with printing paper, and thus exposed on warm days to the action of the air and the diffused sunlight, or, if the temperature should be below 10°, they are dried by artificial heat not exceeding 40°, but carefully screened from steam and smoke. The drying is finished, and the so treated substance called *air-dried*, when it does no longer incur loss of weight.

With the exception of certain succulent fruits, which on drying undergo a partial decomposition, all vegetable substances are submitted to analysis air-dried.

To estimate the water still retained by the air-dried material, a small sample of the latter is properly comminuted as described under C., and intimately mixed. Two to five grams of the substance thus treated are then put into a platinum crucible, and the latter placed into an *air-bath made of copper*, and provided with a thermometer. The whole apparatus is now heated to a temperature of 110° to 120° by means of a small spirit (or gas) lamp. (For particulars see Div. III., 1.)

Vegetable substances in the natural state less than those educts, obtained in the course of the analysis, and which, on account of their liability to decompose under the influence of heat, atmospheric air, or moisture, cannot be dried in the air-bath, must have the last traces of water removed under the jar of the *air-pump* with the simultaneous application of water-absorbing substances, such as quicklime, fused chloride of calcium or concentrated sulphuric acid, any of which are placed close to the substance in question, under the receiver of the air-pump.

## C.—Comminuting Apparatus.

Thick or woody roots, stems or large branches, are at first thinly split with an *axe;* the single pieces are then cut up transversely on the *cutting-board* and separated by means of a *sieve of wire gauze*, with apertures not larger than one-sixth of an inch; the coarser parts are further comminuted in the *stamping-box* by means of *cross-knives* provided with a long handle, and again passed through the sieve. To reduce vegetable parts of this kind to a still finer or powdery state, will generally require much time and labour, but this is essential for a satisfactory final result. Often the operation is greatly facilitated by successively exposing the substance to a gentle heat and then pounding it in a *metal mortar*.

Tough hard woods, after being fastened in a *screw-vice,* may be advantageously rasped by means of a coarse *file.* This is a good but rather tedious process.

Flexible or thin roots are placed at once on the cutting-board; still thinner ones only into the stamping-box, and afterwards, if necessary, into the mortar.

Fresh and succulent roots are split with a knife of horn or of a metal unaffected by the acid juices, first lengthwise, and then transversely.

Woody stems and thick branches are treated like roots; thinner woody stalks are first cut on the board, afterwards stamped in the box; very thin ones go at once into the stamping-box.

Barks may be reduced to powder in the iron mortar, but those with a tough fibrous bast are first treated on the cutting-board.

Fresh herbaceous stalks are bruised in a *stone mortar.* The material of the latter must be possibly pure sandstone; marble would be affected by the always acid juice.

Fresh leaves or flowers are bruised in the stone mortar; dry ones are stamped in the box.

Succulent fruits or coverings of seeds are bruised in the stone mortar, but if large they are previously cut into pieces with the horn knife.

Ordinary dry fruits are pounded in the iron mortar, but require, when tough, to be dried at a very gentle heat.

Seeds are treated like dry fruits; those rich in oil, to prevent the latter from separation, are first broken in the iron mortar and then submitted to a moderate trituration. Should it be possible to separate the pericarps from the seeds, this may be conveniently effected either with a *hammer* or in the iron mortar, and each part is then examined by itself.

### D.—Extracting Apparatus.

Glass and porcelain vessels would answer this purpose best, but very large vessels of those materials are easily broken, and consequently costly. As a substitute the chemist generally employs tin or copper vessels, especially when working with large quantities.

*Glass flasks* are required, of various sizes from the smallest up to those of two litres capacity. The glass of these flasks must be of a uniform thinness, and the edges of their necks must be bent outwards, to allow the insertion of a cork stopper without risk of cracking.

*Infusion-pots* with *lids,* both made of porcelain, may be used for the extraction of substances containing no volatile matters, with either water, diluted acids, or diluted alkalies.

A *tin still* is further required, capable of holding at least 10 litres of water, with *head* and *worm* of the same metal. A still of this kind should never be exposed to the direct fire, but only to the heat of the water-bath, to prevent the metal from fusing and the contents from getting charred. The *water-bath* consists of a *copper boiler*, in which the still is fitted so as to reach nearly to the bottom, while resting rather lightly on the prominent edges of the contracted mouth of the boiler.

The tin still serves for extractions and distillations (*see* F.) on a larger scale.

### E.—STRAINING APPARATUS.

For larger quantities *dishes or deep vessels of porcelain or stone-ware* are used with a *tenaculum* (square framework, constructed of four narrow pieces of wood) of proper size to be placed on the dishes, and carrying on its four prominent points a *linen cloth* or a *linen bag-filter*, through which to strain the substance under investigation. The latter operation is accelerated by occasionally stirring or pressing the contents with a strong *glass rod* or with a *porcelain spatula*, or instead of these a *spatula*, made of *pine* or *beech* and previously cleaned by boiling with water, suffices for most purposes.

Thick, slimy masses let the liquid pass so slowly that it is necessary to strain through a cloth, the meshes of which do not contract by moisture. This is *miller's gauze*, or *bolting cloth of silk*, which is sold in numerous gradations as to the width of the meshes, and may be therefore selected according to circumstances.

After the dripping has ceased, the contents of the cloth or of the bag-filter are submitted to the action of a *screw-press* or of an *hydraulic press*. Those parts of the press, which are in direct contact with the cloth or with the liquid passing from it, must be made of tin.—For pressing smaller quantities, a small *self-acting press*, similar in construction to a bookbinder's press, is to be employed, but modified that it may be fastened to a table, and its sides must be covered with thick plates of glass.

Still smaller quantities are poured into *paper-filters*, spread out in *funnels* of *glass* or *porcelain*, the latter resting on the edges of *glass-jars*, and, if necessary, supported by means of *filter-plates* or other appliances. If the filtering process is slow, the funnel must be covered with a *glass-plate* in order to screen off air and dust. This is not less necessary with very volatile liquids, for instance, alcohol, ether, &c., in order to prevent losses.

More economically, larger quantities or slowly filtering alcoholic liquids are filtered in a *displacement apparatus of glass*, the top of which can be shut nearly air-tight.

For the filtration of liquids, which require warming either because they are then only of sufficient fluidity, or because in lower temperatures the solved parts would become solid, a *water-bath-funnel* is employed, *i.e.* a water-bath of funnel-shape and exactly fitting to the funnel containing the filter, and the water of which is kept boiling by means of a tube inserted at the side, and heated by the flame of a lamp beneath.

## F.—Evaporating Apparatus.

Under this head have to be mentioned *dishes of glass* and of *porcelain*, also of *tin* (see below), *watch-glasses*, *glass beakers* of various sizes.

For accelerating the evaporation at higher temperatures, coals, alcohol or gas, are employed as fuel to act upon the evaporating vessel either directly or divided from it by an *iron plate* or a *dish* of the same metal, either empty or filled with sand *(sand-bath)*, or with water *(water-bath)*.

Should it be necessary to accelerate the evaporation without the application of heat, or under the exclusion of the atmosphere, the *air-pump* is made use of, and additionally the fluid to be evaporated under the receiver is brought in proximity with either concentrated sulphuric acid or anhydrous chloride of calcium or quicklime.

Some liquids leave, when concentrated to a certain degree, a stiff, viscid mass, to dry which entirely in the evaporating vessel is very difficult; but this may be effected with comparative ease by spreading those substances as thinly as possible on *glass plates* or on *porcelain dishes*. Plates of this kind are, among other instances, indispensable for any contents of filters, which may be expected on drying to stick to the paper, and thus prevent the separation without loss.

A kind of evaporation is the process of distillation. For this purpose on a larger scale, the tin-still mentioned already under D. is employed. Distillations of ether or of alcohol may be effected directly from it, but not those of water. For the latter purpose the still has to be furnished with a special contrivance. For instance, in order to obtain the volatile oil from a vegetable (including at the same time the extraction of the substance), a sieve-like perforated disc of tin and furnished, besides, with a larger aperture of about one inch diameter, is inserted into the still, within one to two inches from the bottom of the vessel. On this disc the vegetable substance, moistened previously with water, but not so as to form a pulp, is spread out, and is subjected to the steam of water coming from the copper boiler, and conducted by means of a tin tube, terminating under the false bottom, into the substance to be distilled. A mechanism of this kind, besides other implements intended for digesting, evaporating, drying, &c., purposes (dishes of tin, &c.), also a *refrigerator*, together with the tubes required for supplying the cold, and for removing the hot water, form the well-known *Beindorf's apparatus*.

Distillations on a smaller scale are carried on in *retorts of glass,* the vapours produced are condensed in a *Goebel's refrigerator,* and collected in glass receivers.

## G.—MISCELLANEOUS APPARATUS.

Under this category I shall comprise the rest of the utensils, required for a phyto-chemical laboratory, but without any further description.

*Aræometers* for light and for heavy liquids.

*Barometers.*

*Bladders.*

*Blowpipe.*

*Caoutchouc* in tubes and in thin plates.

*Caoutchouc cement,* from heated indiarubber and kaolin.

*Coal pincers.*

*Coal* prepared for cutting glass.

*Crucibles* of porcelain and of platinum.

*Florentine glass bottle.*

*Funnel-tubes.*

*Glass bottles* with or without glass stoppers.

*Glass jars,* graduated.

*Glass jars,* to be fitted air-tight on glass plates.

*Glass rods.*

*Glass tubes* (including capillary tubes for determining the fusing points).

*Glazed paper.*

*Lamps* of glass and of brass.

*Microscope.*

*Parchment-paper.*

*Pipettes.*

*Plated iron* and *platinum foil.*

*Pycnometer.*

*Rings* with iron arms.

*Scissors* of various sizes.

*Separation-funnel.*

*Stands* of wood (for test tubes, retorts, thermometers, &c.).

*Stoppers* of glass, caoutchouc, cork.

*Stoves* of clay and of iron.

*Test-paper* of litmus and of turmeric.

*Test-tubes.*

*Thermometers.*

*Thread* of various thickness.

*Tongues* of iron and of brass.

*Washing bottle* for washing precipitates.

*Wires* of iron and of platinum.

# DIVISION II.

## CHEMICALS REQUIRED FOR PHYTO-CHEMICAL ANALYSES.

———•———

THESE chemicals may be divided into—

    A. Absorbents.
    B. Solvents.
    C. Reagents.

### A.—ABSORBENTS.

These are principally: concentrated sulphuric acid, anhydrous chloride of calcium and quicklime.

They are used for drying a solid or 'liquid substance at ordinary temperature and under exclusion of the air, by placing them in a glass beaker close to the substance under the receiver of an air-pump, and by working the latter occasionally after intervals of two to three hours.

Every one of the three substances named has its special value. For quickness of action anhydrous lime is unsurpassed, and especially so, when it has been reduced to about the size of peas; but its power is soon exhausted, as it absorbs only 1 eq. water, equal to $\frac{1}{3}$ its own weight. Next in quickness acts concentrated sulphuric acid, which absorbs up to 3 eq. water, increasing its weight by one half. The slowest action is exercised by anhydrous chloride of calcium, but though the least powerful yet it absorbs 6 eq. water or nearly its own weight, and besides it takes up another quantity of water required to convert the compound Ca Cl + 6 HO into the liquid state.

By taking into account these different qualities, the choice of the absorbent cannot be difficult. Should it be required to conduct the drying process as quickly as possible, quicklime is resorted to; if less pressed for time, sulphuric acid; and if time is of no consequence, chloride of calcium. [There may be other considerations to guide in the choice of these absorbents.]

Besides these there are many other hygroscopic substances, but none exceeds in activity and cheapness the above three.

After they have done their service as desiccating agents, the chloride of calcium solution may be evaporated and fused again. The aqueous sulphuric acid may be utilised as such; and the hydrate of lime, as left, would be of no further value.

## B.—SOLVENTS.

These are principally *ether, alcohol* and *water*; also *diluted acids* (especially hydro-chloric and sulphuric), *diluted alkalies* (potash or soda-ley and liquor of ammonia); less frequently used are *benzol, chloroform, wood-spirit, sulphide of carbon, petroleum, oil of turpentine,* &c.

*Ether* withdraws from the plants, almost without exception, easily and completely chlorophyll, wax, fixed and volatile oils, free acids; also certain alkaloïds, pigments, indifferent bitter ingredients, resins; sparingly or not at all saccharine substances. But even such compounds as are insoluble in pure ether, as humus substances, salts of anorganic acids, proteins, &c., dissolve, though in exceedingly small quantities, in ether containing traces of water or alcohol, and retard thereby the isolation of the first-named substances in their pure state.

This inconvenience might be avoided entirely or for the greatest part by using ether of 0·720 sp. gr. (at 15°), which is entirely free from water and alcohol.

Of greater consequence still than with ether is the strength of the *alcohol* (spirit of wine, ethylalcohol) employed as a solving agent. Absolute or anhydrous alcohol of 0·792 sp. gr. (at 15°) dissolves like ether readily chlorophyll, volatile oils, free acids, alkaloïds, indifferent bitter substances, resins and pigments; less easily wax, fats and saccharine matters; sparingly humus substances, salts of anorganic acids and proteins.

Yet as it is troublesome to prepare and to keep alcohol in its anhydrous state, it is reserved for special cases, and in its stead an alcohol of 0·815 sp. gr. is generally in use, the solving power of which is nearly the same in regard to wax and fats, and is sometimes even greater towards the other substances mentioned. Unfortunately it also dissolves to some extent gum and similar matters, which have afterwards to be removed by special processes.

It is not advisable to use weaker alcohol as a solvent, as this would prevent complete isolation of various constituents.

Dilute alcohol is employed in some cases for withdrawing resins from fats, or for separating different resins from each other.

For extracting purposes, spring or any other ordinary water should not be used but always *distilled water*, or, in exceptional cases, fresh rain-water, but then only such as has been caught directly from the sky, and not by means of spouts.

Water is also employed to withdraw from the ethereous, alcoholic, &c., extracts, gum, protein substances in their soluble form, acids, most of the alkalies and alkaline earths in combination with organic acids, most of the salts of the alkaloïds, some alkaloïds in the free state, many bitter substances, some other indifferent matters, pigments; while chlorophyll, wax, fats, resins, certain pigments are not dissolved by water; ethereal oils, certain alkaloïds, bitter principles, other indifferent substances, some pigments dissolve in it only sparingly.

Acids are not used as solvents in the concentrated state, but as addition to water, chiefly in order to dissolve pectin and oxalate of lime. As a rule the *diluted acids* (*hydrochloric, sulphuric*) are only applied after the vegetable substance has been exhausted successively by ether, alcohol and water.

In cases where the search for alkoloïds and their isolation without loss is to be instituted, water mixed with only 2 to 3 per cent. of the concentrated acids is employed, but even in this diluted state the acid acts not only as a solvent but also altering and decomposing, for instance, on glucosids, and even on some alkaloïds.

*Acetic* acid of about 20 per cent. is employed for separating oxalate, phosphate, and sulphate of lead, which are insoluble in the reagent, from the compounds of lead with other organic acids, which are then re-precipitated from their solutions on addition of a base.

The *fixed caustic alkalies* likewise are only used in a very diluted state, and are always preceded by ether, alcohol, water, and acids. Their usefulness is very limited, because in most cases and by their agency only those protein substances are dissolved which are insoluble in water, while again the partial decomposition of these is very difficult to avoid.

Another inconvenience connected with these alkalies consists in the conversion of so-called extractive substances, and of such as have not been dissolved by the other extracting agents into partly humus-like dark-coloured products, which greatly impede the further process of the analysis. The only important use of diluted alkalies is to soften the vegetable fibre to such an extent as to facilitate its purification.

*Liquor of ammonia* as a solvent is only employed for treating complex substances after they have been withdrawn from the plant, in order to remove one or other of the constituents, especially resins, some of which are soluble in liquor of ammonia, others not.

All the other above-mentioned chemicals—*benzol, chloroform, wood-spirit, sulphide of carbon, petroleum, oil of turpentine, &c.*—

are, like liquor of ammonia, never used for directly extracting the plant, but only for separating mixed substances. In some cases they have proved very useful in separating a substance from tenaciously attached colouring particles, the latter being insoluble, or nearly so, in those liquids, while, on the contrary, they prove excellent solvents for alkaloïds, resins, &c. The solvent may be finally removed, either by evaporating (chloroform, wood-spirit, sulphide of carbon); or the alkaloïds are withdrawn from their respective solutions in benzol, petroleum, or oil of turpentine by shaking with water acidulated with hydrochloric acid, as evaporation at a high temperature would be impracticable.

## C.—REAGENTS.

### A.—*Examination of the dry substance.*

The first trial consists in exposing the respective substance to the action of *heat*, when is to be observed whether fusion, either total or partial volatilisation, or carbonisation takes place.

In the first case, *i.e.*, with a fusible substance, is to be determined the *fusing-point*, by inserting small pieces of about the size of a millet-seed into a capillary tube, or with fats or waxy matters by allowing them in the melted state to rise into such a tube, and to congeal therein; then put this tube, together with a thermometer reaching at least to 300°, into a test-tube, the latter into a glass-flask, this on an iron plate, and heat the latter by means of a spirit-lamp. The heating has to be conducted very slowly and gradually, so as to enable a careful observation of the mercury rising from degree to degree. The height of the mercury is read off and noted down as soon as the fat begins to melt, without being entirely fluid. The latter precaution is necessary, because in the transition from the solid into the liquid state heat is absorbed and becomes latent, causing either a stoppage of the mercury, or even a retrogressive movement. — Every estimation of the fusing-point has to be made at least twice, and if the two experiments do not harmonise, oftener.

After complete fusion and with increased heating, the substance sublimates, if *volatile*; if not, decomposition takes place, indicated by the black colour assumed by the substance, and by the emission of empyreumatic gases. Yet, it sometimes happens, that substances sublimate or decompose without having been fused first. In a capillary tube this cannot be observed satisfactorily, because the liquid sample in it rises mechanically under the influence of the heat without having been sublimated at all, and because the tube is too narrow for inserting slips of test-paper.—Repeat, therefore, the experiment with a fresh sample, in a glass tube of at least 1-24th inch diameter, and closed at the bottom. The *non-volatility* is now characterised by a discolouration (blackening), and by the evolution of empyreumatic vapours, which must be tested as regards their acid or alkaline reaction by both litmus and turmeric paper, inserted into the upper part of the tube. A brown colouration of the turmeric paper indicates with certainty a nitrogenised substance (proteinoïd), and therewith is always connected an unpleasant odour like that of burnt horn or feathers (so-called horny odour). If the turmeric paper remains unaltered, then the substance contains either no nitrogen or only a little of it, and in this case the litmus paper assumes a red colour.

After the substance has been tried on its behaviour towards heat, it is next exposed to the action of *mineral acids*, and first to that of *concentrated sulphuric acid*, which may effect alterations of colour, solutions, or odours. Warming the substance with it would only be advisable, if no perceptible alteration takes place in the cold; else a carbonisation would be the result. The solution thus obtained becomes sometimes turbid on addition of water, as it may throw down the substance altered or unaltered.—*Diluted* with 5 to 10 parts water, the sulphuric acid serves as reagent of glucosids, as it possesses the property of separating these after some time on warming, and to convert the carbo-hydrate contained in the glucosid into sugar, which is recognised by the formation of red suboxyd of copper on warming, after the acid has been neutralised by excess of soda, and mixed with dissolved alkaline tartarate of copper.—In a still more diluted state (1 : 50 or 100) the sulphuric acid may be used for indicating the alkaloïd-nature of a substance, inasmuch as the latter dissolves more readily in it than in pure water.

*Nitric acid*, in a moderately concentrated state (from 1·3 specific gravity upwards), acts almost always decomposing (oxydising) on dry substances, under the simultaneous appearance of yellow-brown fumes, accompanied, in most instances, by a colouration of the acid, by a change of the colour of the substance, or by a solution of the latter. Exposed to the heat, even a weaker acid shows already oxydising effects, and evolves brown-yellow vapours. Very diluted, it may be used instead of sulphuric acid; in great dilution, for indicating the alkaloïd nature of a substance.

*Concentrated hydrocholoric acid* (of 1·16 to 1·20 specific gravity) acts least energetically of these three mineral acids, but produces sometimes a characteristic colouration of the substance, or gives a specific odour after a solution has been effected, and therefore it should be used in every case.—Diluted with 5 to 6 parts water it may be used, like sulphuric acid, for the discovery of glucosids; and still more diluted, it may indicate the alkoloïd-nature of a substance, though in both these cases it is not superior to sulphuric acid, and is therefore used less frequently.

After the acids, *caustic alkalies* must be applied; *liquor of ammonia*, of 0·960 specific gravity, *potash* or *soda-ley*, of 1·15 specific gravity. They are destined for solving purposes, some-times aided by a gentle heat. Should the substance have been dissolved in any of the three alkalies, and should diluted acids have proved unable to dissolve the same substance, then yet has to be tried, whether the alkaline solution on over-saturating with an acid will throw down the substance again or not. In the first case no alteration has been effected by the alkali; in the latter case a change has taken place. It may also happen that the alkaline solution becomes turbid on diluting with water, or even on adding a fresh portion of the alkali. In the latter case what has been separated is not the original body, but a compound of it with the alkali, and insoluble in alkaline liquids. In dissolving nitrogenised, especially protein substances in fixed alkalies, appli-cation of heat must be avoided entirely or as much as possible, in order to prevent decomposition, mostly connected with the evolu-tion of ammonia.

The other liquids, the influence of which on the dry substance may be tried also, and which act only as solvents, are *ether, alcohol, water, benzol, chloroform, wood-spirit, sulphide of carbon, petroleum, oil of turpentine;* but the resulting solutions are only available for testing with other chemicals after the solvents, as ether, alcohol, benzol, &c., have been removed and are replaced by water. Their further treatment must be referred to the following sub-division :—

*B.—Examination of the Substance in Solution.*

Under this head are comprised not only the liquids obtained in the course of the analysis, as well as the solutions of the solid substances obtained in the same way, and dissolved either in water or in diluted acids or in alkalies, but also the extracts obtained by means of ether, alcohol, water, diluted acids and alkalies. The three last-named solutions are submitted directly to the action of reagents, the alcoholic or ethereous solutions only, after the respective solutions have been removed by evaporation or distillation and replaced by water. By this change of the ether or alcohol by water the resulting aqueous solution will seldom be clear, but mostly turbidified by substances insoluble in water, to be separated by filtering, and to be submitted to a special treat-ment (specified in Division III.) Such solutions, obtained by treatment with water from the alcoholic and ethereous extracts, often retain dissolved, through the influence of other constituents, small quantities of resinous and other substances, which are insoluble by themselves in pure water.

The reagents used for testing the aqueous solutions of any vegetable matter and brought into the order in which they might be employed most conveniently, are as follows:—

*Blue litmus-paper.*—A red colouration of it indicates free acids or acid salts. Extracts of vegetables will always produce this effect.

*Yellow turmeric-paper.*—A brown colouration of it would indicate basic salts, but such an effect has scarcely ever been obtained with vegetable extracts.

*Red litmus-paper* may be dispensed with in most cases, yet it happens sometimes that an alkaline reaction is recognised more easily with this paper than with turmeric-paper.

All other tinged papers, as those of dahlia, violets, &c., are inferior to the above three.

*Ether* (of 0·720 specific gravity) frequently produces turbidity even in small quantity, but it is best to add as much as the watery liquid is able to dissolve, or until a small quantity of ether floats on the aqueous solution after the mixture has been well agitated for some time. The turbidity may be produced by gum, by salts of anorganic bases with anorganic or organic acids, by protein substances, humus-like matters, &c.

*Alcohol* (of 0·815 specific gravity) usually effects no alteration when added in small quantity; therefore more of it is required than of ether, or on the average a quantity equal in bulk to the aqueous liquid. A turbidity, obtained hereby, is mostly caused by gum.

*Liquor of ammonia* (of 0·960 specific gravity) imparts clearness to liquids which possess an opalescent appearance (*see above*) from traces of resins. Coloured liquids become darker with it. But ammonia may also produce precipitates; should any of them disappear on the addition of alcohol, it is with tolerable certainty to be concluded that an alkaloïd is present. The same conclusion is justified when a further addition of ammonia acts redissolvingly on the precipitate, and therefore the precaution should be used to apply the reagents only successively drop by drop, because sudden large additions would redissolve any otherwise visible precipitate, and prevent its being noticed.

Should the precipitate, obtained with ammonia, disappear neither with alcohol nor with an excess of ammonia, it may be either alumina or phosphate of lime, or phosphate of ammonia-magnesia or oxalate of lime. The first of these precipitates is flocky and soluble in potash-ley, also in acetic acid when newly precipitated; the second one is also flocky, and when newly precipitated, soluble in acetic acid, but not in potash-ley; the third behaves towards acetic acid and potash-ley like the second, but is distinguished by its crystalline appearance; the fourth precipitate is of a finely pulverulent form and insoluble either in potash-ley or in acetic acid.

Ammonia is also employed for saturating liquids, especially those which have been precipitated with acetate of lead, in order to render them alkaline and so to produce another precipitate.

*Carbonate of ammonia* (1 part in 9 parts water) acts similarly to liquor of ammonia, but less energetically. Its principal use in phyto-chemical operations is to remove completely any excess of lead left from the precipitation with acetate of lead, and in this respect it is superior to sulphuret of hydrogen, commonly recommended, on account of greater simplicity and ease.

*Leys of potash or of soda* (of 1·330 specific gravity) produce in general the same appearances as ammonia, but act more decidedly. They impart to coloured liquids a still darker colour.

A precipitate, produced by potash or soda, and soluble in an excess of the reagent, may be alumina partly or entirely.

In the presence of ammonia-salts the ammonia is displaced by potash or soda, and may be recognised either by the smell or by the fumes obtained with acetic acid, and tested by holding a glass rod moistened with the acid over the surface of the liquid.

*Carbonates of potash or of soda* (1 part in 9 parts water) possess in general the properties of potash or soda-ley, but in a less degree, and cannot dissolve alumina. They may be used for removing an excess of oxyd of lead, when there is reason to avoid the use of carbonate of ammonia.

*Solution of baryta* (1 part Ba O + 9 HO in 19 parts water) behaves towards coloured liquids similar to the alkalies; it also produces precipitates which may be either alkaloïds, from which the acid is withdrawn, or components of baryta with sulphuric acid, phosphoric acid, organic acids, or with substances acting like acids, as pigments, resins, &c.

*Lime-water* (1 part Ca O in 700 parts water) acts like solution of baryta, but less decidedly; it does not precipitate sulphuric acid but oxalic acid (which is not precipitable by diluted solution of baryta), oxalate of lime being only soluble in mineral acids.— Of the four organic acids, malic, citric, tartaric, and racemic, the first is not precipitated by lime-water, either cold or hot; the second is precipitated only in the heat, and becomes clear again on cooling; the third yields cold a precipitate soluble in chloride of ammonium; the fourth yields also a precipitate in the cold, but insoluble in chloride of ammonium.

*Chloride of ammonium* (1 part in 9 parts water) serves for distinguishing tartaric from racemic acid. (*See* Lime-water).

*Chloride of barium* (1 part in 9 parts water). A precipitate, obtained by means of it, contains certain anorganic acids. If it is insoluble in hydrochloric acid, then it contains sulphuric acid; if soluble, wholly or partially, then the soluble portion contains phosphoric acid.

*Chloride of calcium* (1 part in 9 parts water) is used for detecting a few acids through the formation of precipitates. The latter, if newly precipitated, soluble in acetic acid, may contain phosphoric acid; if not soluble, it contains oxalic acid.

*Acetate of lime* (1 part in 9 parts water) is used for testing oxalic acid instead of chloride of calcium, should the presence of chlorides in the liquid have to be avoided. From liquids, containing acetates only, or free acetic acid, the oxalic acid can be removed entirely by means of the above reagent.

*Chloride of iron* [perchloride of iron] (1 part in 9 parts water) indicates the various tannic acids by yielding blue or green precipitates of various gradations of colour, or similarly coloured liquids; though a green tinge, produced by chloride of iron, must not be taken as satisfactory proof of the presence of tannin. Gum arabic also is precipitated by the reagent.

*Sub-sulphate of iron*, or green vitriol (1 part in 9 parts water), is sometimes used for testing tannic .acids instead of chloride of iron, as the colours of the precipitates, produced by these two iron salts, differ sometimes considerably.

*Glue* (1 part isinglass in 100 parts alcohol of 40%) is the test of the true tannic acids, with which it yields insoluble or hardly soluble, and mostly grey and flocky, precipitates. Should the solution of glue prove too thick, it must be heated gently and shortly before use (not diluted with water or with alcohol).

To remove tannic acid from liquids which contain other acids isinglass is used, in small narrow strips previously soaked in cold water.

*Tartarate of antimony and potash* [tartar emetic or *tartarated antimony*] (1 part in 19 parts water) precipitates most of the tannic acids, and may therefore be used for testing these as an accessory reagent.

*Acetate of lead* [sugar of lead] (1 part in 9 parts water), the most important chemical in phyto-analyses used as precipitating and separating agent; it precipitates completely sulphuric and phosphoric acids, and more or less completely most of the organic acids, pigments, protein-substances, and resins. Those anorganic lead-precipitates are insoluble in acetic acid; most of the others are soluble.

*Tribasic acetate of lead* [lead-vinegar or *subacetate of lead*] (of 1·200 specific gravity).—After a vegetable extract has been precipitated by acetate of lead, a new precipitate is generally formed on addition of lead-vinegar, but instead of using at once the latter, it is preferable and more convenient to saturate the acid liquid containing yet acetate of lead, first with ammonia, and then to remove any turbid substances by means of filtering, before this reagent is used.

*Protochloride of tin* [sub-chloride of tin] acts as precipitating agent similar to acetate of lead, but a great disadvantage is that it must be employed in a very acid solution, and even then soon becomes oxydised. It deserves, therefore, no recommendation.

*Sulphate of copper* [blue vitriol] (1 part in 9 parts water) yields, with a few organic acids, precipitates of a characteristic colour, for instance, a green one with salicylous acid ; but may be easily dispensed with.

*Alkaline tartarate of copper* is used as a direct test for grape-sugar (glucose) or fruit-sugar, also for the direct quantitative estimation of grape-sugar and fruit-sugar, and for the indirect one of cane-sugar and starch. To prepare it, dissolve 3·465 grams pure crystallised sulphate of copper in about 40 grams water, add 9 grams pure tartaric acid, and, after it is dissolved, 18 grams hydrate of soda and dilute, after this is also dissolved, with so much water that the whole amounts to 100 cubic centimeters. Ten C. C. of this deep-blue liquid, containing 0·3465 grams sulphate of copper, require for the reduction of the oxyd to sub-oxyd 0·05 grams grape or fruit sugar.

Cane-sugar and starch must be previously converted into grape-sugar by digestion with diluted sulphuric acid. 500 parts grape-sugar, as found above, are equal to 475 parts cane-sugar, or to 450 parts starch.

Liquids which have to be tested with this reagent must be alkaline, or at least neutral; acid liquids have therefore first to be neutralised by potash or soda-ley.

This reagent, when kept ready prepared, must be tried every time before using it on its unimpaired quality by heating it to the boiling point, when it must become neither discoloured nor turbid.

*Tannic acid* [gallotannic acid], (1 part in 9 parts alcohol of 50 per cent.; for precipitating purposes: 1 part in 9 parts water, prepared shortly before use). An important precipitating agent of alkaloïds and of many indifferent bitter substances. As a rule these precipitates do not allow washing with pure water without being decomposed. In order to isolate the substance combined with the tannic adid, the precipitate is spread out on several thicknesses of blotting prper,, supported by a tile or brick, it is then mixed with oxyd of lead or with white of lead and eventually with a little water, is dried at a gentle heat and extracted with alcohol, which leaves undissolved the tannic acid in combination with lead.

*Bi-iodide of potassium* (4 parts iodide of potassium and 3 parts iodine in 93 parts water). Serves principally to distinguish starch, to which it imparts a violet or deep-blue colour. It may also be used with advantage for precipitating alkaloïds which yield precipitates of a more or less kermes-brown colour.

*Sub-cyanide of potassio-platinum* [potassio-platinous cyanide], (1 part in 19 parts water), gives well characterised precipitates with some alkaloïds. It is obtained by dissolving the grey-green subchloride of platinum, obtained by heating the chloride, in an aqueous solution of cyanide of potassium and by crystallising the salt from the solution.

*Iodide of potassio-mercury* (2 parts chloride of mercury, and 5 parts iodide of potassium in 43 parts water) yields with most of the alkaloïds insoluble precipitates.

*Iodide of potassio-bismuth* has been recommended recently as a sensitive precipitant of alkaloïds. For this purpose it is prepared by heating a mixture of 32 parts sulphide of bismuth, and 47½ parts iodine, until under evolution of sulphur iodide of bismuth has sublimated. Treat the sublimate with a concentrated solution of iodide of potassium hot as long as the latter dissolves anything, decant the solution from the insoluble portion and mix with an equal volume of a concentrated solution of iodide of potassium.

*Phosphate of soda*, the common crystallised salt (1 part in 14 parts water), precipitates some alkaloïds; it may also be used instead of sulphate of soda, for removing lead from liquids, if the introduction of sulphuric acid into the liquid has to be avoided.

*Phospho-molybdate of soda* (6 parts molybdic acid, 12 parts crystallised carbonate of soda, and 1 part crystallised phosphate of soda in 31 parts water), mixed with pure nitric acid, until of a pure citron-yellow colour. It precipitates alkaloïds, and is highly important on account of it yielding precipitates with many alkaloïds, which are not precipitated by other reagents. To free these precipitates from other matters which may likewise have been precipitated, washing with water must be avoided on account of their liability to decompose. They are, therefore, treated in the following way. Place the filter with its contents on several thicknesses of blotting paper, and the whole on a new brickbat; leave the precipitate until of a pasty consistence, transfer it then to a porcelain dish and add water so as to form a thin pulp, now add burnt magnesia under continual stirring, until every trace of acid reaction has disappeared and a slightly alkaline reaction towards litmus paper has taken its place, dry with a gentle heat, grind the residue finely and shake with absolute alcohol. Any alkaloïd which has been liberated by the magnesia, passes into the alcohol, remains on evaporating and may now be subjected to a closer investigation.

*Phospho-tungstate of soda* (10 parts tungstic acid, 12 parts crystallised carbonate of soda, and 27 parts water are boiled until the liquid is no longer precipitable by acids, *i.e.*, until the tungstic acid has passed into meta-tungstic acid; add 1 part crystallised

w

phosphate of soda, restore the water lost by evaporation, and filter if necessary). After it has been strongly acidified by nitric acid, it acts similar to phospho-molybdate of soda, and is even more sensitive towards some alkaloïds. In all other respects, what has been said of the preceding reagent is applicable to this one too.

*Picric acid* yields with many alkaloïds yellow and mostly crystalline precipitates.

*Chloride of platinum* [platinic chloride] (1 part in 19 parts water) is one of the most common precipitants of alkaloïds. The precipitates are various shades of yellow, flocky, and distinguishable hereby from those obtained by ammonia or potash, which are pulverulent or crystalline.

*Chloride of mercury* [mercuric chloride] (1 part in 19 parts water). Precipitating reagent of alkaloïds, and forming a white double-compound; but several of which are rather soluble in water, and do therefore not appear in diluted solutions.

Chloride of mercury forms also with protein substances compounds which do not dissolve in water.

*Sub-nitrate of palladium* [palladious nitrate] (1 part in 19 parts water). Precipitant of alkaloïds. The compounds are yellow or brown.

*Sub-nitrate of mercury* [mercurous nitrate] (1 part in 19 parts water under addition of a few drops of nitric acid). As in general vegetable extracts contain compounds of chlorine, a turbidity is nearly always occasioned by this test, and which cannot be removed by nitric acid. It precipitates also most of the organic acids, but the precipitates are mostly decomposed partially on keeping, changing their (as a rule) white colour into a grey one. Even if no precipitate has been obtained, the liquid, containing organic matters, often assumes with this reagent a grey or a darker colour, produced by the reduction of the sub-oxyd of mercury, and indicating eventually gallic acid, which acts strongly reducing on this salt.

*Nitrate of silver* (1 part in 19 parts water) yields with some alkaloïds sparingly or not at all soluble double compounds, which separate as white, flocky precipitates. In other respects, all that has been said of the sub-nitrate of mercury applies also to this test.—As regards its application for the quantitative estimation of formic and of hydrocyanic acids, I refer to these substances in the first division of the first part of this work.

The following substances are not employed for testing, but for separating and purifying:—

*Oxyd of lead*, ground to a subtle powder, serves for removing tannic acid from liquids or from pulpy precipitates. Mix the latter intimately with the oxyd of lead, dry, grind the remnant and withdraw from it the other constituents (usually an indifferent or a basic bitter substance) by means of alcohol. It may be sometimes used also for removing colouring matters, by assimilating those and rendering them insoluble. It serves also for separating fluid from solid fat-acids (*see* Part II., Div. III., *A, a.*)

*White of lead* may be used instead of oxyd of lead, and acts even more decidedly in many cases.

*Acetate of magnesia* (1 part in 9 parts water) serves sometimes for the fractional precipitation of solutions of soaps containing different fat-acids, in order to facilitate their separation.

*Chloride of lime* is used in its clear aqueous solution for bleaching the vegetable fibre (*see* Fibre, p. 82).

*Lime*, slaked with water as to form the finely pulverulent hydrate, serves only for displacing volatile alkaloïds (*see* Div. III., X.).

*Phosphoric acid* is employed for dissolving protein substances (*see* Div. III., IV.) ; also for displacing volatile organic acids (*Ibid*, IX.).

*Sulphate of soda*, Glauber's salt (1 part in 9 parts water), is employed for removing lead from liquids, when carbonates of alkalies have to be avoided; yet, as it does not precipitate the lead completely it is necessary, after filtering, to remove the rest of the lead by means of sulphuret of hydrogen.

*Sulphate of silver* (1 part in 200 parts water) is employed, before distilling the volatile organic acids, to remove hydrochloric acid.

*Sulphuret of hydrogen* serves not so much for freeing the bases from any excess of lead-salts (for which purpose the carbonates or sulphates of alkalies are preferable) as for decomposing newly precipitated and edulcorated lead-precipitates with the view of isolating the acid combined with the lead. The gas obtained from sulphide of iron by means of diluted sulphuric acid has to pass through water in order to get rid of particles carried over mechanically, before it comes in contact with the lead-precipitate.

*Animal charcoal*, as finely ground bone-black purified by hydrochloric acid and heated afterwards to a red heat, is an important means for absorbing pigments, odours and bitter substances, and is even used for isolating the latter pure, by submitting the coal laden with them and after washing with water, to digestion with strong alcohol, which then withdraws the bitter substance or substances.

*Alumina*, in its newly precipitated hydrated state, may often be employed with advantage as separating agent, as it precipitates pigments and many other either little coloured or colourless bodies from their solutions and leaves others dissolved.

# DIVISION III.

## GENERAL SYSTEMATIC COURSE OF PHYTO-CHEMICAL ANALYSIS.

———◆———

ETHER, alcohol, and water, are, as recorded, the most important solvents, respectively extracting agents used in phyto-chemical analysis; acids, alkalies, &c., being of less importance.

In using the above three liquids, it would appear at first of little consequence in what order they are applied to the substance under investigation, as, after all, the parts soluble in them must enter these liquids as extracts; but taking into consideration all circumstances for the purpose, it soon becomes evident that a certain order has to be adhered to. Those special cases are of course excepted where it is required to obtain or to recognise only one constituent, and when the choice of the solvent depends already on the nature of the particular constituent, as for example, when ether is required for extracting a fixed oil, and water for gum. Still, if a thorough investigation of the chemical constitution of plants should be intended, then I would recommend to follow the practice adopted by me since many years,— to treat the substance first with ether, then with alcohol, and last with water, the reasons for which treatment I shall give presently.

1. By treating the generality of vegetable substances with ether the latter dissolves fat and wax readily and completely; alcohol dissolves these bodies also, but less readily, and only at the boiling heat, and as it deposits them again on cooling, and as filtering is retarded thereby, it becomes necessary to employ ether for removing the rest of fat or wax. These two substances, therefore, are passing into two different solvents. If the analysis is commenced with alcohol, this inconvenience increases when, as in seeds, the quantity of fat or wax predominates, and it can only be avoided by the previous treatment with ether.

2. Certain alkaloïds, bitter substances or resins, dissolve in alcohol and in ether, others only in alcohol. Now, by commencing

the analysis with alcohol the whole of them is dissolved, while by commencing with ether a separation of two different groups is effected at the outset.

3. Not less advantageous is the previous treatment with ether and then with alcohol for the subsequent one with water, whereas, if the commencement be made with water, inconveniences of various kinds are incurred. Amongst these stand out in the first line the difficulty of filtering, caused sometimes, it is true, by a considerable amount of gum, but also and often alone by finely suspended particles of resins and fats. These could, indeed, easily be removed by a few drops of dissolved acetate of lead, but this would disturb the course of the analysis by introducing acetic acid and by otherwise complicating the process. Since the water is to be employed not only cold but also hot, fats and resins present would be liable to undergo alterations, and greater still would be the difficulty with amylaceous substances, since the formation of a paste would make straining impossible, and to overcome this difficulty by means of adding an acid, in order to convert the paste of starch into sugar, would involve troubles of another kind, for instance, the breaking up of glucosids or resins, or the displacement of volatile acids.

Before commencing the analysis the question must be answered, how much of the material has to be worked upon, and how much may be available. If it consists of whole herbs or of parts of these as roots, barks, leaves, flowers or seeds, no more than 100 grammes of the substance should be employed, even if there be no scarcity of material. With very costly or not easily procurable substances a less amount must suffice, and with gummous or resinous exudations or with excretions of other kind it is possible to operate even upon 20 grammes, since the number of constituents of these bodies is comparatively small.

After the analysis, completed according to I. to VII., has given a clear idea of the chemical constitution of the substance under investigation, then those constituents which have been obtained in too small a quantity for a thorough investigation, such as alkaloïds, bitter substances, volatile oils, volatile acids, must be prepared from considerably larger quantities of the raw material, according to the methods indicated under VIII. to X.

## I.—QUANTITATIVE ESTIMATION OF THE WATER.

Weigh off from the substance in the possibly finest state of comminution 2 to 5 grammes, according to its bulk, into a platinum crucible, the weight of which and of its lid having been before carefully determined ; place the crucible uncovered into a metallic air-bath furnished with a centigrade thermometer, put the whole on a little clay-stove, and heat by means of a very small gas-flame or of a spirit-lamp to 120°, and keep at this temperature for about an hour. After a few trials the operator will soon become accustomed to control the temperature at nearly 120°, by placing the flame at the requisite distance from the air-bath; and by using for drying operations of this kind the same stove, the same lamp, and the same length of flame, no new trials for regulating the degree of heat will be required afterwards. A difference of a few degrees above or below 120° is of no consequence; but the mercury should never fall below 115° nor rise above 125°. After the substance has been thus heated for an hour, remove the flame, cover the crucible with the lid, shut the air-bath, take out the crucible when quite cold, place on a balance and determine the loss of weight.

Afterwards the same process of heating and weighing is once more to be repeated; if the second result agrees exactly with the first, or if it only differs by a few milligrammes, no third trial is required; else the heating process has to be repeated a third time or more, until with two following operations the same result be obtained.

Lastly the loss of weight is calculated for 100 parts, and the result is registered as the amount of percentage of water of the substance.

## II.—TREATMENT WITH ETHER.

Place 10 grammes (or less, *see above*) of the air-dried and possibly fine reduced substance into a glass flask of $\frac{1}{2}$ to $\frac{2}{3}$ litres capacity; add of ether of 0·720 until after thoroughly soaking the substance is covered $\frac{1}{2}$ or 1 inch high by the liquid, secure the flask with a cork and shake occasionally, taking care lest portions of the contents should get into the neck of the vessel. Having left it to macerate for four days, lift the cork a little, warm the flask in a water-bath or in any other manner to a temperature below the boiling-heat of ether (36°) for a few hours, remove the mattrass, and again macerate for two days.

Now transfer the contents of the flask (should the ether contain a great quantity of oil, resin, &c., pour out first the liquid so far as it easily can be removed, and after it has passed the filter, add the remaining substance with the aid of small quantities of fresh ether), into a displacement apparatus, rinse the flask repeatedly with ether, and wash the substance with this and with small quantities of ether until the latter assumes not only no colour, but leaves, on evaporating a few drops on a watch-glass, no residue of any consequence.

To push the washing to this degree, requires mostly much ether and also time, but it is necessary in order to prevent (1) that any portions of the constituents, soluble in ether, should be lost;

(2) that any portions of the constituents, soluble in alcohol *and* in ether, should be transferred to the alcoholic solution. The loss of time involved by performing this first part of the operation will be balanced by the simplification of the analysis, and as to the ether, it may easily be recovered by distillation.

Throw the substance, exhausted by ether, into a wide porcelain dish, spread out thinly, expose to the open air, stir assiduously with a porcelain spatula, drive off the rest of ether at a very gentle heat, keep at ordinary temperature for another day, weigh the whole, mix uniformly, estimate with 2 to 5 grammes of the substance, by drying at 120°, the amount of hygroscopic water, and calculate from this in centesimal proportions the weight of the *whole* substance in the anhydrous state. By adding to this weight the amount of water, found under I., and by deducting the sum from 100, the rest will be equal to the weight of the substances dissolved by ether. If, for instance, the amount of water under I. be $10°/_o$, and that of the substance exhausted by ether and in the anhydrous state $78°/_o$, then the ether will have dissolved 12%, because $100 - 10 - 78 = 12$.

The ether of the united tinctures—the colour, taste, and reaction of which have preliminarily to be ascertained—is either distilled off, if it amounts to at least $\frac{1}{2}$ litre, or it is left to evaporate in an airy place, and lastly, with aid of a very gentle heat, in a glass beaker of known weight; and the residue is tested as below.

For recovering the ether choose a tubulated retort, which would hold at least double the quantity, or fill to one-half and add successively of the ether to keep at the same level. The tubulus of this retort must be as wide as possible, and fixed in such a way as to allow the contents to be poured out to the last drop. After filling the retort with the ethereous solution, plunge into the liquid a glass rod, of such a length as to protrude about one inch out of the liquid, place the retort in a water bath, adapt to it a Goebel's refrigerator and distil at least $\frac{7}{8}$ of its contents. Pour what is left in the retort into a glass beaker (weighed), rinse the retort with small portions of the distillate, and keep the beaker with the liquid and a glass rod in it in a moderately warm place. After the ether has nearly evaporated, add to the remnant 10 grammes of distilled water, warm until the ethereous smell has completely disappeared, and let cool.

The contents of the beaker, now, will consist either of two different strata of liquids, *A* and *B*, and of a more or less solid, tough, plastic mass, *C*, or of the latter and only one kind of liquid. In the first case, the upper liquid *A* is a fat, this, for instance, in analysis of seeds largely present; has only one liquid been obtained, then remains all that is said under *A* as a matter of course unnoticed.

*A.—Examination of the upper liquid or of the fatty stratum.*

After the stratum of fat has so far congealed on cooling as to be removable by a spatula, it is thrown into a porcelain dish. Add to it an equal volume of water, warm until fused entirely, stir assiduously, and keep warm half an hour, let cool, perforate the hard layer of fat, pour the submatant aqueous liquid into the beaker containing the lower aqueous stratum *B* and the solid mass *C*, and repeat the operation once more with fresh water, or as often as the water shows an acid reaction.

Should the fat in ordinary temperature or in a cool place prove fluid, or of too soft consistence as to allow its removal by means of the spatula, then the whole mixture—warmed if necessary—is poured into a high, narrow, glass jar. Pour, after the two layers have completely separated, the oil stratum into a porcelain dish—the last portions of it by means of a pipette—pour back the aqueous liquid into the beaker, mix the oil in the dish with an equal volume of water, warm gently under stirring for half an hour, pour back into the jar, separate as before the oil from the water, and repeat, should the water exhibit an acid reaction, the operation a third time.

Now digest the fat, freed from all matters soluble in water, with three times its weight alcohol of 70% (of 0·890 specific gravity), in order to remove any resin present. The result in this case is only approximate, but fortunately the simultaneous occurrence of much fat and of much resin is very rare. Any small portions of resin contained in the fat pass completely into the alcohol of 70%, and traces of fat, dissolved by the latter solvent when hot, separate on cooling. Leave the alcoholic solution in a cold place for one day, filter, decolourise with animal charcoal if necessary, filter again, and bring to dryness at a moderate heat.

A remnant, obtained hereby, is a *resin* (or a glucosid), generally in very small quantity; it is to be tested regarding its external characteristics, as fusibility, solubility in benzol, chloroform, woodspirit, sulphide of carbon, petroleum, oil of turpentine, alkalies, and in concentrated mineral acids. If it is a glucosid it will form sugar, when heated with diluted sulphuric acid (1 part acid and 10 parts water) for a quarter of an hour, and the presence of sugar in the liquid will be recognised by the reduction of alkaline tartarate of copper to red sub-oxyd, when heated together for a short time, and after previous neutralisation with carbonate of baryta and filtration. The sugar may also be recognised, and with greater certainty, by its sweet taste, after the acid liquid has been neutralised by carbonate of baryta, filtered and evaporated to dryness. In this case the properties of the other product, obtained by the breaking up of the glucosid, have also to be determined.

The *fat*, freed from all resin, has now to be weighed, and to be studied in respect to its external characteristics, colour, smell, and taste; of fats solid at ordinary temperature the fusing-point is to determined (*see* Div. II., *C*, *a*), of fats liquid at ordinary temperature the freezing-point, and of both the specific gravities in the liquid state, and the temperature when thus examined ; try also the solubility in ether, alcohol of 100 and of 90 %, benzol, chloroform, wood-spirit, sulphide of carbon, petroleum, and in oil of turpentine at ordinary temperature and when heated; spread also thin layers of the fat on a glass-plate, and see whether it will dry or not after being left for not less than a fortnight at a mean temperature ; and lastly, try its behaviour towards concentrated mineral acids and caustic alkalies.

Should a saponification have been effected by means of any fixed alkali, then from a weighed quantity of the fat a soap is to be formed (with an alkali free from chlorides and nitrates), and to be decomposed by adding an excess of dilute sulphuric acid and digesting at a gentle heat (below the boiling-point of water), then allow the mass to stand cold, perforate the hard layer of fat-acids (*a*), pour off the acid aqueous liquid (*b*), wash repeatedly by adding fresh water, digest, set aside into the cold, perforate, and pour off the water, and examine the fat-acids according to *a*, and the united acid liquids according to *b*.

(*a*) *The mass of fat-acids* obtained, is always a mixture of one liquid and of one or several solid fat-acids. After determining the fusing point of this mass, add one and a-half times its weight pure white of lead, triturated with water to a fine pulp, digest for a few hours under frequent stirring and at about 100°, transfer the lead-soap to a wide-mouthed bottle, add about five times its weight ether, secure the bottle with a cork stopper, shake with care lest any of the contents should get into the neck, filter after 24 hours into a larger bottle, wash the remaining portion of solid fat-acids combined with lead with ether as long as it dissolves anything, mix the united ethereous filtrates with hydrochloric acid and shake for a few minutes. After the chloride of lead has subsided, try if a sample of the ethereous liquid, mixed in a test-tube with sulphuret of hydrogen, and well shaken, will assume a brown or a black colour; if so, add again hydrochloric acid, shake, let subside, and test again with sulphuret of hydrogen. Filter the ethereous liquid, after the whole of the lead has been precipitated, into a glass beaker, and drive off the ether by exposure to the open air and the rest by means of the air-pump.

The fat-acid remaining in the glass beaker is now probably either *oleic acid*, obtained from non-drying oils, or *linoleic acid*, which is the liquid constituent of most drying oils. Should the fat-acid differ from either oleic or linoleic acid, its peculiar properties have to be further investigated; besides, now the atomic weight has to be estimated, and the elementary analysis of the acid is to be performed.

The *estimation of the atomic weight* is accomplished by heating the fat-acid for some time with half its weight of crystallised carbonate of soda, and with five times its weight of water. Evaporate the saponaceous mass until it be equal to double the weight of the fat-acid employed, treat this with alcohol of 70%, filter, precipitate the filtrate with an aqueous solution of acetate of lead, wash the precipitate, consisting of either oleate or linoleate of lead, by decantation, dry at 110°, and incinerate about 1 gramme carefully in a weighed porcelain crucible. Mix the contents of the crucible, now consisting of oxyd of lead and of metallic lead, after every trace of coal has disappeared, with a few drops of nitric acid of 1·200, allow to dry carefully at a gentle heat; raise to a red-heat and weigh.

One hundred parts oleate of lead leave 29·018 oxyd of lead.

One hundred parts linoleate of lead leave 31·472 oxyd of lead.

Decompose the portion of the lead-soap insoluble in ether with diluted hydrochloric acid warm, fuse the fat-acid obtained repeatedly with warm water until free from lead (tested by sulphuret of hydrogen), dry at 100°, and determine the fusing-point. Now, dissolve in five times its weight hot alcohol of 90%, keep cold for two days, collect what has crystallised, press, drive off the last traces of alcohol, and determine again the fusing-point. If this coincides with the one obtained before, then the second mass of fat-acid (the lead-compound of which is insoluble in ether) is a single compound; if, contrarily, the second fusing-point is higher, then the crystals have to be recrystallised until a product of a constant fusing-point be obtained. Now, compare this fusing-point with those of the different solid fat-acids (*lauric, myristic, palmitic, stearic,* and other acids), and see if it agrees with any of those (commonly with palmitic acid), when the identity with the latter will be evident. To make sure it is advisable to determine, at least by one experiment, the atomic weight (*see above*) also, and should this not harmonise, then an elementary analysis has to decide the question. The two latter alterations are indispensable whenever the fusing-point does not coincide with that of any of the fat-acids known as yet. I need scarcely remark that all the other properties of such an acid have also to be investigated.

If the lead-compound, insoluble in ether and separated from the mixture of fat-acids, possesses a lower fusing-point than the portion crystallised from the alcoholic solution, then at least one other fat-acid is present. To isolate it: mix the alcoholic mother-leys, allow to evaporate slowly, to stand cold during the night, collect every morning what has crystallised and determine the fusing-points. Those crystals formed first are likely to belong to the fat-acid already found before, and must therefore be at once removed; those which form afterwards are kept separately, and only such portions the fusing-points of which are the same should be united. Now crystallise the portions of the lowest fusing-point in hot alcohol and proceed as before, *i.e.*, evaporate the mother-ley slowly and examine the crystals obtained each morning. Recrystallise again the portions of the lowest fusing-point until finally an acid is obtained which does not alter its fusing-point on recrystallising. Compare the fusing-point of this second acid likewise with those of the fat-acids known, and if found different from any of them, assume it to be a new acid and determine its properties, chemical composition, and atomic weight.

These operations require very much time and yield a satisfactory result only with an adequate quantity of raw material. Sometimes the separation of the solid fat-acids is rendered quicker and more effectual by converting them, as is done with the liquid fats, into soda-soaps, and by precipitating the alcoholic solution with acetate of magnesia, but in such a manner that (after the quantity of the acetate of magnesia required for precipitating the whole of the soap has been found previously by testing with a small sample of the liquid) the precipitation is effected in at least three distinct equal parts successively; the precipitate obtained from each portion is then washed separately, decomposed with hydrochloric acid, and the fat-acid thus separated is treated as already previously explained.

Sometimes mixtures occur of more than two fat-acids; to analyse which the quantity obtained from 100 grammes of the vegetable substance would prove inadequate, and therefore more of the fat has to be obtained first, or the analysis in this respect can not be proceeded with.

(*b*) Distil the acid liquids, separated from the fat-acids, in a retort furnished with a refrigerator and receiver, until about three-quarters of the whole have passed over. It is the purpose of this operation to find out if any *volatile fat-acids* soluble in water be present (as *formic, acetic, propionic, butyric, valerianic, caproic, caprylic, capric*). If the distillate proves neutral, then no such acid is present; should it show acid reaction, then mix it with some carbonate of baryta previously triturated to an impalpable pulp with a little of the distillate, and evaporate. Should, during the evaporation, the used carbonate of baryta be completely dissolved, then a new quantity ought to be added, for only an excess of it gives the certainty that nothing of the volatile acids is lost. After the liquid has been brought to a certain degree of concentration, and when found of neutral reaction, it has to be filtered from the excess of carbonate of baryta.

The carbonate of baryta remaining on the filter, after it has been washed with warm water, may possibly contain a salt of a new volatile fat-acid, which with baryta might form an insoluble combination. This must be tested preliminarily by heating a sample of the dried remnant in a glass tube closed at the bottom; if it turns black, such a new acid is present. To isolate it, triturate the above remnant with ten times its weight phosphoric acid of 1·08 specific gravity ($10^o/_o$ $PO_5$), place this into a retort, and submit it to distillation, when the volatile acid will pass over, the physical and chemical properties of which must be investigated.

The liquid filtered off from the carbonate of baryta, and the water used for washing it, are now poured off from any crystals which may have formed. Evaporate to half its volume, keep cold for a few days, separate from any crystals, evaporate and repeat these operations several times in order to effect the separation of the three acids—*capric, caprylic,* and *caproic.* These acids are distinguished by the different degree of solubility of their baryta-compounds, for the caprate of baryta dissolves in 200, the caprylate in $106\frac{1}{2}$, and the caproate in $12\frac{1}{2}$ parts cold water. Every crop of crystals is to be collected separately and tested respecting the properties of the respective acid according to the instructions given in the first part of this work. The baryta-compounds of the *other five volatile acids* can not be separated by crystallisation, for their solubility does not much differ (the formate of baryta dissolves in 4, the butyrate in $2\frac{1}{2}$, the valerianate in 2, the acetate and the propionate in about one part cold water). Nor is it possible to effect a separation by means of alcohol, as the latter dissolves them sparingly or not at all. The only course left is to test separately on each of these five acids according to the instructions given in the first part of this work.

Besides these five acids an entirely *new acid* might be present, the properties of which have to be determined by a special investigation.

The weight of the whole of these acids is found by heating to a red heat the baryta-compound, dried at $100^c$, until completely incinerated, by sprinkling with carbonate of ammonia, heating again, weighing the carbonate of baryta, calculating the amount of pure baryta, and by deducting the latter value from the weight of the baryta-compound employed. The rest represents the acid or the mixture of acids. Or precipitate the baryta-compound hot with diluted sulphuric acid, add a little nitric acid, in order to facilitate filtering, collect the sulphate of baryta, determine its weight, calculate from it the pure baryta and deduct its weight as above.

(c) In the saponification of fats *glycerin* is always obtained besides fat-acids. To convince yourself of its presence, saturate the acid contents of the retort, left after the distillation of the volatile fat-acids, with carbonate of soda, evaporate on the water-bath nearly to dryness, triturate the salty mass to a fine powder, shake with absolute alcohol, filter, and let the filtrate evaporate in a glass beaker. A remaining syrup of sweet taste is glycerin (page 94).

*B.—Examination of the lower or aqueous liquid.*

Separate this liquid, mixed with the water used for washing the fat, from the sediment present almost in every case, edulcorate the sediment with small quantities of water (if the sediment be of a viscous nature, under warming, then cooling and pouring off), until the water assumes no longer an acid reaction; concentrate the whole of the liquids to about 100 grammes, pour into a glass jar and leave to rest for a few days. A slight *resinous sediment* will have formed, which has to be filtered off.

After the liquid has been examined in respect to its physical properties (colour, smell, taste, reaction towards litmus-paper), allow a small portion of it to evaporate to a small bulk on a watch-glass of considerable size and transfer to a cold place (in hot weather to a cellar or cooled by ice). If crystals have formed after one to two days, the rest of the liquid is also evaporated and kept cold. Collect the crystals in a filter, rinse with a little water, spread the filter on several thicknesses of blotting-paper, and remove the whole of mother-ley by changing the blotting-paper several times. Afterwards dissolve the crystals in the least quantity of hot water, let cool, collect the crystals formed after one or two days, and dry as before on paper. The substance is now so pure that its properties can be recognised.

As a preliminary treatment this body is to be examined on the presence of nitrogen, by exposing a sample of the size of a lentil with double its weight of sodium, in a dry test-tube to a temperature slowly increasing to a red-heat. After the whole has cooled down again add water, shake, filter, mix the filtrate with a few drops of a stale solution of subsulphate of iron, and, after a good shaking, with hydrochloric acid in excess. In the presence of nitrogen blue flocks will be obtained either immediately or after some time, if not, the substance is free from nitrogen.

(a) *The substance contains nitrogen.* Most likely an *alkaloïd*, combined with an acid, or free.*

Examine the physical properties of the substance, its behaviour in the heat to indifferent solvents (including volatile and fixed oils), to alkalies, to concentrated and diluted acids; the behaviour of its aqueous solution to the hydrates and to the carbonates of alkalies, to the special tests of alkaloïds (tannic acid, chloride of gold, bi-iodide of potassium, subcyanide of potassio-platinum, iodide of potassio-mercury, iodide of potassio-bismuth, phosphate of soda, phospho-molybdate of soda, phospho-tungstate of soda, picric acid, chloride of mercury, chloride of platinum, subnitrate of palladium, nitrate of silver). The solution has also to be tested

---

* That this may happen notwithstanding the acid reaction may be exemplified by caffein, which, though very rich in nitrogen, is of such a weak basicity as to crystallise pure and uncombined from acid solutions.

on sulphuric and hydrochloric acids, as in their presence the alkaloïd would be in combination with any of them.

If the alkaloïd be present in the state of a salt, precipitable by an alkali, and soluble in an excess of the latter, it can only be obtained by cautiously adding of the alkali until the liquid just begins to exhibit an alkaline reaction towards litmus-paper. If, on the contrary, it is precipitated permanently, the same precaution need not be taken, and the alkali may be added in a slight excess. Collect the precipitate, after it has completely subsided, in a filter, wash with water and dry at ordinary temperature or at a very gentle heat. It has to be tested as indicated in the preceding paragraph.

If the alkaloïd be present as a salt and not precipitable by alkalies, it is most effectually isolated in the following way. Convert the chloride into the sulphate by precipitating its solution with just a sufficient quantity of sulphate of silver, filter off the chloride of silver, mix the filtrate intimately with carbonate of baryta equal in weight to the alkaloïd salt employed, digest for one day at ordinary temperature, and filter. If the alkaloïd be soluble in water, it will now be present pure in the filtrate; if it be insoluble in water, it remains in the filter mixed with the sulphate of baryta, from which it may be obtained after drying by means of alcohol.

Should the alkaloïd prove new and unknown, its elementary analysis, determination of atomic weight, and the preparation and examination of some of its salts are required.

If the material should be insufficient for these experiments, the continuation and conclusion must be deferred to sections IX. or X., when larger quantities of raw material are taken in hand.

(*b*) *The substance is non-nitrogenised and is indifferent.* It belongs to this category if, besides being non-nitrogenised, it has no acid taste and no or only slightly acid reaction. Test like the substances under *a*, on its physical properties, behaviour in the heat towards indifferent solvents, to alkalies, to acids, and to tannin.

Of great importance is treating with dilute sulphuric acid, as it shows if the substance is a *glucosid* or not. For this purpose the acid must be diluted with ten times its weight of water. Digest the body with it for an hour at a temperature of about 100°, and let cool down. If the substance remains undissolved, or if, in consequence of the treatment, a new body is formed, ocular inspection will show if it be the original body or a product of it. The whole must be filtered and the contents of the filter washed until all the acid is removed. From the united filtrates or, should the whole have been dissolved, from the liquid as it is, take a small sample and over-saturate with caustic soda, add a few drops of the alkaline copper solution; heat and proceed as indicated under *A* with a resin.

If sugar has been detected by the reduction of copper, the properties and constitution of the other product obtained by the breaking up of the glucosid have to be investigated. For this purpose decompose the greater portion of the remaining glucosid by means of digesting with dilute sulphuric acid, collect what has separated in a filter, wash and dry. Should, on the contrary, the product be soluble in water, the acid liquid is saturated with carbonate of soda and evaporated to dryness. Extract the dry mass with alcohol of 95%, filter off the sulphate of soda and let evaporate. The product of decomposition of the glucosid will separate slowly either in crystals or in a pulverulent or any other form, while the sugar remains in the mother-ley. Should the product be so soluble in alcohol as not to be separable from the sugar by crystallising, it may be obtained by any of the three following methods:—(1) If it be insoluble in pure water, evaporate the alcoholic tincture and remove the sugar by means of water. (2) If it be soluble in pure water, add a little yeast, destroy the sugar by fermentation, filter off the yeast, and evaporate. (3) If it be soluble in ether, shake the dry mass obtained on evaporating the alcoholic solution with ether, and evaporate the solvent.

It may also happen that by treating with dilute sulphuric acid another decomposition takes place with the glucosid (as, for instance, salicin breaks up first into saligenin, and this again becomes decomposed instantly into saliretin). In order to ascertain this, the substance has to be submitted to the action of milder reducing agents, as, for instance, yeast, synaptase (purified emulsin), and the product, if there be one, is compared with the one obtained by means of dilute sulphuric acid. The method adopted for this purpose may be the same as indicated for the production of saligenin from salicin, viz., by digesting the substance with one-tenth synaptase and with water sufficient for a solution for about twelve hours, and at a temperature not exceeding 40°, by shaking with ether and by evaporating the latter. But, should the product be insoluble in ether, chloroform, benzol, or any other liquid must be tried which does not dissolve sugar, and therefore alcohol must be left out of consideration.

(c) *The substance is non-nitrogenised, and is an acid.* It belongs to this category, if it has not only an acid reaction, but also an acid taste—the least frequent of the three cases *a*, *b*, *c*. Should the acid prove quite new or imperfectly investigated, its properties and composition have to be ascertained first. Afterwards saturate the acid with potash, soda, or ammonia, and test the solutions thus obtained with salts of the earthy and of the heavy metals; or if the latter yield no precipitates (by forming soluble compounds with the acid in question) transfer the acid to these other basic bodies, and investigate carefully the salts hereby produced.

(d) *The mother-leys* of *a* or *b* or c, together with the liquids obtained by recrystallising (containing always traces of the crystalline body) or *the original aqueous solution of the ethereous extract* (if no crystalline body has been obtained from it by evaporating) is concentrated to about 50 grammes and tested with the following chemicals. For every experiment only 10 to 15 drops are employed, and the reagent is only added drop by drop.

*Liquor of ammonia.*—It effects either a darker tinge without cloudiness or it produces turbidity. The first reaction is common with all organic coloured matters under the influence of alkalies; the latter indicates with tolerable certainty the presence of an alkaloïd, and if such a body has been obtained previously by crystallisation, and should have proved precipitable by ammonia, the precipitate obtained from the mother-ley is only a rest of it.

If ammonia has effected a precipitate in the sample, the whole liquid is treated in the same way; the precipitate is collected on a filter, washed with water, and examined as under *B, a.* Evaporate the filtered liquid and the water used in washing to the original bulk, in order to drive off the ammonia, and proceed further as below. Proceed likewise and at once if ammonia has produced no turbidity in the sample.

*Carbonate of ammonia.*—All that has been said about liquor of ammonia applies also to this test.

*Leys of potash or of soda.*—They effect usually a darker colouration than ammonia. A turbidity produced by any of them may be an alkaloïd, or lime, or magnesia, inasmuch as small quantities of the salts of these bases pass always into the ethereous extract of vegetable substances. If it be lime or magnesia, the cloudiness does not disappear after shaking the sample with twice its volume of alcohol of 90 or 95 $\%$.

The precipitate may also be a mixture of an alkaloïd and of alkaline earths. In this case its quantity is diminished by the alcohol. To make sure of it, filter the alcoholic liquid, evaporate almost to dryness, redissolve in a little water with aid of one or two drops of hydrochloric acid, and add potash-ley. Should the liquid remain clear now, no alkaloïd was present; if a turbidness is produced, an alkaloïd is present, and this one not precipitable by ammonia (as it would have been indicated before by that same reagent).

If the precipitate produced by the fixed caustic alkali has been proved an alkaloïd or a mixture of it, the whole liquid must be treated with the alkali. Collect the precipitate in a filter, shake with alcohol, if only partly soluble in it, filter, evaporate, and examine as under *B, a.* The liquid separated from the precipitate, and the water used for washing, are saturated with acetic acid, are then concentrated to the former volume and examined as below. If the ley has had no precipitating effect, proceed at once to the next test.

If a peculiar penetrating odour should be evolved by the caustic alkali, a *volatile alkaloïd* may be present; the examination of which must be deferred to section X.

*Carbonate of potash or of soda.*—In the main, all that has been said of the preceding test applies to this also.

*Solution of baryta.*—It causes, like the alkalies, darker tints in the liquids and eventually it precipitates. The precipates may be not only alkaloïds and alkaline earths, but also compounds of the reagent with sulphuric, phosphoric, organic acids, and with acid-like bodies as pigments and resins. A closer investigation of such a precipitate is not advisable here on account of its complex nature.

Of more importance than solution of baryta is *lime-water*. It is similar in its effects to the alkalies, though not so energetic, and produces either a darker colour or turbidness, the latter being also indicative of a series of acids the compounds of which with lime are insoluble or sparingly soluble in water. Consequently a turbidness, produced immediately after adding the test, may prove the presence, irrespective of phosphoric acid, of oxalic, tartaric, tannic, and other acids, and is oxalic acid, if the turbidness does not disappear with acetic acid.—Filter off the precipitate, obtained by lime-water in excess, and heat the filtrate to the boiling point; a cloudiness, which disappears on cooling, indicates citric acid. If neither cold nor hot a turbidity is obtained; acids may be present which yield such compounds with lime as are soluble in water, as for instance malic, quinic, lactic acids, and which will come under notice in the course of the analysis.

*Chloride of calcium.*—A turbidness obtained by it indicates oxalic acid, if not removable by acetic acid. If it dissolves in the latter, it may have been occasioned by phosphoric acid, but in this case the acetic acid has to be added immediately, as else the precipitate would become crystalline and almost insoluble in acetic acid.

*Chloride of iron.*—The alterations effected by this test consist mostly in the production of various colours (with or without turbidness), to recognise which the mostly dark liquid must be diluted so far as to be of only a slightly yellowish colour. A green colour, obtained with the test, indicates *iron-greening*, a blue or violet colour an *iron-blueing tannic acid*. If these acids are not present in too small quantities, coloured precipitates are also obtained, though sometimes of such a mixed colour as to appear brown instead of green, and grey instead of blue.

*Gallic acid* yields, with chloride of iron, a reaction similar to that of *gallotannic acid*, but the *gallate* of iron is abundantly soluble in acetic acid, in the hydrates, and in the carbonates of alkalies, this being not the case with *gallotannate* of iron.

*Sulphate of iron.*—It often produces, with tannic acids, gradations of colour different from those obtained by the chloride of
iron, and has to be used with the same precautions.

*Glue* serves to confirm the presence of the *true tannic acids*, by
producing a dirty, flocky precipitate. Some substances, generally
accepted as iron-greening tannic acids, produce no turbidness
with glue. They are, therefore, either no true tannic acids, or
such as yield with glue compounds soluble in water.

*Tartarated antimony.*—It precipitates some only of the tannic
acids, and serves therefore to distinguish such as differ in this
respect.

*Acetate of lead.*—It produces almost always a flocky, more or less
considerable precipitate, which is sometimes light, sometimes dark,
but in most cases of a dirty brownish or earth-colour, while the
liquid becomes light and eventually clear as water. The precipitate
may contain, besides traces of phosphoric and sulphuric acids, the
*oxalic* and *tannic acids* of the ethereous extracts, small portions of
*resin* and of acids yielding insoluble or sparingly soluble compounds
with lead. No sulphate, phosphate, or oxalate was present in the
precipitate if it dissolves completely in acetic acid.

*Alkaline copper-solution* (sodio-cupric tartarate).—This test requires an alkaline or at least neutral condition of the sample under
trial; an acid condition must therefore be previously removed by
one or two drops of potash or soda-leys. After adding one to two
drops of the reagent, heat to the boiling-point; a yellowish and
afterwards red turbidness of sub-oxyd of copper indicates *sugar;*
but as other matters (tannic, gallic acid, &c.), have a similar
reducing power, this experiment is not decisive for the presence of
sugar, unless corroborated by a sweet taste. But as the latter is
not easily perceived in the presence of other matters which might
have passed into the ethereous extract, and even less so with very
small quantities of it, the presence of sugar cannot here be decided
on, and must be referred to experiment in the course of the
analysis. Should more than mere traces of sugar be contained in
the substance under trial, most of it will pass into the alcoholic
extract (III.)

*Tannic acid.*—It precipitates most of the *alkaloïds*, but also
many indifferent bodies; it therefore serves only to give collateral
evidence for the presence of the one or other of these bodies.

---

After the application of the above tests has proved insufficient
for indicating the presence of an alkaloïd, try again small quantities
of the liquid with the special alkaloïd-tests, mentioned under *B, a*
(page 301), such as yield sparingly or not at all soluble precipitates
with alkaloïds. If the result with all of them be positive, the pre-

sence of an *alkaloïd* need not be doubted any longer ; but even if the result be only positive with some of them, an alkaloïd might be present. The behaviour to caustic alkalies has shown in some degree, if it be volatile, and the closer examination is carried on according to section X. As to non-volatile alkaloïds, see below, under *h*.

(*e*) After having tried the tests under *d*, proceed to the discovery of the *organic acids* contained in the liquid *B*. For this purpose precipitate the remaining aqueous liquid (to which the sample treated with acetate of lead may be added) with acetate of lead in excess, collect the precipitate, after it has subsided, on a filter, and edulcorate with water, as long as it assumes an acid reaction towards litmus-paper. (Testing with sulphuretted hydrogen in order to find out when washing is finished, is inadmissible, because most of these lead-precipitates are not quite insoluble in water). Should the precipitate begin to decompose during washing, (if the water running off becomes suddenly milky), the washing must be finished, even if the acid reaction has not yet disappeared. The filtrate and the water used for washing are reserved for *f*.

From the washed precipitate take with a glass-rod a sample of about the size of two peas, put the rod into a test-tube, add 4 to 5 grams water, and heat to the boiling-point. A clear solution obtained hereby shows that only acids are present, the lead-compounds of which are soluble in hot water, as, for instance, *malic acid*, and the precipitate has then only to be examined under *α*. (If a complete solution can be expected, has already been ascertained by the preliminary reactions under *d*, for a complete solution could not have been effected in the presence of sulphuric, phosphoric, oxalic, tannic, citric, and of many other acids.) If, on the contrary, the liquid remains turbid, it is filtered boiling hot, without washing, and the filtrate is left to cool. If no sediment (crystalline or pulverulent) is obtained, the liquid is concentrated and again left to stand cold. If no sediment has formed now, proceed to *B*, since the precipitate does not contain any organic acids such as yield lead-compounds soluble in hot water.

On the other hand, if a sediment has been obtained in the above way, the whole precipitate is washed off the filter into a spacious porcelain-dish under addition of water ten times the volume of the moist precipitate. Heat under continual stirring with a glass-rod, or with a porcelain-spatula, for a quarter of an hour; filter hot and wash with hot water. The filtrate is now examined under *α*, and the insoluble remnant under *β*.

*α*. Free the aqueous solution from the sediment produced in the cold, evaporate to a small bulk, collect the solids which may have separated on a filter, wash with a little cold water and dry. The mere external appearance of the substance will teach already to some extent, whether it is of a complex constitution or not. In

x 2

the latter and more frequent case it may be *malate* of lead, a supposition which will be confirmed by incineration (page 298), when malate of lead will leave $56.91\%$ oxyd of lead. Should the result be at variance with this supposition, and even be confirmed by a second incineration, the elementary analysis is resorted to and the properties of the acid must be investigated. To perform this, triturate a portion of the lead-compound with water to an impalpable pulp, wash it into a glass-jar, impregnate the milky liquid under continual stirring with well-washed sulphuret of hydrogen, allow the excess of the latter to evaporate at the air, filter off the sulphide of lead, after it has completely subsided (which takes sometimes a very long time, and may be accelerated by gently heating), evaporate a part of the filtrate at a very gentle heat (to prevent etherification of the alcohol) and at last under the receiver of the air-pump, and employ the other part of the filtrate for testing with solution of baryta, lime-water, &c., and for the preparation of some other salts, while the dry acid is tried on its behaviour in the heat, &c.

If there is reason to suppose that the water has withdrawn more than one compound from the lead precipitate, these compounds must be separated from each other by crystallization, and examined separately in the above manner.

$\beta$. The small portion of the *lead precipitate, left undissolved by hot water*, is tried moist by shaking with acetic acid of about $20\%$. If it prove *soluble* and completely so when heated, the further examination is confined to No. 1. If it remains *turbid*, it must be filtered and saturated with ammonia; should it now remain clear, then nothing has been dissolved by the acetic acid, and it has then to be examined under No. 2.

If, on the other hand, a turbidity has been produced by ammonia in the acid filtrate, the whole of the precipitate is transferred from the filter into a beaker by means of a horn-knife. Then mix with acetic acid to a thin pulp, cover the vessel with a glass plate, agitate the contents assiduously and apply, if necessary, a gentle heat. Filter after about an hour and wash with water until the latter passes off nearly void of acid reaction, and examine separately the *solution in acetic acid* under 1, and the *undissolved remnant* under 2.

1. Add to the *solution in acetic acid*, carefully and under continual stirring, ammonia in such a quantity as to leave the acid very slightly in excess (an excess of ammonia must be corrected eventually by a few drops of acetic acid.) This causes the part $(\beta)$ of the precipitate, dissolved in the acid, to separate again. Let subside, collect on a filter, wash completely with water, spread half of the precipitate thinly on a glass or porcelain plate, dry at the air or at a very gentle heat, triturate and reserve for further use.

Suspend in a glass-jar the other half of the washed precipitate in about five times its volume of absolute alcohol, treat with sulphuret of hydrogen in excess, allow the latter to evaporate at the open air, filter off the sulphide of lead, and evaporate the liquid with only a gentle heat, and at last in the vacuum. After the alcohol has been driven off, dilute with water half of the remaining liquid, to be used for testing purposes, and evaporate the other half to *dryness*.

The *dried remnant*, if consisting only of a yellowish varnish of a pure, astringent taste (often followed by a slightly bitter flavour), is most likely only a *tannic acid*, which then has to be characterised in its aqueous solution by the proper tests of the preceding paragraph (chloride of iron, glue, tartarated antimony). The dry lead-compound serves for the elementary analysis and for the estimation of the atomic weight.

In the absence of tannic acids, the dried body appears often in crystals and with a more or less acid taste, and has then to be tested on *citric acid*, &c., by means of lime-water. Acids of a peculiar odour, as *benzoic acid, cinnamic acid,* &c., betray themselves to some extent by this property.

A taste, at once astringent and acid, and a not entirely amorphous state, indicate a mixture of tannic with one or more acids of a different kind. A thorough examination of the latter is only possible, after the tannic acid has been removed. This is done most effectually and without fear of contamination by cutting isinglass into short, narrow strips, leaving these to soak in a porcelain-dish with water, until converted into a swollen-up jelly-like mass, adding the acid liquid in question and keeping the whole at ordinary temperature and under stirring, until a sample of the liquid on examination proves free from tannic acid. The complete absorption of the tannic acid by isinglass is effected slowly and may take several days. Heat must be avoided entirely, as likely to convert the glue into its soluble modification. After the process is finished, the liquid has to be filtered, and contains now the other acid or acids, the nature of which has to be examined. If necessary, the dry part of the lead-precipitate is also decomposed in the same way, in order to obtain more of the acid, and this ought to be done in all cases where two or more require investigation.

2. The portion of the lead-precipitate which remains undissolved after the successive treatment with hot water and acetic acid (or the whole lead-precipitate, if not affected by hot water or by acetic acid) may contain of mineral acids, phosphoric or sulphuric; of organic acids, oxalic acid, and appears, as a rule, of a grey-brown colour, from traces of humus-like or of other colouring matters.

To test on oxalic acid, heat the moist precipitate to the boiling-point with a solution of carbonate of soda, boil for a quarter to half an hour, filter, saturate the filtrate with acetic acid in excess, filter again if necessary, and add lime-water or acetate of lime. In the presence of oxalic acid a considerable turbidity is immediately produced. To determine the quantity of oxalic acid, precipitate the acid liquid with a sufficient quantity of acetate of lime and convert the precipitate by heat into carbonate of lime. One hundred parts carbonate of lime represent 72 parts oxalic acid.

(*f*) Add to the liquid, separated by filtering from the lead precipitate under *e*, and to the first portions of water used for washing the same, slowly and under continual stirring, liquor of ammonia, to a very slightly acid condition. (Any excess of ammonia must be again corrected by acetic acid). The precipitate obtained hereby is of less bulk than the first, but lighter and usually of pale-yellow colour. It contains no sulphuric, phosphoric, or oxalic acids, but may contain either traces of other acids precipitable in acid solutions by acetate of lead, or acids precipitable by acetate of lead only from neutral solutions; or acids of both kinds. Its examination may therefore serve either for completing the analysis of that portion of the lead-precipitate which is soluble in acetic acid (*e*, *β*, 1), or for the discovery of new acids. But, before proceeding any further, add a little acetate of lead, until a precipitate is no longer produced.

Collect the precipitate after subsiding on a filter, wash, spread a part of it on glass or porcelain and let dry; suspend the other part in absolute alcohol and decompose with sulphuret of hydrogen. After the sulphide of lead has subsided, and when the liquid has become clear and has lost the smell of sulphuret of hydrogen, it is filtered and the filtrate is cautiously evaporated. Test the remaining liquid, freed from alcohol, on such acids as have been found under *e*, *β*, 1, and observe any discrepancies. Should the precipitate contain only one acid, the dried portion of it serves for determining its constitution.

(*g*) Mix the liquid, separated from the precipitate, with the first portion of the water used for washing the same, and add subacetate of lead. The white or yellowish-white precipitate obtained thereby contains no tannic acids, and probably of other acids only a few; no gum or sugar, which, though also precipitable by subacetate of lead, do not pass into the ethereous extract, or only to a very slight extent.

Let the above precipitate subside, collect in a filter, and wash out. The washing has to be interrupted as soon as decomposition sets in; which is recognised by the milky appearance of the water passing through the filter. Now, suspend the precipitate in a glass-jar in about ten times its volume of water, decompose with sulphuret of hydrogen, keep the whole at the air until the excess of the gas has disappeared and the liquid has become clear, filter, concentrate the filtrate, try its physical and chemical properties, bring slowly to dryness, and in the first place see if it agrees in properties with *quinic acid* (page 183).

(*h*) Free the liquid separated from the precipitate *g* from oxyd of lead by precipitating with carbonate of ammonia, and evaporate to a syrup over a water-bath, in order to drive off most of the acetic acid and of the ammonia. Of acids this syrup may contain especially lactic and quinic acids, but probably, besides, matters of an *alkaloïdal*, or of an indifferent nature (*bitter substances, sugar*).

To recognise *lactic acid*, mix a part of the syrup with an equal volume of a cold saturated solution of acetate of zinc, and allow the mixture to stand in a cool place for one or two days. If no crystals have formed after this period, no lactic acid is present; in the other case, a crystalline mass, usually in the form of a crust and consisting of lactate of zinc, is obtained, which is further to be examined regarding its amount of water of crystallisation and of oxyd of zinc (*see* Lactic acid, p. 116).

*Quinic acid* is recognised by boiling a large quantity of the syrup with milk of lime until the ammonia is driven off, and then proceeding as indicated under Quinic acid (p. 183).

Should the syrup have a bitter taste, try if its aqueous solution with tannic acid will produce a precipitate. (This has been tried already on page 301, but with a weaker solution, and may with a better prospect of success be repeated now on a more concentrated solution). Should a precipitate arise in the sample by tannic acid, then the whole rest of the syrup is precipitated likewise. Wash the precipitate as well as possible (p. 288), mix well with oxyd of lead or with white of lead, dry in a gentle heat and extract with alcohol. The tannic acid remains in combination with oxyd of lead, while the respective substance passes pure into the alcohol, remains after the evaporation of the solvent, and may then be examined regarding its properties.

On the contrary, if no turbidity is produced by tannic acid, digest the aqueous solution of the syrup with a great quantity of animal charcoal over the water-bath for several hours, collect the coal on a filter, wash with cold water, transfer it moist, but freed from superfluous water, into ten times the weight of the dry coal employed, alcohol of 95%, heat and boil for a quarter of an hour; filter hot, wash with hot alcohol, and evaporate the united alcoholic filtrates with a very gentle heat. Should an amorphous mass remain without any signs of crystallisation, try if by treating with ether a separation, purification, &c., can be effected. The closer investigation will show whether the substance is of a basic or of an indifferent nature, and it must be treated accordingly as indicated before.

Keep the syrup for at least a week in the cold. Anything which will have separated after this period, either of a pulverulent or of a granular or of a crystalline form, is collected in a funnel, the neck of which is loosely covered by a glass stopper. Wash the sediment, after the liquid has run off, with a little cold water, and submit it to a series of experiments comprising its basic as well as its acid constituents. It will prove most probably a salt, and perhaps acetate of ammonia, the two constituents of which have been supplied by the testing chemicals.

### C. —Examination of the solid mass.

It may contain *wax, resin, fat, chlorophyll*. After the water has been removed by gently heating for some time, determine its weight and note down its external characters. Next let three times its weight alcohol of $70\,^{\circ}/_{\circ}$ act on it warm for half an hour, keep cold for one day, filter and wash the remaining part with a little alcohol of the same strength as before.

(*a*) Shake a sample of the filtrate, which may be possibly coloured by chlorophyll, with animal charcoal for some time and treat, if the colour has become lighter, the whole filtrate in the same way, filter and evaporate. The remnant is a *resin*, and has to be tested as indicated on page 296.

(*b*) Free the undissolved portion of *C* from alcohol at a gentle heat, weigh again (to ascertain the weight of the dissolved resin), and treat with ten times its weight alcohol of $90\,^{\circ}/_{\circ}$ hot. Usually a complete solution will be effected; any remaining body is *fat—fat-oil*—which is rinsed repeatedly with small quantities of warm alcohol of $90\,^{\circ}/_{\circ}$.

*a.* The undissolved *fat-oil* is generally of little amount, and allows only a few experiments concerning its taste and smell, its capability of drying when exposed in thin layers to the air, and its solubility in different solvents. If an oil has been obtained already under *A*, the oil of *C* is always identical with it and needs no further examination.

*β.* The hot filtered alcoholic solution of the substance *C* deposits most of the *wax* on cooling. Let the whole stand in the cold for one day, collect the deposit on a weighed filter, wash with cold alcohol of $90^{\circ}/_{\circ}$, dry at a mean temperature, weigh and determine, as far as possible, its physical properties including fusing-point. Generally, the quantity obtained is insufficient for a thorough investigation, but if there be a sufficiency of the wax, the latter is purified by repeatedly dissolving in hot alcohol; afterwards its constitution is determined by treating with potash-ley, or if necessary, by fusing it with caustic potash, and by examining the products obtained.

*γ.* The alcoholic liquid, left after the separation of the wax, may retain only traces of fat and wax, and deserves no further consideration.

### III.—Treatment with Alcohol.

Return the substance, exhausted by ether, to the dry flask used before, pour on it as much alcohol of $95°/_o$ as will cover the well-soaked substance one-half to one inch high, secure the flask with a cork, perforated lengthwise by a canal of about 1 millimeter in width, digest for three days at a temperature not exceeding 70°, let stand cold for one day, transfer the contents to a displacement apparatus, and wash thoroughly with alcohol of the same strength. But, should the alcohol have formed a deposit on the top of the substance, after having been kept cold for one day, then the flask must be warmed for half-an-hour before filtering and the washing is performed also with warm alcohol.

The contents of the displacement apparatus, after being washed, are spread out thinly in a shallow porcelain-dish and dried, at last with a very gentle heat. Weigh the substance, after it has been left cold for one day, determine the amount of hygroscopic water of a sample of 2 or 5 grams and calculate from this the weight of the whole substance in the dry state. The latter value deducted from that found under II. represents the weight of the portion dissolved by alcohol.

Concentrate the solution under addition of the alcohol, used for washing the insoluble part, by evaporation in a weighed beaker at the open air, and finally at a gentle heat. Larger quantities, consisting of more than half a litre, are distilled in a retort until seven-eighths have passed over. Then pour the remnant into a beaker, rinse the retort a few times with a little alcohol, and gently heat the solution. Mix the residue, after the odour of the alcohol has almost disappeared, with 10 grams water, heat until the last traces of alcohol are driven off, and let cool.

The contents of the beaker will now consist of a liquid and of a sediment, which in some cases may be very slight or even wanting altogether. Before separating the solid from the liquid portion, try if an addition of more water will produce any turbidity. If this should be the case—which generally takes place in the presence of much resinous matter—the liquid must be mixed with an equal volume of water, warmed for a few minutes and left to cool. It is afterwards tried again with more water, and, if it becomes turbid again, the same process is repeated as before, and so on, until a liquid be obtained which remains clear by mixing with water. The liquid, together with the whole sediment, is now brought on a weighed filter, washed thoroughly with cold water, and left to dry; while the solution is examined as follows:—

*A.—Examination of the aqueous solution of the alcoholic extract.*

Try the united filtrates in regard to taste, odour, colour, &c., and reaction towards litmus-paper, evaporate a small sample to a small bulk on a watch-glass, and keep cold, in order to facilitate the formation into crystals. (In concentrating the liquid, traces of resin separate eventually from which the liquid is previously freed by decanting or filtering). In case of any crystals having formed, the whole filtrate is concentrated and kept cold for a few days (in summer time in the cellar, and, better still, surrounded by ice). Proceed with the crystals obtained, as indicated under II, *B.*

### *B.—Examination of the portion of the alcoholic extract insoluble in water.*

Wax, fat and chlorophyll, having passed into the ethereous solution, the examination, in this case, is entirely confined to resins, a considerable number of which are insoluble in ether, but soluble in alcohol. If the vegetable substance under trial has been rich in resins, most of them, as a rule, will be found in this part of the extract.

After noting down the physical properties of the substance and its behaviour in the heat, try if you can free it from any dyeing matters, by dissolving a little of it in alcohol under addition of animal charcoal, and digesting for some time. If no alteration of the colour is observable, the colour belongs to the resin itself and cannot be destroyed by this means; in the other case the whole substance is treated in the same manner and again brought to dryness.

Whether the resin, so obtained and eventually decolourised, be a mixture of several resins or not, is now determined by treating successively with cold as well as boiling alcohol of 70, 80, 90, 95, and 100 °/₀, and by comparing with each other the single portions, obtained after evaporating.

Every individual resin is afterwards examined in regard to its fusing-point, its solubility (in amylic alcohol, benzol, chloroform, sulphide of carbon, petroleum, oil of turpentine, concentrated sulphuric acid), elementary composition, atomic weight (calculated from the compounds with oxyd of lead), and by digesting it with dilute sulphuric acid, in order to ascertain if it be a glucosid or not.

### IV.—TREATMENT WITH COLD WATER.

Re-transfer the substance, exhausted by ether and alcohol, to the flask, as before, add distilled water sufficient to form a thin pulp, and keep at the ordinary temperature and under assiduous shaking for six days. In hot weather the flask with its contents, in order to prevent fermentation, is kept either in a cool cellar or in water kept cold by artificial means. After this, expose the whole to a temperature not exceeding 30° for one day, and filter in a displacement apparatus.

This latter operation, though comparatively easy with ethereous and alcoholic liquids, often engenders great difficulties with aqueous extracts on account of the presence of colloïdal matters, as albumin and gum. It is therefore better to first try a small quantity, and if filtering is found practicable, to give the whole into the displacement apparatus and to wash with water, until the latter passes off clear. Any decomposing effect of heat has to be avoided by performing the operation in a cool place.

If, on the contrary, filtering through paper should prove incomplete and slow, it is necessary to strain the mass through miller's gauze (silk bolting cloth), not through linen, calico, or flannel, because the meshes of these latter fabrics would soon become filled up by swelling, and so prevent the liquid from passing through. Spread a circular piece of the silk gauze (preferably of No. 10 to 13) over a wide glass jar, press it in the middle slightly into the vessel, secure the part of the gauze resting on the edges by means of a wide porcelain ring, pour the contents of the flask into the bag-like cavity, and rinse the flask with a little water. After the liquid has ceased dripping, take hold of the cloth, tie it together close over the contents with pack-thread, and press out as much liquid as possible. Or use a small bookbinder's press, the side wedges (those parts that come in contact with the liquid) of which are covered with glass or porcelain plates. Again, mix the pressed-out substance in a porcelain dish with water to a pulp, strain and press, and repeat these operations once more. Mix the whole of the liquids well together, and allow them to subside in a cool place.

An amount of starch or inulin, surpassing mere traces, may be separated from the vegetable substance in a merely mechanical way, i.e., by kneading. A partial separation is even effected in straining through gauze, but may be considerably improved in case of any starch or inulin having been observed in the first liquid by kneading the substance after the second pressing, and after the gauze-bag containing it has been tied as before, under water and for a considerable time, adding the milky water to the other liquids and kneading again under fresh water until the substance is entirely exhausted. From the united liquids, starch or inulin subside gradually. Filter, or, if impracticable, decant the clear liquid, collect the sediment of starch or inulin on a filter, dry at 110°, and weigh.

By filtering the aqueous extract in the displacement apparatus, starch or inulin are, of course, left in the remnant. The presence of starch is easily ascertained by shaking a small quantity of the substance with solution of iodine, when it will assume a violet or blue colour. The presence of inulin can only be ascertained by kneading. Starch and inulin have, as yet, not been observed in the some plant; and, whereas the former occurs in widely different plants, the latter has been found as yet only in the order of Compositæ, at least to any considerable extent.

Starch or inulin having been observed in the substance left after filtering and washing, this substance is tied up in gauze and kneaded. Collect the white powder deposited from the united liquid on a filter, dry at 110°, and weigh it. The liquid is thrown away, as the substance before kneaded had been exhausted already by cold water.

In order to find out the form and size of the starch or inulin obtained, they have to be examined under a microscope magnifying at least 400 diameters; after this, try their behaviour to warm water in order to determine their solubility or the temperature required for forming a paste. Any heterogeneous substance that has been kneaded out together with starch or inulin will remain on dissolving in water; as, for instance, oxalate of lime, which, on burning, yields carbonate of lime, the weight of which gives the weight of the oxalate of lime according to 625 $(Ca\ O + CO_2)$ $=1025\ (Ca\ O + C_2\ O_3 + {}_2HO)$. The weight of the oxalate of lime found in this way must be subtracted from that of the starch or inulin.

After the substance has been exhausted in the above manner by cold and lukewarm water, and eventually by kneading, determine its weight whilst still moist. Now take 5 grams of it, dry at 100°, weigh, calculate from it the weight of the whole residue, conceived dry, and subtract the latter weight from that of the substance, exhausted by ether and alcohol. The rest represents the weight of the substances including eventually starch or inulin; the weight of which, as found above, has to be subtracted from the rest in order to give the weight of the matters soluble in water.

Heat the aqueous solution, obtained clear by filtering and subsiding, in a porcelain dish to the boiling-point; boil for about a quarter of an hour and allow to cool. Any turbidness or flocky precipitate is produced by albumin. Let subside, collect on a weighed filter, dry at 110°, and weigh. The albumin obtained is mostly of a dark colour, and, consequently, impure; but the colouring matter is hardly removable and is of too little amount to impair the numerical result.

Concentrate the filtered liquid to about 100 grams at a temperature not exceeding 70° to 80° (if, as usually, a little albumin should separate again, it must be collected and added to that obtained before); evaporate a small sample of it on a watch-glass to a thicker consistence, and allow both portions to stand cold. In case of crystals having formed in either of these liquids—a rare occurrence—evaporate the whole liquid to a small bulk, and keep in the cold. Collect what will have crystallised on a filter, wash with a little cold water, purify by recrystallising in hot water, and examine according to II., *B*, *a* or *b* or *c*.

The remaining liquid, mixed with the water that has been used for washing and evaporated again to the original weight of 100 grams, or the original aqueous solution reduced to the same weight—after no crystals have been obtained from it—is examined in small quantities with the reagents indicated under II., *B, d.*

After these experiments have been made, precipitate and treat the remaining liquid as under II., *B, e, f, g*, successively with acetate of lead, ammonia and subacetate of lead, and examine the precipitates as indicated there.

The precipitate (*e*), obtained by acetate of lead, may in this case contain, besides the substances indicated there, gum and a protein-substance not identical with albumin. These substances will remain with the sulphide of lead as insoluble in alcohol. To ascertain their presence, wash the sulphide of lead well with alcohol, dry at 100°, triturate and shake with cold water. This dissolves the gum readily, and leaves behind the protein-substance rendered insoluble in water by drying and heating. The presence of gum in the cold water will be recognised by its forming on evaporation an amorphous tasteless varnish.

After this, digest the sulphide of lead with phosphoric acid of 1·04, warm, filter, edulcorate the remaining sulphide of lead with water, and neutralise the whole filtrate carefully by means of an alkali. The protein-substance is thrown down; collect it on a filter, wash, dry at 110°, and determine its weight. The quantity of the substance obtained hereby is insufficient for a closer investigation.

If no gum or protein-substance has been found in the lead-precipitate *e*, the lead-precipitates *f* and *g* have to be tried for it, and this is done with the precipitate *f* in the same manner as with precipitate *e*, whereas the precipitate *g*, its decomposition not being effected under alcohol but under water, is treated in the following way:—Boil the liquid, filtered off from the sulphide of lead and containing the gum, and probably the protein-substance, for a few minutes, when the protein-substance, if present, will separate; filter, evaporate to a small bulk, and shake a sample of the liquid with alcohol of 90%. If the mixture remains clear, no gum or protein-substance is present; if it becomes turbid, precipitate the whole liquid with alcohol, wash the sediment with alcohol, dry at 100°, and treat with cold water. A clear solution contains only gum; any insoluble portion is protein-substance. As the sulphide of lead might in this case as well contain protein-substance, it must be treated as above, with phosphoric acid, &c.

The liquid, after separating from the precipitate, produced by lead-vinegar, is treated according to II., *B, h.*

## V.—Treatment with Boiling Water.

Mix the moist substance, exhausted by ether, alcohol and cold water, in a deep porcelain dish with so much water as to form a very thin pulp, heat under continual stirring to the boiling-point, let boil for an hour, filter, or if necessary, strain through gauze in the manner indicated under IV., and wash the remnant with boiling water.

Determine the weight of the remaining substance in the supposed dry state according to IV., and subtract from the weight of the substance found under IV., the rest representing the weight of the matters dissolved in the boiling water.

Concentrate the united liquids to about 100 grams, and try in small portions with the tests indicated under II., B, d; in most cases slight or no reactions will be obtained.

If no inulin has been obtained under IV., examine on starch, by mixing a portion of the liquid with solution of iodine. A violet or blue colouration indicates starch, which is either a remnant from that obtained by kneading under IV., or it represents, if no starch has been obtained there, the whole amount of starch contained in the vegetable substance.

If no reaction of starch is obtained, a small amount of inulin might have been present in the vegetable substance, but is now converted into gum by the boiling water.

Any precipitate, produced by lead-salts, must be examined according to II., III. and IV.

## VI.—Treatment with Diluted Hydrochloric Acid.

The substance exhausted by ether, alcohol and water, contains as a rule only constituents of minor interest, and in most cases the analysis may now be considered as finished. Considering, however, that the object of this work is to investigate every constituent of vegetables, though it be of minor interest, I now resort to those solving agents which, if employed in the beginning, would have caused alterations and decompositions highly detrimental to the analysis, but which now may be used with comparative impunity.

Give the substance left in V., and still moist, into a beaker (weighed before), make up with water to 500 grams, add 20 grams pure hydrochloric acid of 1.12 specific gravity, cover the beaker with a glass-plate and expose to a moderate heat for two days. Filter and wash the substance on a filter with water, until it passes off free from acid reaction.

After washing the substance, determine the weight of the dry substance as under IV., subtract from the weight under V. and note down the rest as the weight of the matters dissolved in the acid water.

(*a*) Saturate the acid extract, mixed with the water used for washing, with ammonia in excess. Generally only a darker colour is produced; should a turbidness be obtained, try, with a small quantity, if it becomes clear again by immediately over-saturating with acetic acid.* If this is not the case, oxalate of lime is present. Allow the liquid to become perfectly clear—sometimes (with barks, &c.) a jelly-like mass of pectic acid is obtained by over saturating with ammonia, which prevents the precipitate from subsiding, but may easily be dissolved in a gentle heat—collect the precipitate on a filter, wash with water and redissolve in dilute hydrochloric acid. Over-saturate the solution, if necessary after filtering, with ammonia and immediately afterwards with acetic acid, collect the deposit of pure oxalate of lime, dry at 100° and weigh it. This lime-compound contains, if no oxalic acid has been previously found, the whole of the oxalic acid contained in the original vegetable substance. In 100 parts of it $= Ca\,O + C_2 O_3 + 2\,H\,O$ are contained 43.90 parts oxalic acid. The oxalate of lime may also be converted into carbonate of lime by heating to a moderate red heat in an open crucible, and the oxalic acid may be calculated from the latter salt; 100 parts $Ca\,O + C\,O_2$ correspond to 72 parts oxalic acid.

(*b*) Unite the liquid that has been separated from the precipitate effected by ammonia, with the liquid that has been separated from the pure oxalate of lime, over-saturate the mixture, if necessary, with ammonia, allow the precipitate to subside, collect on a filter, wash and dry. It consists chiefly of phosphate of lime and phosphate of ammonio-magnesia, but may eventually contain traces of alkaloïds, met with in the preceding sections, or possibly an alkaloïd that has not been affected by extracting with ether, alcohol and water. Submit it therefore to the influence of warm alcohol of 90%, filter and evaporate the filtrate. Any residue, obtained hereby, has to be examined according to II., *B, a*.

(*c*) Evaporate the liquid, left in *b*, to about 100 grams, in order to remove the excess of ammonia and of water, and try with a few sensitive alkaloïd tests which are without influence on ammonia-salts, viz., with tannic acid, chloride of gold, bi-iodide of potassium, iodide of potassio-mercury, iodide of potassio-bismuth, phosphate of soda, chloride of mercury, nitrate of palladium. In case of any turbidness, obtained, an alkaloïd may be present.

Usually, for economy's sake, tannic acid is employed for precipitating the whole liquid, if an alkaloïd is supposed to be present. Triturate the precipitate, obtained by tannic acid or by any other of the above tests, after washing and while still moist, intimately

---

* The acetic acid must be added immediately after the ammonia, because phosphate of lime and phosphate of ammonio-magnesia are only soluble in acetic acid when newly precipitated.

with an excess of burnt magnesia (or finely pulverised oxyd of lead or white of lead), dry in a gentle heat, pulverise, treat with cold water (in order to remove alkali-salts, chloride of magnesium, iodide of potassium, &c.), boil with alcohol of $90\%$, evaporate the alcoholic liquid and submit to examination what has remained.

(d) The remaining liquid may probably contain traces of dyeing matters, of gum and of resin, the further examination of which is perfectly valueless and involves only loss of time.

## VII.—TREATMENT WITH DILUTED POTASH-LEY.

Bring the substance remaining after the treatment with hydrochloric acid, and while still in a moist state, back into the beaker, add water enough to make the whole up to 500 grams, dissolve in the mixture 10 grams hydrate of potash (or of soda), digest for two days in a temperature of $50°$ to $60°$, let cool, filter and wash the residue thoroughly.

Dry 5 grams of the residue at $110°$, weigh, calculate the weight of the whole residue conceived dry, and subtract this weight from the one obtained under VI., the rest representing the weight of the matters dissolved by the ley.

Should the mass prove too thick for filtering, it must be diluted previously, and is, if necessary, clarified by subsiding.

(a) The alkaline liquid is generally more or less brown, sometimes black-brown. By over-saturating with hydrochloric acid, a brown, flocky precipitate is invariably obtained, which contains so-called humic acids, but which may also contain protein-substances. The two substances are only incompletely separable by liquor of ammonia, the humic acids being more soluble in it than the protein-substances. For the quantitative estimation of the two bodies it is sufficient to collect the precipitate produced by hydrochloric acid, to wash, to dry at $120°$, and to determine the amount of nitrogen by heating with soda-lime. By multiplying with $6\frac{1}{2}$ the weight of the nitrogen obtained, the quantity of the protein-substances is determined, and by subtracting the latter weight from that of the precipitate, the weight of the humic acids is obtained.

(b) The liquid obtained by filtering in a, contains small quantities of humic acids as well as of protein-substances, the closer investigation of which is impracticable.

(c) The residue left after the treatment with potash-ley, and the dry weight of which has been calculated, is generally accounted for as vegetable fibre or cellulose, though this name is not correct, as this fibre contains not only colouring matters but also mineral compounds. Regarding the complete removal of the first and the estimation of the latter impurities, see under Fibre, p. 82.

## VIII.—DISTILLATION WITH WATER.

In the preceding analytical course the volatile oils have found no place, though their presence might have been conspicuous sometimes, especially in the ethereous extract. For their production and examination another sample of the substance must be employed, but not less than 1 kilogram, or much more on account of the comparatively small amount of volatile oil contained in most plants.

Besides volatile oils there are other ingredients of plants that have been obtained in the preceding analysis in too small a quantity or not at all, and the discovery of which becomes easier with a larger amount of raw material.

The production of these two classes of bodies may be combined in the following way:—

(*a*) Macerate the substance with water for one or two days and distil, preferably on the water-bath and by means of steam, in a Beindorf's apparatus, described on p. 277, and provided with a good refrigerator. From what will be seen afterwards the weight of the empty tin still (without head or other accessories) must be determined and engraved on the tin. As recipient, a Florentine flask or a similar contrivance is employed.

The distilling water dissolves invariably certain quantities of oil. Either more oil is present than requisite for saturating the water, or there is less of it. In the first case the oil separates on or under the water, according to its density. To clarify the at first usually turbid oil allow to stand undisturbed for a few days, and prevent the solidifying of some oils by keeping them in a moderately warm place. After it has become clear, and if it floats on the water, remove it by means of a thin cotton wick, used as a syphon; if it be heavier than water, pour off the latter and remove the rest by means of the same syphon.

If no volatile oil has been separated in the distillate, *i.e.*, if the water has not been saturated with it, a saturation or separation may be obtained in most cases by submitting the distillate to another distillation with a fresh quantity of raw material so as to increase the amount of oil in the same quantity of water. Sometimes this process (cohobation) must be repeated a third time. Should it be desirable to obtain the oil without distilling two or three times, the aqueous distillate must be shaken with one-fifth its volume of ether for half-an-hour. Allow the mixture to become clear, pour off the ether and let evaporate spontaneously in a beaker. The volatile oil, mixed with a few drops of water, remains, though only in very small quantity.

In order to enable an investigation of more than the external characters of a volatile oil (colour, smell, taste, density) at least 50

Y

grams of it are required. The examination has to be conducted as indicated under Essential Oils, p. 77.

(*b*) The aqueous distillate contains, besides small quantities of volatile oil, almost invariably some volatile organic acid, and has therefore an acid reaction. If the examination of these acids be required the distillate must not be treated with ether, as the latter removes both oil and acid.

Test first on hydrocyanic acid, and employ in case of its presence a weighed quantity of the distillate for its quantitative estimation. Both is done according to the instruction given under "Hydrocyanic acid," p. 107.

Now mix the rest of the distillate with carbonate of baryta, rubbed down so as to form a fine milk, and evaporate on the water-bath, driving away together with the water any hydrocyanic acid present. After the liquid has been reduced to 50 grams, filter off the excess of carbonate of baryta and evaporate slowly to dryness. As the residue will be very small it can only be examined in regard to the more common volatile acids occurring in distillates, as formic acid, acetic acid, and as regards the lower members of the series of fat-acids, as propionic, butyric, valerianic acids, while a closer investigation is deferred to section IX.

After the distillation has been finished, remove the tube conducting the steam, the head and the false bottom of the still, place the latter on the balance and determine the weight of the water by subtracting from the whole weight the weight of the still and of the raw material employed, add as much pure water as to make the whole up to twice the quantity of the raw material, add alcohol of 90%, equal in weight to the whole of the water, mix the whole assiduously with a spatula of beech-wood, replace the still once more into the hot water of the boiler, refit the head and continue firing for the rest of the day—the distillation ought to be commenced in the morning—but not so strong as to make the alcohol pass over. After the apparatus has been left undisturbed for one day, the contents are strained, preferably, by means of a strong linen bag-filter, fastened to a tenacle, under assiduously stirring the mass in the filter with a spatula. After dripping has ceased, submit the remnant to the strongest possible pressure, put back into the still, mix with alcohol of 45% to a pulp, strain, press and repeat the same operation once more. Clarify the united alcoholic liquids by subsiding and filtering, distil off the alcohol completely in the tin still, remove the head and allow the contents of the still to cool down as slowly as possible, in order to remove the last traces of alcohol (for this purpose the still is left suspended in the hot water of the boiler).

The contents of the still are now either a clear liquid or—more frequently—a fat is floating on the surface and a resin has subsided. Pour off from the latter and remove the fat by filtering through a wet filter.

All these three substances—resin, fat, aqueous liquid—together with the remaining substance, have to be examined separately.

(*a*) The resinous mass is usually of a dark colour. Wash with water and try if it contains an alkaloïd by triturating it assiduously with water containing 1-20th of its weight hydrochloric acid of 1.12 specific gravity, keeping at ordinary temperature for a few hours (heat applied with the acid is liable to split up the resin), filter the acid watery solution and wash the resin well with water.

*a.* Evaporate the acid liquid at a very gentle heat, try any crystalline or non-crystalline residue with the proper tests (II., *B*, *a*) and see, if it be a new or a known alkaloïd.

*β.* The resinous mass, remaining in *a*, is compared with the resin obtained previously (II., *A* and III., *B*), and the knowledge of the latter is, if necessary, completed by means of the resin now obtained.

(*b*) Submit the fat to the same treatment with hydrochloric acid and water as the resin, examine the acid liquid for alkaloïds, compare the fat with the fat obtained previously (II., *A*, *B*), and complete its investigation by means of the new material.

(*c*) Over-saturate a small sample of the aqueous solution of invariably acid reaction with ammonia. If a precipitate is obtained hereby, precipitate the whole liquid with ammonia. If no precipitate has been obtained with ammonia, try in the same way successively carbonate of ammonia, ley of potash (or of soda) and carbonates and bicarbonates of potash or of soda.

*a.* Let the precipitate, obtained by means of the above tests, subside, wash, dry, triturate, digest with alcohol of 90°/₀, warm, filter, evaporate the alcohol and examine any remnant on alkaloïds (II., *B*, *a*).

*β.* Precipitate the liquid remaining in *c*, after it has been made slightly acid with acetic acid, or the original clear liquid, in case no precipitate has been produced by alkalies, with acetate of lead completely and proceed according to II., *B*, *e*. The chief object is in this case the investigation of non-volatile organic acids and the completion of the former investigations.

(*d*) The remaining mass, exhausted by alcohol of 45%, contains most of the gum, and may be used for determining this substance if the former investigations (IV.) have been without a satisfactory result. For this purpose spread it on shallow dishes to drive off the alcohol, mix with cold water, strain after some time, press, let the liquid subside, decant, evaporate to a small bulk, and throw down the gum by means of alcohol, &c., according to the instructions given under "Gum," p. 99.

## IX.—DISTILLATION WITH ACID WATER.

In distilling vegetable substances with water, volatile acids are obtained in the distillate, but generally in such small quantities (except hydrocyanic acid) that it is often impossible, especially in a mixture, to recognise their nature. In order to obtain larger quantities of these volatile acids, the distillation with acid water is resorted to, while employing at least 1 kilogram of fresh raw material, or larger quantities if required.

Usually sulphuric acid is used for this purpose, although phosphoric acid is preferable on account of its non-volatility, and because it does not become decomposed towards the end of the distillation, while sulphuric acid is reduced to sulphurous acid, which passes over and adulterates the distillate. The volatile anorganic acids occurring in plants (hydrochloric and nitric acids) are likewise obtained by distilling with sulphuric or phosphoric acid, but the hydrochloric acid may easily be removed from the distillate, and the presence of nitric acid is of less consequence.

As the volatile acids (*i.e.*, those volatile by the steam of water, consequently not benzoic, cinnamic, oxalic, &c., acids) are always contained in plants as compounds soluble in water, the distillation of the aqueous extract is preferable to the distillation of the whole vegetable substance. Prepare, therefore, first an aqueous extract by mixing the finely-comminuted substance with four to six times its weight of pure water; keeping the whole for one day at a gentle heat (in the tin still of Beindorf's apparatus); straining, pressing, adding 50 to 60 grams phosphoric acid to every kilogram of the liquid; filtering (filtering before the addition of the acid would have been difficult or impossible; should it prove difficult still, subsiding instead of filtering must be resorted to) and adding a solution of sulphate of silver, as long as a precipitate is produced. Allow the chloride of silver to subside, filter, pour the filtrate into a glass retort, and distil two-thirds by means of a good refrigerator.

Pour the distillate into a porcelain-dish, add an adequate quantity (for every kilogram of raw material, about 10 grams) of carbonate of baryta, rubbed down to the finest powder with a portion of the distillate, and evaporate under assiduous stirring with a glass-rod on the water-bath. Any hydrocyanic acid present in the distillate evaporates, while the other volatile organic acids combine with the baryta, and remain in the liquid. Should the whole of the carbonate of baryta be dissolved in evaporating, a fresh portion of it must be added in order to prevent the loss of volatile acids. After the liquid has reached a certain concentration, and is no longer of acid reaction, it is filtered, and the contents of the filter are edulcorated with water.

The examination of the contents of the filter, and of the filtrate, is conducted according to II., *A, b*; but, besides the low acids of the fat-acid series named there, other acids, as angelic, salicylous, &c., acids, must not be neglected, or even the investigation of as yet unknown acids.

The discovery and estimation of (with water) non-volatile acids is not effected with the liquid remaining from the distillation of the volatile acids, on account of the great amount of phosphoric acid contained in it; but, with the liquid *c, β* of the preceding (VIII.) section; whereas the acid remnant, if necessary, may be employed for the investigation of volatile alkaloïds (X.)

## X.—DISTILLATION WITH ALKALINE WATER.

For a thorough investigation of volatile alkaloïds, at least 1 kilogram of the dried raw material is required. Similar to volatile acids, their separation is effected best by distilling the aqueous extract of the respective vegetable substance with an alkali, for the reason given before, and in order to prevent rising of the liquid and evolution of much ammonia. The latter is obtained, indeed, even in distilling the extract, because no vegetable extract is free from ammonia-salts or from other nitrogenised compounds, but in much less quantity.

Extract the substance by a warm digestion with four to six times its quantity of water (which, in presence of tannic acid, is mixed with 1-25th its weight hydrochloric acid of 1·12 specific gravity), press, pour the whole liquid into a copper-boiler (well adapted for this purpose is the boiler of the Beindorf's apparatus); saturate with slaked lime, add as much lime, slaked and mixed with water to the finest pulp, as to effect an excess of 50 grams quick-lime to every kilogram of raw material, and distil by means of a good refrigerator, until the distilling water is void or nearly void of alkaline reaction.

In case of any scarcity of raw material, the investigation of volatile alkaloïds may be effected by means of the acid residue left from the distillation under IX., but modified in such a way as to use caustic soda instead of quicklime, in order to prevent the formation of insoluble phosphate of lime.

Saturate the whole distillate exactly with diluted sulphuric acid, evaporate on the water-bath, and after due concentration in a weighed dish under the receiver of an air-pump, until no further loss of weight be observed. Weigh the salty remnant, triturate if necessary, shake in a flask with absolute alcohol; collect the insoluble portion (sulphate of ammonia) on a filter, wash with absolute alcohol, dry by means of the air-pump, and weigh. By subtracting the latter weight from that of the whole remnant, the quantity of the sulphate of the alkaloïd is obtained; and by deter-

Z

mining in the alcoholic solution, and after the alcohol has been driven off, the quantity of sulphuric acid by means of a salt of baryta, and by subtracting this weight from that of the sulphate of the alkaloïd, the rest represents the quantity of the pure alkaloïd.

The isolation of the alkaloïd is effected by distilling the sulphate with soda-ley, shaking the distillate with ether, decanting the ethereous liquid and evaporating by means of the air-pump.

The investigation of the physical and chemical properties of the composition and constitution of the volatile alkaloïd is the object of further experiments.

## Table of Comparison between Celsius' and Fahrenheit's Thermometric Scales.

| C. | F. | C. | F. | C. | F. |
|---|---|---|---|---|---|
| – 20° C. | = – 4° F. | ÷ 90° C. | = + 194° F. | ÷ 200° C. | ÷ 392° F. |
| – 15° | = + 5° | 95° | = 203° | 205° | 401° |
| – 10° | = + 14° | 100° | = 212° | 210° | `410° |
| – 5° | = + 23° | 105° | = 221° | 215° | 419° |
| 0° | = + 32° | 110° | = 230° | 220° | 428° |
| + 5° | = + 41° | 115° | = 239° | 225° | 437° |
| 10° | = 50° | 120° | = 248° | 230° | 446° |
| 15° | = 59° | 125° | = 257° | 235° | 455° |
| 20° | = 68° | 130° | = 266° | 240° | 464° |
| 25° | = 77° | 135° | = 275° | 245° | 473° |
| 30° | = 86° | 140° | 284° | 250° | 482° |
| 35° | = 95° | 145° | = 293° | 255° | 491° |
| 40° | 104° | 150° | = 302° | 260° | 500° |
| 45° | 113° | 155° | 311° | 265° | 509° |
| 50° | 122° | 160° | = 320° | 270° | 518° |
| 55° | 131° | 165° | = 329° | 275° | 527° |
| 60° | 140° | 170° | = 338° | 280° | 536° |
| 65° | 149° | 175° | = 347° | 285° | 545° |
| 70° | 158° | 180° | 356° | 290° | 554° |
| 75° | 167° | 185° | 365° | 295° | 563° |
| 80° | 176° | 190° | = 374° | 300° | 572° |
| 85° | 185° | 195° | 383° | | |

## Table of Comparison between Baume's Scale and the Specific Gravity of Alcohol.

| | | | | | | | |
|---|---|---|---|---|---|---|---|
| 10 degrees B. | = | 1.000 sp. gr. | | 27 degrees B. | — | 0.896 sp. gr. | |
| 11 | ,, | = 0.993 | | 28 | ,, | = 0.890 | |
| 12 | ,, | = 0.986 | | 29 | ,, | = 0.885 | |
| 13 | ,, | = 0.980 | | 30 | ,, | = 0.880 | |
| 14 | ,, | = 0.973 | | 31 | ,, | = 0.874 | |
| 15 | ,, | = 0.967 | | 32 | ,, | = 0.869 | |
| 16 | ,, | = 0.960 | | 33 | ,, | = 0.864 | |
| 17 | ,, | = 0.954 | | 34 | ,, | = 0.859 | |
| 18 | ,, | = 0.948 | | 35 | ,, | = 0.854 | |
| 19 | ,, | = 0.942 | | 36 | ,, | = 0.849 | |
| 20 | ,, | = 0.936 | | 37 | ,, | = 0.844 | |
| 21 | ,, | = 0.930 | | 38 | ,, | = 0.839 | |
| 22 | ,, | = 0.924 | | 39 | ,, | = 0.834 | |
| 23 | ,, | = 0.918 | | 40 | ,, | = 0.830 | |
| 24 | ,, | = 0.913 | | 41 | ,, | = 0.825 | |
| 25 | ,, | = 0.907 | | 42 | ,, | = 0.820 | |
| 26 | ,, | = 0.901 | | 43 | ,, | = 0.816 | |

## Table Showing the Specific Gravity of Alcohol of Different Percentage by Weight at 60° F.

| | | | | | |
|---|---|---|---|---|---|
| 5% = 0.9914 sp. gr. | 40% = 0.9396 sp. gr. | | 75% = 0.8603 sp. gr. |
| 10 = 0.9841 | 45 = 0.9292 | | 80 = 0.8483 |
| 15 = 0.9778 | 50 = 0.9184 | | 85 = 0.8357 |
| 20 = 0.9716 | 55 = 0.9069 | | 90 = 0.8228 |
| 25 = 0.9652 | 60 = 0.8956 | | 95 = 0.8089 |
| 30 = 0.9578 | 65 = 0.8840 | | 100 = 0.7938 |
| 35 = 0.9490 | 70 = 0.8721 | | |

## Table showing the Specific Gravity of Alcohol of Different Percentage by Volume at 60° F.

| | | | | | |
|---|---|---|---|---|---|
| 5% = 0.9928 sp. gr. | 40% = 0.9519 sp. gr. | | 75% = 0.8773 sp. gr. |
| 10 = 0.9866 | 45 = 0.9435 | | 80 = 0.8639 |
| 15 = 0.9811 | 50 = 0.9343 | | 85 = 0.8496 |
| 20 = 0.9760 | 55 = 0.9242 | | 90 = 0.8340 |
| 25 = 0.9709 | 60 = 0.9134 | | 95 = 0.8164 |
| 30 = 0.9655 | 65 = 0.9021 | | 100 = 0.7946 |
| 35 = 0.9592 | 70 = 0.8900 | | |

## Table of Comparison between the Specific Gravity of Alcohol and its Percentage Over or Under Proof.

| | | | | | |
|---|---|---|---|---|---|
| 0.8156 sp. gr. | = 67% over proof. | | 0.9225 sp. gr. | = | 2% under proof |
| 0.8199 | = 65 ,, | | 0.9248 | = 4 | ,, |
| 0.8221 | = 64 ,, | | 0.9270 | = 6 | ,, |
| 0.8259 | = 62 ,, | | 0.9295 | = 8 | ,, |
| 0.8298 | = 60 ,, | | 0.9318 | = 10 | ,, |
| 0.8336 | = 58 ,, | | 0.9340 | = 12 | ,, |
| 0.8376 | = 56 ,, | | 0.9362 | = 14 | ,, |
| 0.8415 | = 54 ,, | | 0.9384 | = 16 | ,, |
| 0.8450 | = 52 ,, | | 0.9409 | = 18 | ,, |
| 0.8484 | = 50 ,, | | 0.9435 | = 20 | ,, |
| 0.8516 | = 48 ,, | | 0.9446 | = 22 | ,, |
| 0.8550 | = 46 ,, | | 0.9465 | = 24 | ,, |
| 0.8582 | = 44 ,, | | 0.9485 | = 26 | ,, |
| 0.8615 | = 42 ,, | | 0.9503 | = 28 | ,, |
| 0.8646 | = 40 ,, | | 0.9521 | = 30 | ,, |
| 0.8678 | = 38 ,, | | 0.9540 | = 32 | ,, |
| 0.8708 | = 36 ,, | | 0.9559 | = 34 | ,, |
| 0.8738 | = 34 ,, | | 0.9572 | = 36 | ,, |
| 0.8769 | = 32 ,, | | 0.9587 | = 38 | ,, |
| 0.8797 | = 30 ,, | | 0.9602 | = 40 | ,, |
| 0.8825 | = 28 ,, | | 0.9617 | = 42 | ,, |
| 0.8854 | = 26 ,, | | 0.9632 | = 44 | ,, |
| 0.8883 | = 24 ,, | | 0.9645 | 46 | ,, |
| 0.8910 | = 22 ,, | | 0.9658 | 48 | ,, |
| 0.8938 | = 20 ,, | | 0.9672 | = 50 | ,, |
| 0.8966 | 18 ,, | | 0.9702 | 55 | ,, |
| 0.8994 | 16 ,, | | 0.9732 | = 60 | ,, |
| 0.9022 | = 14 ,, | | 0.9760 | = 65 | ,, |
| 0.9049 | = 12 ,, | | 0.9789 | = 70 | ,, |
| 0.9075 | 10 ,, | | 0.9819 | = 75 | ,, |
| 0.9100 | 8 ,, | | 0.9851 | = 80 | ,, |
| 0.9125 | = 6 ,, | | 0.9885 | = 85 | ,, |
| 0.9151 | = 4 ,, | | 0.9921 | = 90 | ,, |
| 0.9176 | = 2 ,, | | 0.9959 | = 95 | ,, |
| 0.9200 | = proof spirit | | 1.0000 | = 100 | ,, |

## Table of Comparison between French Metrical and English Weights.

| | | | | | | |
|---|---|---|---|---|---|---|
| 1 Milligramme | = 0.015432 grs. av. | | 1 Gramme | = | 15.432 grs. av. | |
| 1 Centigramme | 0.154323 ,, ,, | | 2 ,, | | 30.865 ,, ,, | |
| 1 Decigramme | 1.543234 ,, ,, | | 3 ,, | = | 46.297 ,, ,, | |
| 1 Gramme | 15.432348 ,, ,, | | 4 ,, | | 61.729 ,, ,, | |
| 1 Kilogramme | = 15432.348 ,, ,, | | 5 ,, | | 77.162 ,, ,, | |
| | — 35.2739 ozs. ,, | | 6 ,, | = | 92.594 ,, ,, | |
| | = 2.2046 lbs. ,, | | 7 ,, | = | 108.026 ,, ,, | |
| | | | 8 ,, | = | 123.459 ,, ,, | |
| | | | 9 ,, | = | 138.891 ,, ,, | |

## Table of Comparison between Cubic Centimeters and English Cubic Inches.

| | | |
|---|---|---|
| 1 Cub. Centimeter = 0.061024 cub. in. | 6 Cub. Centimeter = 0.366144 cub. in. |
| 2 „ „ = 0.122048 „ | 7 „ „ = 0.427168 „ |
| 3 „ „ = 0.183072 „ | 8 „ „ = 0.488192 „ |
| 4 „ „ = 0.244096 „ | 9 „ „ = 0.549216 „ |
| 5 „ „ = 0.305120 „ | |

## Table of Comparison between Litres and Fluid Ounces.

| | |
|---|---|
| 1 Litre = 35.2754 fluid ozs. | 6 Litres = 211.6524 fluid ozs. |
| 2 „ = 70.5508 „ „ | 7 „ = 246.9278 „ |
| 3 „ = 105.8262 „ „ | 8 „ = 282.2032 „ |
| 4 „ = 141.1016 „ „ | 9 „ = 317.4786 „ |
| 5 „ = 176.3770 „ „ | |

## Table of Atomic and Molecular Weights of the Principal Elementary Bodies.

| | Molecular Weights. | Atomic Weights. | | Molecular Weights. | Atomic Weights. |
|---|---|---|---|---|---|
| Aluminium = Al. = | 27.5 | 13.75 | Manganese = Mn. = | 55 | 27.5 |
| Antimony = Sb. = | 122 | 122 | Mercury = Hg. = | 200 | 100 |
| Arsenic = As. = | 75 | 75 | Molybde- | | |
| Barium = Ba. = | 137 | 68.5 | num = Mo. = | 96 | 48 |
| Bismuth = Bi. = | 210 | 105 | Nickel = Ni. = | 58.8 | 29.4 |
| Boron = B. = | 11 | 11 | Nitrogen = N. = | 14 | 14 |
| Bromine = Br. = | 80 | 80 | Oxygen = O. = | 16 | 8 |
| Cadmium = Cd. = | 112 | 56 | Palladium = Pd. = | 106 | 53 |
| Calcium = Ca. = | 40 | 20 | Phosphorus = P. = | 31 | 31 |
| Carbon = C. = | 12 | 6 | Platinum = Pt. = | 197·4 | 98.7 |
| Chlorine = Cl. | 35.5 | 35.5 | Potassium = K. = | 39.1 | 39.1 |
| Chromium = Cr. = | 52.2 | 26.1 | Silicium = Si. = | 28 | 28 |
| Cobalt = Co. | 58.8 | 29.4 | Silver = Ag. = | 108 | 108 |
| Copper = Cu. = | 63.4 | 31.7 | Sodium = Na. = | 23 | 23 |
| Fluorine = F. | 19 | 19 | Strontium = Sr. = | 87 | 43.5 |
| Gold = Au. = | 197 | 197 | Sulphur = S. = | 32 | 16 |
| Hydrogen = H. = | 1 | 1 | Tin = Sn. = | 118 | 59 |
| Iodine = I. = | 127 | 127 | Tungsten or | | |
| Iron = Fe. = | 56 | 28 | Wolfram = W. = | 184 | 92 |
| Lead = Pb. = | 207 | 103.5 | Uranium = U. = | 120 | 60 |
| Lithium = L. = | 7 | 7 | Zinc = Zn. = | 65.2 | 32.6 |
| Magnesium = Mg. = | 24 | 12 | | | |

# OMISSIONS AND ERRATA.

——◆——

**Aconitin.**—Dr. C. R. A. Wright has shown that the use of alcohol acidulated with a mineral acid, for exhausting the root of Aconite, causes an alteration of the alkaloïds originally present; hence it is recommended, according to Duquesnel's method, to percolate by alcohol acidulated with tartaric acid, to evaporate at a low temperature, or, better still, in vacuo, to crystallise from ether after the separation of the base by sodium-carbonate, and to purify by conversion into a crystalline salt, for which purpose the hydrobromide is well fitted. Dr. Wright's formula for pure Aconitin$=C_{33}H_{43}NO_{12}$.—(Blackett).

**Alstonin.**—Alkaloïd of the bark of Alstonia constricta, F. v. M. Obtained by treating the alcoholic extract with water and a little hydrochloric acid, adding to the filtered solution a small excess of ammonia, dissolving the separated flocks in ether, evaporating the ethereous solution, and purifying the remaining A. by dissolving again in dilute acid and repeating the above process.—Orange yellow, brittle, pellucid mass, of very bitter taste, melts below 100°, and is carbonised in higher temperatures; dissolves easily in alcohol, ether, and dilute acids, sparingly in water. All its solutions in the dilute state exhibit a strong blue fluorescence, which is not affected by acids or alkalies. Its alcoholic solution has a slightly alkaline reaction. Alstonin combines with acids, but does not completely neutralise them. Hydrochloric and other strong acids, also alkalies, decompose it partly on evaporation in the water-bath to a dark-coloured acid substance. The hydrochloride of A. gives precipitates with the chlorides of platinum and mercury, the iodides of potassio-mercury and of potassio-bismuth, bi-iodide of potassium, the phospho-molybdate and the phospho-tungstate of soda, bichromate of potash, picric acid, and by the alkalies and alkaline carbonates. Tannic acid does not precipitate the hydrochloride, but does so the acetate and the pure base. Concentrated nitric acid dissolves A. with crimson colour, yellow on warming; sulphuric acid reddish brown, afterwards dirty green; hydrochloric only effects a yellowish solution. Alstonin differs from Ditamin chiefly by its behaviour towards concentrated acids, and by its fluorescence, which has not been recorded of the other alkaloïd.—(F. von Mueller and L. Rummel.)

**Chrysophanic Acid** contained to the extent of up to 84°/₀ in the Goo- or Bahia- or Arariba- or Araroba-powder, obtained from the medulla of the wood of Centrolobium robustum, Mart., and C. tomentosum, Benth.; large Brazilian trees (Silva-Lima, Neumann, Voigt.)

**Duboisin.**—Volatile alkaloïd of the leaves and twigs of Duboisia myoporoides, R. Br., and probably identical with the Piturin found by Staiger in Duboisia Hopwoodii, F. v. M. Prepared like Nicotin.—Yellowish oily liquid, lighter than water, of a strong narcotic odour, resembling that of nicotin and also of cantharides, of a very strong alkaline reaction, neutralises the acids completely; dissolves in any quantity of water, alcohol, and ether; throws down ferrous oxyd from subsulphate of iron; dissolves colourless in concentrated acids. Its hydrochloride in a weak aqueous solution is precipitable by bi-iodide of potassium, the iodides of potassio-mercury and of potassio-bismuth, and by tannic acid, not by other alkaloïd reagents. Nicotin, to which Duboisin resembles, is distinguished from the latter by its specific gravity, its less powerful odour, and by its hydrochloride in a diluted aqueous solution being precipitated by phosphomolybdate of soda, picric acid, and chloride of platinum.—(F. von Mueller and L. Rummel.)

**Erythrophlaein** (see page 229).—This alkaloïd has also been obtained from the bark of Erythrophlaeum Laboucherii, F. v. M. The E. crystallises in needle-shaped crystals, which are often united in tufts and arranged in the form of an oblique cross; it is almost tasteless, not bitter, either by itself or united with acids; its concentrated solution in aqueous alcohol is precipitable on addition of absolute alcohol; its salts are likewise precipitated by phospho-molybdate and phospho-tungstate of soda; it appears to be nearly related to Laburnin. It occurs in the dry bark to the extent of about 2%. This alkaloïd may be sought for also in a species recently rendered known from the Seychelles.—(F. von Mueller and L. Rummel.)

**Pilocarpin,** $C_{23} H_{34} N_4 O_4$, according to Kingsett, *Journ. Chemic. Soc.*, October, 1876. This alkaloïd, the only one of the Jaborandi, on distillation with caustic potass yields trimethylamin. To obtain Pilocarpin the Jaborandi is exhausted with alcohol and tartaric acid by a process similar to that followed by Duquesnel for isolating aconitin.—(C. R. Blackett).

**Pyrocatechin** (Oxyphenal or Oxyphenic acid) is formed by the dry distillation of Catechin, and is also obtainable from wood-vinegar. It is a white crystalline mass, melts at 112° C., and volatilises even at lower temperatures. It has a bitter taste, but is scarcely of acid reaction. In contact with hydrochloric acid it tinges fir-woods violet; it dissolves in water, alcohol, and ether. The aqueous solution deposits a white precipitate with lead-acetate, and colours ferric salts dark-green; nitric acid decomposes it into oxalic acid and a small portion of a yellow nitro-compound (Watts). For its direct occurrence in plants see page 182.

**Tannic Acid** (its estimation).—Exhaust with boiling water, filter, add a solution of chrome-alum as long as a sediment is formed, filter after a few hours, wash out, dry at 100° C., and weigh; then incinerate the dry precipitate and weigh again the remaining oxyd of chromium; the first weight minus the latter one gives the amount of tannic acid.—(F. v. Mueller and L. Rummel). The active tannic principle in oak-bark seems as yet not strictly chemically defined.

P. 269, add, before Coniferæ: 3, *Gymnospermæ*.

Many substances, of which as yet the chemical formula has not been carefully ascertained, or of which well-marked characteristics are not yet on record (such as Podophyllum-resin, oil of Erigeron Canadensis, &c.), are omitted in this work for the present.

P. 40, l. 18.—Instead of Fibrin read Fibre.

P. 68, l. 11.—Strike out "alkaloïd."

P. 91, l. 20, 21, 25, 27.—Instead of "Gingkoic, Gingko, Gingkoate" read "Ginkgoic, Ginkgo, Ginkgoate."

P. 149, l. 14.—Instead of "Eucalyptin" read "Eucalypten."

P. 171, l. 29.—Instead of Pholobaphen read Phlobaphen.

P. 174, l. 11.—Instead of "ceropic" read "ceropinic."

P. 226, l. 18.—Instead of "acids" read "acrids."

P. 229, l. 2.—Instead of "$C_{23} H_{26} N_2 O_4$" read "$C_{46} H_{26} N_2 O_8$."

P. 231, l. 10.—Instead of "$C_{22} H_{23} NO_8$" read $C_{44} H_{23} NO_{16}$."

P. 272.—Read Hule-Caoutchouc instead of Ulé Caoutchouc.

M'Carron, Bird & Co., Printers, 37 Flinders Lane West, Melbourne.

www.ingramcontent.com/pod-product-compliance
Lightning Source LLC
Chambersburg PA
CBHW021451210326
41599CB00012B/1021